Geoinformation Technology and Artificial Intelligence

Geoinformation Technology and Artificial Intelligence

Edited by Kaylin Davies

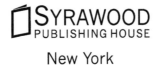

SYRAWOOD
PUBLISHING HOUSE
New York

Published by Syrawood Publishing House,
750 Third Avenue, 9th Floor,
New York, NY 10017, USA
www.syrawoodpublishinghouse.com

Geoinformation Technology and Artificial Intelligence
Edited by Kaylin Davies

International Standard Book Number: 978-1-64740-399-7 (Hardback)

Cataloging-in-publication Data

Geoinformation technology and artificial intelligence / edited by Kaylin Davies.
    p. cm.
Includes bibliographical references and index.
ISBN 978-1-64740-399-7
1. Geographic information systems. 2. Artificial intelligence--Geophysical applications.
3. Geophysics--Data processing. 4. Information storage and retrieval systems--Geography.
I. Davies, Kaylin.
G70.212 .G47 2023
910.285--dc23

TABLE OF CONTENTS

Permissions

List of Contributors

Index

PREFACE

I am honored to present to you this unique book which encompasses the most up-to-date data in the field. I was extremely pleased to get this opportunity of editing the work of experts from across the globe. I have also written papers in this field and researched the various aspects revolving around the progress of the discipline. I have tried to unify my knowledge along with that of stalwarts from every corner of the world, to produce a text which not only benefits the readers but also facilitates the growth of the field.

Geoinformation technology refers to a science that deals with the acquisition, storage, management, analysis and delivery of geographic and spatially referenced information. It comprises an interactive site for deeper study of raster data on greenhouse gas emissions as well as a website for spatial analysis and visualization of vector data. Artificial intelligence (AI) is a field of computer science that focuses on developing and managing technology that can learn to make decisions and perform actions autonomously on behalf of humans. It uses logic-based techniques and advanced analysis including machine learning to assist and automate decisions, interpret events, and take actions. The combination of geoinformation technologies and AI technologies has various applications such as exposure modeling in environmental epidemiology and automating cyber-physical systems. This book elucidates the concepts and innovative models around prospective developments with respect to the application of artificial intelligence in the field of geoinformation technology. It will serve as a valuable source of reference for graduate and postgraduate students.

Finally, I would like to thank all the contributing authors for their valuable time and contributions. This book would not have been possible without their efforts. I would also like to thank my friends and family for their constant support.

Editor

1

A Comparative Study of the AHP and TOPSIS Techniques for Dam Site Selection using GIS

Ali Jozaghi [1,*], Babak Alizadeh [1], Mohsen Hatami [2](ID), Ian Flood [2], Mohammad Khorrami [3](ID), Nastaran Khodaei [4] and Erfan Ghasemi Tousi [5]

[1] Department of Civil Engineering, University of Texas at Arlington, P.O. Box 19019, Arlington, TX 76019, USA; babak.alizadeh@mavs.uta.edu
[2] M. E. Rinker, Sr. School of Construction Management, University of Florida, P.O. Box 115703, Gainesville, FL 32611, USA; mohsen.hatami@ufl.edu (M.H.); flood@ufl.edu (I.F.)
[3] Department of Civil Engineering, Ferdowsi University of Mashhad, Mashhad 9177948974, Iran; mohammad.khorrami@mail.um.ac.ir
[4] College of Architecture and Planning, University of Colorado Denver, P.O. Box 173364, Denver, CO 80217, USA; Nastaran.khodaei@ucdenver.edu
[5] Department of Civil and Architectural Engineering and Mechanics, University of Arizona, P.O. Box 210072, Tucson, AZ 85721, USA; erfang@email.arizona.edu
* Correspondence: ali.jozaghi@mavs.uta.edu

Abstract: The application of multiple criteria decision-making (MCDM) techniques in real-life problems has increased in recent years. The need to build advanced decision models with higher capabilities that can support decision-making in a broad spectrum of applications, promotes the integration of MCDM techniques with applicable systems, including artificial intelligence, and Geographic Information Systems (GIS). The Analytic Hierarchy Process (AHP) and Technique for Order of Preference by Similarity to Ideal Solution (TOPSIS) are among the most widely adopted MCDM techniques capable of resolving water resources challenges. A critical problem associated with water resource management is dam site selection. This paper presents a comparative analysis of TOPSIS and AHP in the context of decision-making using GIS for dam site selection. The comparison was made based on geographic and water quality criteria. The geographical criteria are geology, land use, sediment, erosion, slope, groundwater, and discharge. The water quality criteria include Soluble Sodium Percentage, Total Dissolved Solid, Potential of Hydrogen, and Electrical Conductivity of water. A ratio estimation procedure was used to determine the weights of these criteria. Both methods were applied for selection of optimal sites for dams in the Sistan and Baluchestan province, Iran. The results show that the TOPSIS method is better suited to the problem of dam site selection for this study area. Actual locations of dams constructed in the area were used to verify the results of both methods.

Keywords: dam site selection; water resources; MCDM; topographical conditions; morphological conditions; TOPSIS; AHP; Geographic Information System; Sistan and Baluchestan

1. Introduction

The economic merits of dams outweigh the detriments and costs, thus providing a good reason for construction of dams around the world. Selection of the best location for construction a dam is one of the most complex and controversial decisions in water supply management [1]. Just as an optimal site selection can enhance the security of the reservoir in a region and groundwater regeneration, a bad site selection can undermine them. A well-selected site will not only provide direct benefits, but its careful design can also provide the additional benefit of a recreational area surrounding the reservoir. Conversely, a poorly selected site could cause detrimental influences such as negative biophysical, socio-economic, and geopolitical impacts, often through the loss of ecosystem services provided by fully functioning aquatic systems [2–5]. Thus, for selection of dam sites, it is necessary to conduct a precise study over the interest area considering factors affecting this selection. However, this procedure is costly and time consuming. With advances in computing and information technologies, determining competitive solutions in terms of cost, time, and a variety of other objective variables is greatly facilitated.

A powerful tool that plays a noticeable role in this process is the Geographical Information System (GIS) and its applications in hydrology. GIS, which can integrate large layers of data, is a computer-based system handling the attribute data as well as spatial data where geographical information is a key characteristic [6–8]. The use of such technology as GIS in this sector compared to buildings and other infrastructure is still behind and this has still remained as a challenge in the lifecycle management of dams. In addition to GIS, Multiple Criteria Decision-making (MCDM) helps decision-makers select between alternative solutions, in this case sites for construction of dams, where there are many criteria [9]. Two of the most widely used MCDM techniques are known as Analytic Hierarchy Process (AHP) [10] and Technique for Order of Preference by Similarity to Ideal Solution (TOPSIS) [11–14]. The integration of MCDM and other data analysis tools such as GIS have been commonly used by past researchers, particularly in dam site selection studies [15–18]. It should be noted that GIS is powerfully used in dam site selection [19–21]; however, its ability in a certain region may differ from other locations. The AHP approach is an effective tool for system analysis and solves decision problems by reducing complex decisions to a series of pairwise comparisons. AHP is an effective multiple criteria decision-making technique that has been used to solve decision problems in a variety of fields [22,23]. Furthermore, AHP includes an effective technique for checking the consistency of the decision-maker's evaluations, thus decreasing the bias in the decision-making process [24]. The TOPSIS method is based on the idea that the best alternative has the shortest distance from the ideal solution and the farthest distance from the negative ideal solution. The ideal solution is assumed to be an alternative that has the best values for all considered criteria whereas the negative ideal solution is identified with a hypothetical alternative that has the worst criteria values [11]. Many papers have been published in the field of dam site selection using MCDM techniques for choosing a feasible location for building a dam. However, the authors are not aware of any comparative study between the TOPSIS and AHP techniques for solving dam site selection problems.

A wide range of studies have been conducted regarding risk or performance analysis on multiple watersheds, dams or tunnels all around the world with a diverse hydro-climatological regimes [25–33]. Anane et al. (2012) carried out a study to prioritize the best locations for irrigation with treated wastewater (TWW) in Tunisia [34]. Potential feasible locations were identified based on resources conflicts, cost effectiveness, land suitability, social acceptance and environment factor. Several researchers have applied fuzzy systems on decision-making methods [35–39]. Using fuzzy AHP combined with GIS, they were able to map and prioritize the appropriate sites for different purposes. Reliable data and advanced technologies are two necessary elements for efficient management of transportation networks and their transit corridors [40–42]. Soil properties and groundwater level should be investigated in design phase of the road transportation network to have a high-quality subgrade. As such, there have been a significant progress in development of the standards for data transfer between different software packages and technologies [43]. Choudhary and Shankar (2012) considered location selection for thermal power plants as a multi-criteria decision problem and applied a STEEP-fuzzy AHP-TOPSIS framework to rank the alternative locations and select the most suitable for a thermal power plant in India [44]. Their study considered five major criteria, including social, technical, economic, environmental, and political. The weight of qualitative and quantitative criteria that impact the location selection process was determined by the fuzzy AHP. The alternative locations were ranked based on their overall performance using the TOPSIS technique. Kim et al. (2012) used the TOPSIS method in a fuzzy environment to rank the best out of ten sites for treated wastewater (TWW) in a South Korean urban watershed [35]. They considered four main criteria, including technical, social, economic and environmental criteria. The uncertainty of weighting values and input data were considered using triangular fuzzy numbers, and data were collected using individual interviews. The results showed that the fuzzy TOPSIS approach provided different rankings of the best sites for TWW compared to a traditional MCDM technique, highlighting the need for fuzzy-based MCDM. In 2012, a study was conducted by Minatour et al. in the western part of Iran using the AHP technique to help choose an appropriate location for an earth dam [45]. In this study, nine categories of criteria and eleven sub criteria were considered based on their ranked importance for locating the earth dam site. The criteria used were: topographic conditions, economic development, health of dam site, river flow regime, annual yield, volume of reservoir, annual volume of sediment, probable maximum flood, average annual evaporation, access to materials and facilities, overall cost, water quality, dam body and reservoir, probable dam break, environmental, social, and political. According to the criteria in the dam site selection, using the experiences and opinions of experts, and existing information and data, four feasible alternatives were proposed and prioritized using the AHP technique. The most optimal site was selected and then approved by the experts involved in the project [45]. In another study, Sengül et al. (2014) developed a model using the Fuzzy TOPSIS method to rank renewable energy supply systems in Turkey [36]. The Interval Shannon's Entropy methodology was used to determine weight values for the criteria. Criteria data were obtained from the literature and from official sites in Turkey. The results of this research offer energy policymakers and decision-makers optimal alternative solutions for resource allocation in Turkey.

Zyoud et al. (2016) used AHP and TOPSIS methods within a fuzzy environment to create a framework for water loss management in developing countries [37]. They proposed a hierarchical structure of the decision problem that consisted of four levels: overall objective; main criteria; evaluation criteria; and options. In this study, the weightings of criteria were determined by fuzzy AHP, and fuzzy TOPSIS was also used to rank options in terms of their potential to meet the overall objective based on the evaluations and preferences of decision-makers. It was assumed that the most important option was a pressure management and control strategy. Additionally, employing advanced techniques and establishment of district metered areas were determined to be the second and third most important, respectively. Moreover, based on sensitivity analysis results, the strongest and weakest options were less sensitive to changes in the weights of the evaluation criteria. Özcan et al. (2017) applied the AHP and TOPSIS methods for maintenance strategy selection in hydroelectric power plants in Turkey [46]. In their study, a combined AHP-TOPSIS methodology was used for choosing the most critical equipment. Nine items of equipment of critical importance for hydroelectric power plants were determined. A Goal Programming (GP) model was proposed to obtain maintenance strategy combinations for the equipment. The results showed that there was an improvement of 77.1% in failure frequency of the power plant resulting from employing the wrong maintenance strategy on critical equipment compared to the period when the model was not used. Esavi et al. (2012) compared the AHP and Fuzzy-AHP methods for underground dam site selection in a hydrologic catchment in Iran [47]. They concluded that the Fuzzy-AHP shows higher capability and more flexibility in selecting the appropriate underground dam location. An intercomparison between AHP and TOPSIS techniques was carried out by Önüt and Soner (2018) to select an optimal transshipment site in Istanbul, Turkey [48]. They used fuzzy sets to account for uncertainties in different criteria and they derived criteria weights based on a pair-wise comparison using the AHP method. Mulliner et al. (2016) conducted a comprehensive analysis of five different MCDM techniques, including TOPSIS and AHP, to assess sustainable housing affordability using different economic, social and environmental criteria in Liverpool, UK [49]. Their results show that overall ranking of alternatives varies from method to method, and there is no perfect technique for this problem. Hence, where possible, applying a selection of different methods to the same problem is ideal. Balioti et al. (2018) applied the AHP and TOPSIS methods with fuzzy logic to select the optimal type of spillway in a dam in Northern Greece [50]. They concluded that these tools are required to account for additional parameters beyond technical and construction costs. Table 1 shows a quick review of recent research study on dam site selection using AHP and TOPSIS techniques.

This paper presents a study based on the selection of suitable dam sites in southern parts of the Sistan and Baluchestan province, Iran. Therefore, a comparative study of two MCDM models (TOPSIS and AHP) is applied in the GIS environment to determine the proposed locations for dam construction. For this purpose, based on the literature and similar research experiences, several criteria were used in the selection procedure. Therefore, the main objectives of the present study are as follows:

- To propose several factors affecting the selection of dam site in Sistan and Balouchestan Province, Iran, based on previous experiences around the world.
- To integrate MCDM and GIS in the study area for the purpose of dam site selection.
- To consider topographical and morphological conditions in dam site selection.
- To compare the outputs of TOPSIS and AHP applications in dam site selection.
- To help the decision-makers in the construction of new dams in the interest area.

This paper is organized as follows. At first, the area of the available dam sites is introduced. Then, the data and methods for choosing the best locations for construction of dams is thoroughly explained. Afterwards, results of implementing the TOPSIS and AHP techniques are discussed. Finally, the conclusions and recommendations for the future researches is provided.

Table 1. A review on recent researches on dam site selection using AHP and TOPSIS.

Reference	Year	Site Selection for	Applied Method
[47]	2012	Underground Dam	AHP
[45]	2013	Dam	AHP
[51]	2013	Dam	TOPSIS
[52]	2013	Underground dam	AHP
[53]	2014	Subsurface dams	AHP
[54]	2015	Small underground dams	AHP
[55]	2015	Dam	AHP
[21]	2016	Dam	AHP
[56]	2017	Dam	AHP
[57]	2018	Dam	AHP
[1]	2018	Dam	TOPSIS

2. Methodology

2.1. Study Area

Sistan and Baluchestan Province covers an area of 181,785 km^2 and, with a population of 2.5 million, is the second-largest province of Iran. Nonetheless, it is one of the most thinly populated provinces in the country. Sistan and Baluchestan Province includes a desert and semi-desert climate and the average annual precipitation in this province is about 110 mm and its average temperature is from 22 to 37 °C. This region is variable in terms of areas and heights. Also, because of the proximity of the province to the Lut desert (Dasht-e-Lut), the province has a warm and arid climate with relatively moderate winter. Temperature variations overnight, as well as high annual fluctuations and low relative humidity, and annual rainfall are much less than the average rainfall in the country [58].

As shown in Figure 1, the study area is located in southern part of the province, covered an area around 32,000 km^2, and bounded by Pakistan and Makran Sea. It includes four cities namely Chabahar, Sarbaz, Konarak, and Nikshahr. There are four perennial rivers, Sarbaz with an average flow rate of 7 m^3/s, Kajoo with average flow rate 5 m^3/s, Shi Kalak with average flow rate 0.3 m^3/s and Kahir with average flow rate 5.5 m^3/s. The study area also contains five large existing dams, Kahir Dam (reservoir capacity: 314×106 m^3), Zirdan Dam (reservoir capacity: 207×106 m^3), Shi Kalak Dam (reservoir capacity: 10.5×106 m^3), Land Dam (reservoir capacity: 10×106 m^3) and Pishin Dam (reservoir capacity: 167×106 m^3). In terms of climate, Sistan and Balouchestan province can be considered a desert. There is virtually no rainfall all year long in Sistan and Baluchestan. These regions are arid and semi-arid and continuity of multi-year droughts and low level of precipitation during recent years has caused water shortages. Köppen and Geiger have classified this location as BWh [59].

Figure 1. The location of the study area (blue area).

2.2. Data Description

Based on the literature and the experts' opinions, we considered the following criteria: Soluble Sodium Percentage (SSP), Total Dissolved Solid (TDS), Potential of Hydrogen (PH), Electrical Conductivity (EC) of water, Geology, Land use, Sediment, Erosion, Groundwater and Discharge.

The presence of suitable geological conditions at the dam site is one of the most important parameters for locating the dams. For this reason, it is necessary to exclude sites that are ineligible in terms of geology. To do this, we can use assign the least possible weight to areas that are suffer from inappropriate geological conditions for dam construction. In addition, the rapid increase of population and the related constructions, and deforestations have caused severe erosion of soil in most of the interest watersheds. The corresponding erosion in the rivers exacerbates the accumulation of sediment in water reservoirs. Therefore, it is necessary to remove the sites that are inappropriate in term of erosion before the analysis steps are performed. For this purpose, areas with very low erosion are selected as the best regions, and the worst area is allocated to the worst erosion rate.

The purpose of this research is to find the best places to construct a dam for drinking and irrigation of agricultural areas. For this reason, water quality data should be weighed for this purpose and weighed. In terms of agriculture, the most important qualitative criteria in agriculture classification are salinity and the amount of sodium. Because these two do not only affect the growth of the plant, it also determines the degree of water eligibility for irrigation and its effect on soil permeability. The salinity is measured using the standard of electrical conductivity and sodium with the standard solution of sodium solution. Therefore, we collected the following data: SSP, TDS, PH, and EC of water in the interest area.

2.3. Choosing Appropriate Sites Based on Topography and Valley Shape

A dam is naturally connected to an environment. Accordingly, the morphology of the river valley plays a crucial role in the choice of a dam site. If we ignore the use of reservoirs for recreational purposes, the main criterion influencing the dam location is reservoir capacity; thus, the best site will be a narrow site where the valley widens upstream of the proposed dam. The objective is to find sites which are located in a narrow valley and have a suitable volume of water available for irrigation and drinking-water supply.

2.4. Creating a Prediction Surface for Datasets

Visiting every location in a study area to evaluate the height, size, or concentration of a phenomenon is not usually practicable. Instead, scattered sample input point locations can be selected, and a value can be predicted and assigned to all other locations. Creating an elevation surface from a set of sample measurements in which each point represents a location where the elevation has been measured is a typical use for point interpolation. There are many different methods which can be used to interpolate values from a set of data points. All methods rely on the similarity of nearby sample points to create the surface [60]. In this section, a comparative study is made between the three most common spatial interpolation techniques including Inverse Distance Weighted (IDW), Spline, and Kriging. These techniques are applied for each set of data points, including sediment, groundwater, PH, SSP, TDS, EC and discharge. Then, the errors of all aforementioned techniques are analyzed, and the best method is chosen based on the minimum Mean Absolute Deviation (MAD). The mean absolute deviation is the average of all absolute errors and is given by:

$$MAD = n^{-1} \sum_{i-1}^{n} |y_i - \hat{y}_i| \qquad (1)$$

where $\hat{y}_i$ denotes predicted value, y_i is observed value, n is the number of data points [61]. After calculating the MAD for all seven datasets, the best interpolation method is chosen based on the minimum MAD and then a raster surface is created from each dataset. Interventionary studies involving animals or humans, and other studies require ethical approval must list the authority that provided approval and the corresponding ethical approval code.

2.5. TOPSIS

The TOPSIS technique was developed by Hwang and Yoon in 1981, and it is a pragmatic method for dealing with real life MCDM problems [11]. It helps decision-makers and engineers compare and rank a set of alternative decisions. In this method, ranking of alternatives is based on the shortest distance from the ideal solution and the farthest from the negative ideal solution. The TOPSIS procedure comprises the following steps [62]:

• Step 1: Determination of the weight of criteria and construction of the decision matrix

The first and foremost step in the TOPSIS algorithm is creating a decision matrix and determining the weight of criteria. Weighting the criteria, all of which in this study have been determined in GIS layers, is one of the most important and complicated steps of the MCDM methods [63]. In this step, relative weights must be assigned not only to each criterion but to its quantitative and qualitative values based on their importance. Since weighting the criteria is the main step in the decision-making process, a high precision is required to determine weights for each criterion and its values. There are a variety of methods which have been proposed to determine weights in MCDM methods. Tzeng et al. (1998) classified weighting approaches into subjective or objective [64]. The weights in the subjective methods are determined based on preference information of criteria, subjective opinions and decision-maker's knowledge. However, the objective approaches select the weights of criteria based on a mathematical calculation. In this study, the Ratio Estimation procedure, which is a subjective method, is adopted to decide the relative importance of weights of attributes and criteria based on the opinion of experts. In this method, an arbitrary highest weight is assigned to the most important criterion. Correspondingly, smaller weights are assigned to the remaining 8 criteria with lower order until a score is assigned to the least important criterion. The ratio is calculated by dividing each weight to the lowest weight. The ratio, therefore, is equal to w_j/w^*, where w^* is the lowest score assigned to the least important criterion and w_j is the score for the jth criterion. Finally, the weights are normalized by dividing each one by the total [65]. In the next step, a decision matrix $X = (x_{ij})$ is constructed using GIS and the ratio estimation procedure, where $x_{ij} \in \Re$.

- Step 2: Calculation of the normalized decision matrix

Generally, there are benefit attributes and cost attributes in an MCDM problem. To transform various attribute dimensions into non-dimensional units and facilitate inter-attribute comparisons, several known standardized equations are introduced to normalize each attribute value x_{ij} in decision matrix $X = (x_{ij})_{m \times n}$. The following equation is the most frequently used method of calculating the normalized value r_{ij} [66]:

$$R = (r_{ij})_{m \times n} = \begin{matrix} A_1 \\ A_2 \\ \vdots \\ A_m \end{matrix} \begin{pmatrix} u_1 & u_2 & \cdots & u_n \\ r_{11} & r_{12} & \cdots & r_{1n} \\ r_{21} & r_{22} & \cdots & r_{2n} \\ \vdots & \vdots & \vdots & \vdots \\ r_{m1} & r_{m2} & \cdots & r_{mn} \end{pmatrix} \tag{2}$$

where

$$r_{ij} = x_{ij} / \sqrt{\sum_{i=1}^{m} (x_{ij})^2} \tag{3}$$

For benefit attribute $x_{ij}, i \in M, j \in N$

$$r_{ij} = 1 - (x_{ij} / \sqrt{\sum_{i=1}^{m} (x_{ij})^2}) \tag{4}$$

For cost attribute $x_{ij}, i \in M, j \in N$.

All criteria, which have been introduced in the GIS layers, including Soluble Sodium Percentage (SSP), Total Dissolved Solid (TDS), Potential of Hydrogen (PH), Electrical Conductivity (EC) of Water, Geology, Land Use, Sediment, Erosion, Slope, Groundwater and Discharge, are transformed to non-dimensional attributes using Equations (3) and (4).

- Step 3: Calculation of the weighted normalized decision matrix

In the third step the weighted normalized value v_{ij} is calculated by multiplying the normalized decision matrix by the normalized weights of criteria:

$$v_{ij} = w_j \times r_{ij} \tag{5}$$

where $i = 1, \ldots, m; j = 1, \ldots, n, m$ is the number of attribute value in each criterion, n is the number of criteria, and w_j is the normalized weight of the jth criterion, $w_j = \frac{W_j}{\sum_{j=1}^{n} W_j}$ so that $\sum_{j=1}^{n} w_j = 1$, W_j is the original weight assigned to each criterion.

- Step 4: Determination of the positive ideal solutions and negative ideal solutions

The positive ideal solution minimizes the cost criteria and maximizes the benefit criteria; on the contrary, the negative ideal solution maximizes the cost criteria and minimizes the benefit criteria. The equations are as follows:

$$A^+ = [v_1^+, \ldots, v_j^+, \ldots, v_n^+] \tag{6}$$

$$A^- = [v_1^-, \ldots, v_j^-, \ldots, v_n^-] \tag{7}$$

where A^+ denotes the positive ideal solution and A^- denotes the negative ideal solution and

$$\begin{cases} v_j^+ = \max_i \{v_{ij}\} i = 1, 2, \ldots, m \\ v_j^- = \min_i \{v_{ij}\} i = 1, 2, \ldots, m \end{cases} \text{if the } j\text{th criterion is a beneficial criterion}$$

$$\begin{cases} v_j^+ = \min_i \{v_{ij}\} i = 1, 2, \ldots, m \\ v_j^- = \max_i \{v_{ij}\} i = 1, 2, \ldots, m \end{cases} \text{if the } j\text{th criterion is a cost criterion}$$

where v_{ij} denotes the attribute values of each cell for the jth layer.

- Step 5: Calculation of the separation of each alternative from the positive ideal solution and the negative ideal solution

In this step, the separation of each alternative from the positive ideal solution and the negative ideal solution is calculated and then two different GIS layers S_i^+ and S_i^- are created. The equations are as follows.

The separation from the positive ideal solution for each alternative is given as:

$$S_i^+ = \sum_{j=1}^{n} \left| v_{ij} - v_j^+ \right| = \sum_{j=1}^{n} D_{ij}^+ \tag{8}$$

The separation from the negative ideal solution for each alternative is given as:

$$S_i^- = \sum_{j=1}^{n} \left| v_{ij} - v_j^- \right| = \sum_{j=1}^{n} D_{ij}^- \tag{9}$$

- Step 6: Calculation of the relative closeness to the positive ideal solution

The relative closeness of the i-th alternative A_j with respect to the positive ideal solution can be calculated as:

$$C_i^+ = (S_i^-)/(S_i^+ + S_i^-) \tag{10}$$

where $0 \leq C_i^+ \leq 1, i = 1, 2, \ldots, m$.

- Step 7: Determination of the rank of the alternatives according to the relative closeness

In last step, the values of "the relative closeness to the positive ideal solution" layer created in step 6 are determined for sites selected based on topographical conditions. A set of sites can now be ranked by the descending order of the value of C_i^+. The best sites are those that have higher values of C_i^+ and since they are closer to the positive ideal solution, they are preferable and must be chosen [67].

2.6. Analytic Hierarchy Process

The Analytic Hierarchy Process (AHP), developed by Thomas Saaty [10], is an effective method for dealing with complex decision-making. This method helps decision-makers set priorities between alternatives, sub-criteria and criteria in the decision-making process and also helps them to make the best decision [48]. The following are the main steps of the AHP methodology [10]:

- Step 1: Determination of the objective, main-criteria, sub-criteria, alternatives and structure of the hierarchy

This is the main part of the decision-making process, since structuring the decision problem as a hierarchy is fundamental to the process of the AHP. The aim of this study is the rank-ordering of selected sites, based on the mentioned criteria. Figure 2 shows the hierarchy structure for this process.

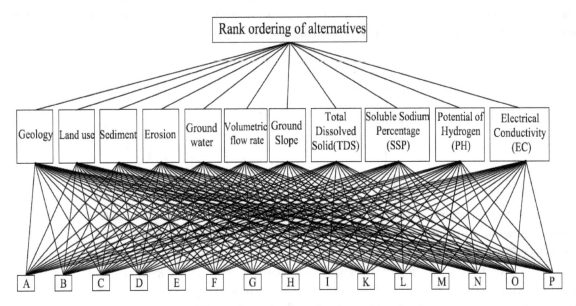

Figure 2. Hierarchy structure for dam site selection.

- Step 2: The pairwise comparison of the criteria with respect to the goal

To calculate the weights for various criteria, the AHP method creates a pairwise comparison matrix A (an $m \times m$ real matrix) as follows:

$$A = \begin{pmatrix} a_{11} & a_{12} & \cdots & a_{1k} & \cdots & a_{1m} \\ a_{21} & a_{22} & \cdots & a_{2k} & \cdots & a_{2m} \\ \vdots & \vdots & \vdots & \vdots & \vdots & \vdots \\ a_{j1} & a_{j2} & \cdots & a_{jk} & \cdots & a_{jm} \\ \vdots & \vdots & \vdots & \vdots & \vdots & \vdots \\ a_{m1} & a_{m2} & \cdots & a_{mk} & \cdots & a_{mm} \end{pmatrix} \tag{11}$$

where a_{jk} represents the importance of the j-th criterion relative to the k-th criterion. The j-th criterion is less important than the k-th criterion provided that $a_{jk} < 1$. In contrast, the j-th criterion is more important than the k-th criterion if $a_{jk} > 1$. As long as two criteria have the same importance, the entry a_{jk} is 1. The pairwise comparison is based on the advice of experts based on questions concerning which criterion is more important with regard to the decision goal, and the relative importance between two criteria is measured according to a numerical scale from 1 to 9 as generated by Saaty (1980) [10]. Once the matrix A is built, the normalization process is implemented simply by (1) calculating the geometric mean of the i-th row, and (ii) normalizing the geometric means of rows in the comparison matrix. This can be represented as:

$$\text{GM}_i = \left[\prod_{j=1}^{n} a_{ij}\right]^{1/n} \begin{array}{l} i = 1, \ldots, m \\ j = 1, \ldots, n \end{array} \tag{12}$$

$$w_i = (GM_i) / (\sum_{i=1}^{m} GM_i) i = 1, \ldots, m \tag{13}$$

Finally, the criteria weight vector w is built.

- Step 3: The pairwise comparison of the alternatives with respect to the criteria

The pairwise comparisons of the alternatives provide matrix S (an $n \times m$ real matrix), where m is the number of criteria and n is the number of alternatives or sites.

$$S = \begin{pmatrix} s_{11} & s_{12} & \cdots & s_{1j} & \cdots & s_{1m} \\ s_{21} & s_{22} & \cdots & s_{2j} & \cdots & s_{2m} \\ \vdots & \vdots & \vdots & \vdots & \vdots & \vdots \\ s_{i1} & s_{i2} & \cdots & s_{ij} & \cdots & s_{im} \\ \vdots & \vdots & \vdots & \vdots & \vdots & \vdots \\ s_{n1} & s_{n2} & \cdots & s_{nj} & \cdots & s_{nm} \end{pmatrix} \tag{14}$$

where s_{ij} represents the value of the i-th alternative with respect to the j-th criterion, n is the number of alternatives, m is the number of criteria. To construct matrix S, first, a pairwise comparison matrix is built for each of the m criteria, called $B^{(j)}$, which is an $n \times n$ real matrix. Each entry $b_{ik}^{(j)}$ of the matrix $B^{(j)}$ represents the evaluation of i-th alternative in comparison to the k-th alternative with respect to the j-th criterion. Once the matrix $B^{(j)}$ is built, the same two-step procedure described for the pairwise comparison matrix A are conducted for the normalization process. Finally, the pairwise comparisons of the alternative matrix S is built.

- Step 4: Calculation of priority vectors

A vector v of global values is created by multiplying S and w, where S is the pairwise comparisons of the alternatives matrix and w is the criteria weight vector.

$$v = S \times w \tag{15}$$

- Step 5: Calculation of the consistency ratio (CR)

In the AHP approach, the pairwise comparisons in a judgment matrix are considered to be adequately consistent if the corresponding CR is less than 10% [10]. To calculate the CR coefficient, a Consistency Index (CI) is defined which is computed as follows:

$$(CI) = (\lambda_{\max} - n) / (n - 1) \tag{16}$$

where λ_{max} is the maximum eigenvalue of the pairwise comparison matrix.

According to Saaty's theorem, for a given positive matrix A, the only positive vector x and only positive constant c that satisfy Ax = cx, is a vector x that is a positive multiple of the Perron vector (principal eigenvector) of A, and the only such c is the Perron value [68] (principal eigenvalue) of A [69]. Saaty calculates λ_{max} by adding the columns of matrix A and then multiplying the resulting vector with the vector x. The CR coefficient is calculated by dividing the CI value by the Random Consistency Index (RCI) as given in (Table 2).

$$CR = CI/RCI \Rightarrow \begin{array}{l} \text{If CR} < 10\%, \text{ achieved data is consistent} \\ \text{If CR} \geq 10\%, \text{ achieved data is inconsistent} \end{array} \qquad (17)$$

Table 2. RCI values for different values of n.

n	1	2	3	4	5	6	7	8
RCI	0.00	0.00	0.52	0.89	1.11	1.25	1.35	1.40
n	9	10	11	12	13	14	15	
RCI	1.45	1.49	1.52	1.54	1.56	1.58	1.59	

- Step 6: Analysis of the AHP scores

If the model is consistent, the best alternative will be chosen based on the matrix calculated in Step 4.

3. Results

3.1. TOPSIS Outputs

To implement TOPSIS in our study, the first task consists of checking the routes of streams using a Digital Elevation Model (DEM) layer and finding locations that are suitable for constructing a dam based on the valley shape. At this step, 16 points (A to P) that meet the aforementioned conditions are chosen. Then, the volume of water that can be stored in the basin is calculated for each selected location based on the crest length. Also, the minimum storage volume and maximum crest length are assumed to be $10 \times 10^6 \text{ m}^3$ and 1.5 km, respectively. Table 3 shows the crest length and storage volume based on the maximum possible height of the dam. Among the introduced points, J does not satisfy the above condition in terms of storage volume. Therefore, we do not consider this point as a potential site for constructing a dam. Figure 3 displays the location of the remaining 15 points in the study area.

Then, the Mean Absolute Deviation (MAD) for the three interpolation methods is calculated. The results for the sediment dataset are shown in Table 4, and it can be found that the exponential kriging provides the lowest MAD and hence can be considered the best interpolation method for the dataset. The applied methods for each criterion are shown in Table 5. All datasets show the lowest MAD in the kriging method, which shows its superiority over the IDW and Spline methods. The weight for each criterion is determined using the TOPSIS algorithm based on expert opinions. Table 6 shows the weights for all the available criteria. As can be seen, volume flow rate and groundwater level are considered the most and least important criteria, respectively.

According to the equations applied in step 4 in TOPSIS algorithm, the positive ideal solution (A^+) and negative ideal solution (A^-) are calculated for all criteria, and then a layer is created for each v_j^+ and v_j^-. The positive ideal solution and negative ideal solution for each criterion are presented in Table 7. The separation of each alternative from the positive ideal solution layer and the separation of each alternative from the negative ideal solution layer are calculated based on Equations (8) and (9), respectively. Figures 4 and 5 show these two separation layers.

Figure 3. The location of potential sites (points A to P) for dam construction in the study area.

Table 3. The selected locations based on topographical and morphological conditions.

Location Name	Reservoir Area (km²)	Storage Volume (m³ × 10⁶)	Maximum Height of Dam (m)	Crest Length (m)
A [1]	4.88	285.53	50	365.0
B	3.00	214.57	90	781.3
C	3.39	24.16	90	766.1
D	12.09	150.93	100	1462.4
E	1.24	37.19	40	1335.9
F	1.42	47.06	60	791.2
G	10.50	919.93	190	1302.4
H [2]	3.39	186.28	60	328.8
I	1.87	49.59	30	1479.8
J	0.36	5.24	30	212.3
K	5.24	266.63	150	1118.3
L [3]	3.94	225.96	80	876.7
M [4]	0.91	10.91	20	140.8
N [5]	7.50	17.21	60	396.8
O	2.53	147.58	90	1488.3
P	35.12	186.03	70	1341.9

[1] Kahir Dam. [2] Zirdan Dam. [3] Land Dam. [4] Shi Kalak Dam. [5] Pishin Dam.

Table 4. Mean Absolute Deviation (MAD) for three spatial interpolation techniques for sediment dataset.

	Methods	MAD
IDW		0.53
Spline	Regularized	0.51
	Tension	0.49
Kriging	Circular	0.37
	Spherical	0.38
	Exponential	0.30
	Gaussian	0.41

Table 5. Different interpolation methods applied for each dataset.

Applied Method	Criterion
Exponential Kriging	Discharge
Circular Kriging	Groundwater
Exponential Kriging	Sediment
Exponential Kriging	PH
Exponential Kriging	SSP
Exponential Kriging	TDS
Exponential Kriging	EC

Table 6. Weight of criteria based on expert opinion.

Criterion	Straight Rank	Ratio Scale	Original Weight	Normalized Weight
Discharge	1	9	9	0.155
Groundwater	6	1	1	0.017
Slope	2	8	8	0.138
Sediment	4	5	5	0.086
Land use	5	3	3	0.052
TDS	4	5	5	0.086
SSP	4	5	5	0.086
PH	4	5	5	0.086
EC	4	5	5	0.086
Geology	3	7	7	0.121
Erosion	4	5	5	0.086
Sum			58	

Table 7. The positive ideal solution and negative ideal solution for each criterion.

Criterion	Positive Ideal Solution	Negative Ideal Solution
Discharge	1.0	0.1
Groundwater	1.0	0.1
Slope	1.0	0.1
Sediment	1.0	0.1
Land use	1.0	0.1
TDS	1.0	0.1
SSP	0.5	0.1
PH	1.0	0.7
EC	0.7	0.1
Geology	1.0	0.1
Erosion	1.0	0.1

Figure 4. Separation of each alternative from the positive ideal solution.

Figure 5. Separation of each alternative from the negative ideal solution.

According to Equation (10), a new layer which indicates the relative closeness to the positive ideal solution is created, as shown in Figure 6. The value of each pixel in this figure indicates how suitable the location is for constructing a dam site. The maximum value shows the best location for the construction, while the minimum value is an indication of the worst location to build a dam. The left-hand side area, which is shown as a brighter color, is a preferable location for dam construction compared to the northern part of the study area.

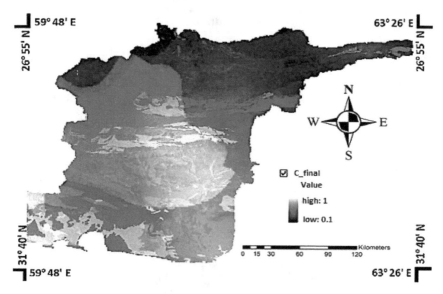

Figure 6. Relative closeness to positive ideal solution.

As mentioned earlier, constructing a dam requires a narrow valley. Therefore, only the 15 points selected are considered as potential dam site locations, and they are ranked based on the C_i^+ value. Table 8 shows the C_i^+ value and ranking of these points.

According to the C_i values calculated by the TOPSIS method, the ranking of five existing dams from the most preferable to the least preferable was Kahir, Shi Kalak, Land, Zirdan and Pishin. If the single best location were to be selected, then Kahir Dam must be chosen, since it has the highest C_i value. As shown in Table 8, four actual dams—Kahir, Shi Kalak, Land and Zirdan—are among the top five locations in the study area. The B location is fourth in this ranking; however, it is not the best site for construction of a dam on the Kahir river, since compared to B, the A location, which represents Kahir Dam, has a higher value of C_i. Analysis results by the TOPSIS method indicate that the Pishin Dam, represented by N in this study, is 11th among the 15 selected alternatives; hence, it can be concluded that this dam does not satisfy the criteria mentioned in this study. However, there are no other options fulfilling the topographical and morphological conditions on that river that can be selected.

Table 8. Results of the rank for 15 selected sites based on the TOPSIS technique.

Site	Value of C_i^+	Ranking	Site	Value of C_i^+	Ranking
A	0.8557	1	F	0.6479	9
M	0.8322	2	I	0.6390	10
L	0.8087	3	N	0.6238	11
B	0.7862	4	E	0.6048	12
H	0.7467	5	D	0.5000	13
C	0.7461	6	O	0.4982	14
K	0.6752	7	P	0.4737	15
G	0.6540	8			

3.2. AHP Outputs

The pairwise comparison matrix of criteria (Matrix A) is created using the AHP method to calculate the weights for each criterion. As can be seen in Table 9, discharge, slope and geology are the most important criteria, as they have the highest weights, and groundwater is the least important one, among others. The pairwise comparisons of the alternatives (Matrix S) are built based on selected sites and criteria (see step 3 in AHP algorithm for more details). Table 10 shows the S matrix. By multiplying the alternative matrix (Table 10) and weight vector (Table 9), the priority vector v, which assigns values to each site based on AHP method, is created, as shown in Table 11.

Using Equation (16) and Table 12, CI and RCI values for each matrix are obtained, respectively. Finally, using Equation (17), CR values are calculated, as shown in Table 13. CR values are less than 10%, an indication of the consistency of all pairwise comparison matrices (see step 5 in AHP algorithm for more details).

Ranking of the dams according to the calculations based on the AHP technique are as follows: Land Dam, Shi Kalak Dam, Kahir Dam, Zirdan Dam and Pishin Dam. The best actual dam based on the AHP method is the Land Dam, as presented in Table 13. Since there are two other alternatives, *B* and *C*, on the Kahir river that have a higher score than Kahir Dam, *A*, it can be inferred that the results achieved from the AHP technique, unlike TOPSIS, are not sufficiently accurate, but nevertheless, results from the AHP method show that the Pishin Dam is 11th among the 15 selected alternatives.

Table 9. The criteria weight vector w.

Criteria	Q [1]	GW [2]	Slope	Sediment	Land Use	TDS	SSP	PH	EC	Geo [3]	Erosion	w
Q	1.00	9	1.13	1.8	3.00	1.8	1.8	1.8	1.8	1.29	1.8	0.155
GW	0.11	1	0.13	0.2	0.33	0.2	0.2	0.2	0.2	0.14	0.2	0.017
Slope	0.89	8	1.00	1.6	2.67	1.6	1.6	1.6	1.6	1.14	1.6	0.138
Sediment	0.56	5	0.63	1.0	1.67	1.0	1.0	1.0	1.0	0.71	1.0	0.086
Land use	0.33	3	0.38	0.6	1.00	0.6	0.6	0.6	0.6	0.43	0.6	0.052
TDS	0.56	5	0.63	1.0	1.67	1.0	1.0	1.0	1.0	0.71	1.0	0.086
SSP	0.56	5	0.63	1.0	1.67	1.0	1.0	1.0	1.0	0.71	1.0	0.086
PH	0.56	5	0.63	1.0	1.67	1.0	1.0	1.0	1.0	0.71	1.0	0.086
EC	0.56	5	0.63	1.0	1.67	1.0	1.0	1.0	1.0	0.71	1.0	0.086
Geo	0.78	7	0.88	1.4	2.33	1.4	1.4	1.4	1.4	1.00	1.4	0.121
Erosion	0.56	5	0.63	1.0	1.67	1.0	1.0	1.0	1.0	0.71	1.0	0.086

[1] Discharge. [2] Groundwater. [3] Geology.

Table 10. The pairwise comparisons of the alternative matrix S.

Alternatives	Discharge	Ground Water	Slope	Sediment	Land Use	TDS	SSP	PH	EC	Geology	Erosion
A	0.106	0.056	0.064	0.061	0.029	0.093	0.053	0.067	0.078	0.034	0.085
B	0.105	0.051	0.066	0.062	0.029	0.009	0.053	0.067	0.078	0.207	0.074
C	0.100	0.050	0.067	0.064	0.057	0.047	0.053	0.067	0.044	0.207	0.074
D	0.032	0.025	0.066	0.072	0.086	0.047	0.053	0.067	0.044	0.034	0.083
E	0.053	0.070	0.067	0.070	0.057	0.009	0.053	0.067	0.078	0.034	0.070
F	0.047	0.068	0.067	0.070	0.086	0.093	0.053	0.067	0.078	0.034	0.068
G	0.055	0.054	0.067	0.069	0.086	0.093	0.053	0.067	0.078	0.034	0.068
H	0.077	0.057	0.067	0.068	0.057	0.093	0.053	0.067	0.078	0.034	0.042
I	0.016	0.120	0.067	0.075	0.086	0.093	0.263	0.067	0.078	0.034	0.090
K	0.046	0.088	0.067	0.076	0.086	0.093	0.053	0.067	0.078	0.034	0.091
L	0.090	0.104	0.067	0.066	0.057	0.093	0.053	0.067	0.078	0.103	0.083
M	0.095	0.064	0.067	0.064	0.086	0.093	0.053	0.067	0.078	0.103	0.053
N	0.112	0.061	0.067	0.037	0.029	0.047	0.053	0.067	0.044	0.034	0.091
O	0.035	0.067	0.066	0.072	0.086	0.047	0.053	0.067	0.044	0.034	0.019
P	0.031	0.064	0.066	0.072	0.086	0.047	0.053	0.067	0.044	0.034	0.010

Table 11. Values of priority vector.

Alternative	Value
A	0.06954
B	0.08232
C	0.08352
D	0.05462
E	0.05568
F	0.06320
G	0.06418
H	0.06383
I	0.07982
K	0.06595
L	0.07843
M	0.07728
N	0.06253
O	0.05018
P	0.04893

Table 12. The consistency ratio (CR).

MATRIX	λ_{max}	CI	RCI	CR
A [1]	11.00850	0.00085	1.52	0.00056
S_1 [2]	15.00100	0.00007	1.59	0.00004
S_2	15.00030	0.00002	1.59	0.00001
S_3	15.00020	0.00001	1.59	0.00001
S_4	15.00010	0.00001	1.59	0.00000
S_5	14.99950	−0.00004	1.59	−0.00002
S_6	15.00000	0.00000	1.59	0.00000
S_7	15.00000	0.00000	1.59	0.00000
S_8	14.93300	−0.00479	1.59	−0.00301
S_9	14.99750	−0.00018	1.59	−0.00011
S_{10}	15.00150	0.00011	1.59	0.00007
S_{11}	14.99920	−0.00006	1.59	−0.00004

[1] Pairwise comparison of the criteria with respect to the goal; [2] Pairwise comparison matrix of the alternatives with respect to the first criterion

Table 13. Results of the ranking for 15 selected sites according to the AHP technique.

Alternative	AHP Score	Rank	Alternative	AHP Score	Rank
C	0.0835	1	H	0.0638	9
B	0.0823	2	F	0.0632	10
I	0.0798	3	N	0.0625	11
L	0.0784	4	E	0.0557	12
M	0.0773	5	D	0.0546	13
A	0.0695	6	O	0.0502	14
K	0.0659	7	P	0.0489	15
G	0.0642	8			

4. Discussion

In the academic literature, there is limited discussion of the dam site selection problem, despite the fact that it represents a very important strategic decision. Generally, dam site selection is performed by traditional methods of decision-making or on the basis of political interests. Cost and resource availability are two important factors in the traditional decision-making approach; accordingly, a location near to a mine or water source may be selected without making a holistic and systematic analysis. In this context, within the scope of the study, a comparative study of the TOPSIS and AHP models using GIS is proposed to determine the most appropriate locations for constructing dams in Sistan and Baluchestan Province, Iran.

The approach considers topographical conditions in addition the factors Soluble Sodium Percentage (SSP), Total Dissolved Solid (TDS), Potential of Hydrogen (PH), Electrical Conductivity (EC) of water, Geology, Land use, Sediment, Erosion, Groundwater and Discharge. Accordingly, the proposed model for the dam site selection problem consists of four basic steps: (1) selecting possible locations based on topographical and morphological conditions; (2) application of the TOPSIS technique using GIS and prioritizing the sites selected in the previous step; (3) application of the AHP method and ranking the sites selected in the first step; and (4) comparing the results of the TOPSIS and AHP models.

It should be noted that dam site selection requires detailed studies and tests; so, to evaluate and compare the outputs, we assumed that the location of existing dams are the best ones. It was used to compare TOPSIS and AHP, as an initial calculation, for the future dam site selections, particularly in the study area. In other words, actual locations of constructed dams in the area were used to verify the results of both methods.

Table 14 compares the outputs of TOPSIS and AHP and shows the same results for the following choices: D, E, G, K, N, O, and P (7 from 15 locations or around 47% similarity). According to the results of the TOPSIS method, four actual dams—Kahir, Shi Kalak, Land and Zirdan, respectively—are the best sites for building a dam in the region. There are three selected sites, A, B and C, on the Kahir river, with the Kahir Dam (A) being the best location among them. The most suitable location among the five sites selected based on the introduced criteria, E, F, G, H and D, is the Zirdan Dam, which is shown by the H character. However, there are no options for the Land Dam, Shi Kalak Dam and Pishin Dam on their rivers. According to the AHP method, the best location in the region for constructing a dam is C, which is located on the Kahir River. In this ranking, four actual dams are not among the top three sites, whereas in the ranking based on the TOPSIS technique, four dams constructed in the area are the best selection. Therefore, it can be concluded that the TOPSIS method may be a more powerful technique compared to the AHP method in the selection of dam sites.

Table 14. Comparison between TOPSIS and AHP outputs.

Alternative	TOPSIS Rank	AHP Rank	Difference in Rank
A	1	6	5
B	4	2	2
C	6	1	5
D	13	13	0
E	12	12	0
F	9	10	1
G	8	8	0
H	5	9	4
I	10	3	7
K	7	7	0
L	3	4	1
M	2	5	3
N	11	11	0
O	14	14	0
P	15	15	0

The difference between the results of TOPSIS and AHP in site selection depends on their strengths and weaknesses, which are thoroughly described in the literature [70–77]. For example, the main advantages of AHP over TOPSIS are its flexibility, its intuitive appeal to decision-makers, and its ability to check inconsistencies. Also, users find the pair wise comparison form of data input convenient and straightforward. However, applying AHP leads to the decision problem being decomposed into several subsystems, and so a substantial number of pair wise comparisons need to be filled. In the case of TOPSIS, the non-linear relations between one-dimensional scores and distance ratios results in the consideration of both negative and positive ideal solutions. In addition, in the TOPSIS framework, we can use variables with different units of measurement. However, TOPSIS in its standard and original form is deterministic and does not include uncertainty in the calculations related to final weightings.

5. Conclusions

This study presents the application of TOPSIS and AHP techniques, integrating GIS, in dam site selection in Sistan and Baluchestan Province, Iran. Based on the review of recent studies on dam site selection using MCDM, it can be found that previous researchers have mostly used AHP and TOPSIS in the application of dam site selection. Therefore, we implemented both methods in a study area in Iran to evaluate their capability for dam site selection in a new local context and to compare the results of both methods. Based on previous experience and the literature, we introduced several factors, including geology, land use, sediment erosion, slope, groundwater, discharge, soluble sodium percentage, total dissolved solid, potential of hydrogen, and electrical conductivity of water. After implementing the techniques in the GIS environment, we found relatively suitable sites for dam construction in the area of interest. Finally, as dam site selection requires detailed studies and tests, to evaluate and compare the outputs, we assumed that the locations of existing dams were the best ones. These were used to compare TOPSIS and AHP, as an initial calculation, for the future dam site selections, particularly in the study area. In other words, actual locations of constructed dams in the area were used to verify the results of both methods.

The main conclusions of the present study can be drawn as follows:

- The results show that the TOPSIS method is better suited to the problem of dam site selection for this study area.
- Using MCDM could provide an overall view for initial calculations to reduce the expenses and to reach a comprehensive study of dam site selection; however, it is necessary to collect accurate in situ data to provide a comprehensive framework. Dam site selection requires detailed studies and tests; therefore, to evaluate and compare the outputs, we assumed that the location of existing

dams were the best ones. These were used to compare TOPSIS and AHP, as an initial calculation, for the future dam site selections, particularly in the study area.

- In the AHP ranking, four actual dams were not among the top three sites, whereas in the ranking based on the TOPSIS technique, four dams constructed in the area were the best selections. Thus, the TOPSIS method may be a more appropriate technique than the AHP method for locating dams based on the criteria mentioned in this study.

Recommendations for future research:

- Comparing the sensitivity of TOPSIS and AHP to different types of dams (e.g., embankment dams, concrete dams).
- Comparing the results of implementing fuzzy TOPSIS and fuzzy AHP with TOPSIS and AHP.
- Collecting more accurate reports of dam site selection to improve and update the criteria and sub-criteria.
- Comparing the sensitivity of TOPSIS and AHP to such factors as the complexity in geological characteristics as one of the main keys in dam site selection study.
- Considering more comprehensive criteria may lead to more accurate results.

Author Contributions: The authors have contributed to this paper at different levels. As the lead author, A.J. conceived and designed the methodology. A.J. along with B.A., M.H. and M.K. contributed equally to manuscript writing and revision. I.F. reviewed the paper and provided helpful comments on improving its quality in all aspects. A.J., B.A., M.H. and E.G.T. contributed equally to analyzing the results of this paper. In addition, finally, N.K., B.A., M.H. and A.J. contributed to the literature review for this paper.

References

1. Noori, A.; Bonakdari, H.; Morovati, K.; Gharabaghi, B. The optimal dam site selection using a group decision-making method through fuzzy TOPSIS model. *Environ. Syst. Decis.* **2018**, *38*, 471–488. [CrossRef]
2. Beck, M.W.; Claassen, A.H.; Hundt, P.J. Environmental and livelihood impacts of dams: Common lessons across development gradients that challenge sustainability. *Int. J. River Basin Manag.* **2012**, *10*, 73–92. [CrossRef]
3. Fearnside, P.M. Environmental and social impacts of hydroelectric dams in Brazilian Amazonia: Implications for the aluminum industry. *World Dev.* **2016**, *77*, 48–65. [CrossRef]
4. Shafieardekani, M.; Hatami, M. Forecasting Land Use Change in suburb by using Time series and Spatial Approach; Evidence from Intermediate Cities of Iran. *Eur. J. Sci. Res.* **2013**, *116*, 199–208.
5. Mulatu, C.; Crosato, A.; Moges, M.; Langendoen, E.; McClain, M. Morphodynamic Trends of the Ribb River, Ethiopia, Prior to Dam Construction. *Geosciences* **2018**, *8*, 255. [CrossRef]
6. Shad, R.; Khorrami, M.; Ghaemi, M. Developing an Iranian green building assessment tool using decision making methods and geographical information system: Case study in Mashhad city. *Renew. Sustain. Energy Rev.* **2017**, *67*, 324–340. [CrossRef]
7. Goodchild, M.F.; Parks, B.O.; Steyaert, L.T. (Eds.) *The State of GIS for Environmental Problem-Solving*; Environmental modeling with GIS; Oxford University Press: New York, NY, USA, 1993; pp. 8–15.
8. Goodchild, M.F. Reimagining the history of GIS. *Ann. GIS* **2018**, *24*, 1–8. [CrossRef]
9. Belton, V.; Stewart, T. *Multiple Criteria Decision Analysis: An Integrated Approach*; Springer Science & Business Media: Berlin, Germany, 2002.
10. Saaty, T. *The Analytic Process: Planning, Priority Setting, Resources Allocation*; Mcgraw: New York, NY, USA, 1980.
11. Hwang, C.-L.; Yoon, K. Methods for multiple attribute decision making. In *Multiple Attribute Decision Making*; Springer: Berlin, Germany, 1981; pp. 58–191. [CrossRef]
12. Tsolaki-Fiaka, S.; Bathrellos, G.D.; Skilodimou, H.D. Multi-Criteria Decision Analysis for an Abandoned Quarry in the Evros Region (NE Greece). *Land* **2018**, *7*, 43. [CrossRef]

13. Rahman, M.A.; Jaumann, L.; Lerche, N.; Renatus, F.; Buchs, A.K.; Gade, R.; Geldermann, J.; Sauter, M. Selection of the best inland waterway structure: A multicriteria decision analysis approach. *Water Resour. Manag.* **2015**, *29*, 2733–2749. [CrossRef]

14. Afshari, A.R.; Vatanparast, M.; Ćoćkalo, D. Application of multi criteria decision making to urban planning: A review. *J. Eng. Manag. Compet.* **2016**, *6*, 46–53. [CrossRef]

15. Mollalo, A.; Alimohammadi, A.; Shirzadi, M.; Malek, M. Geographic information system-based analysis of the spatial and spatio-temporal distribution of zoonotic cutaneous leishmaniasis in Golestan Province, north-east of Iran. *Zoonoses Public Health* **2015**, *62*, 18–28. [CrossRef] [PubMed]

16. Mollalo, A.; Sadeghian, A.; Israel, G.D.; Rashidi, P.; Sofizadeh, A.; Glass, G.E. Machine learning approaches in GIS-based ecological modeling of the sand fly Phlebotomus papatasi, a vector of zoonotic cutaneous leishmaniasis in Golestan province, Iran. *Acta Trop.* **2018**, *188*, 187–194. [CrossRef] [PubMed]

17. Theilen-Willige, B.; Aher, S.P.; Gawali, P.B.; Venkata, L.B. Seismic hazard analysis along Koyna Dam area, western Maharashtra, India: A contribution of remote sensing and GIS. *Geosciences* **2016**, *6*, 20. [CrossRef]

18. Nsanziyera, A.; Rhinane, H.; Oujaa, A.; Mubea, K. GIS and Remote-Sensing Application in Archaeological Site Mapping in the Awsard Area (Morocco). *Geosciences* **2018**, *8*, 207. [CrossRef]

19. Koohbanani, H.; Barati, R.; Yazdani, M.; Sakhdari, S.; Jomemanzari, R. Groundwater recharge by selection of suitable sites for underground dams using a GIS-based fuzzy approach in semi-arid regions. In *Progress in River Engineering & Hydraulic Structures*; International Energy and Environment Foundation: Najaf, Iraq, 2018; pp. 11–32.

20. Jamali, A.A.; Randhir, T.O.; Nosrati, J. Site Suitability Analysis for Subsurface Dams Using Boolean and Fuzzy Logic in Arid Watersheds. *J. Water Resour. Plan. Manag.* **2018**, *144*, 04018047. [CrossRef]

21. Dai, X. Dam Site Selection Using an Integrated Method of AHP and GIS for Decision Making Support in Bortala, Northwest China. Master's Thesis, Lund University, Lund, Sweden, 2016.

22. Tung, S.; Tang, S. A comparison of the Saaty's AHP and modified AHP for right and left eigenvector inconsistency. *Eur. J. Oper. Res.* **1998**, *106*, 123–128. [CrossRef]

23. Khorrami, M. Weighting Energy Efficiency Criteria for a Green Building, Using AHP method, Case study: Green Office Building in Iran. In Proceedings of the 1st National Conference on Green Building, Mashhad, Iran, 14 May 2014.

24. Abdullah, T.; Ali, S.; Al-Ansari, N.; Knutsson, S. Possibility of Groundwater Pollution in Halabja Saidsadiq Hydrogeological Basin, Iraq Using Modified DRASTIC Model Based on AHP and Tritium Isotopes. *Geosciences* **2018**, *8*, 236. [CrossRef]

25. Kaushal, V.; Guleria, S. Geotechnical Investigation of Black Cotton Soils. Available online: https://www.researchgate.net/profile/Vinayak_Kaushal/publication/279192640_Geotechnical_Investigation_of_Black_Cotton_Soils/links/558d959608ae47a3490bcedc.pdf (accessed on 31 October 2018).

26. Hassani, A.N.; Katibeh, H.; Farhadian, H. Numerical analysis of steady-state groundwater inflow into Tabriz line 2 metro tunnel, northwestern Iran, with special consideration of model dimensions. *Bull. Eng. Geol. Environ.* **2016**, *75*, 1617–1627. [CrossRef]

27. Hassani, A.N.; Farhadian, H.; Katibeh, H. A comparative study on evaluation of steady-state groundwater inflow into a circular shallow tunnel. *Tunn. Undergr. Space Technol.* **2018**, *73*, 15–25. [CrossRef]

28. Sadooghi, A.; Kibert, C.; Sadeghi, F.M.; Jafari, S. Thermal performance analysis of a traditional passive cooling system in Dezful, Iran. *Tunn. Undergr. Space Technol.* **2019**, *83*, 291–302. [CrossRef]

29. Piadeh, F.; Moghaddam, M.R.A.; Mardan, S. Present situation of wastewater treatment in the Iranian industrial estates: Recycle and reuse as a solution for achieving goals of eco-industrial parks. *Resour. Conserv. Recycl.* **2014**, *92*, 172–178. [CrossRef]

30. Piadeh, F.; Alavi-Moghaddam, M.R.; Mardan, S. Assessment of sustainability of a hybrid of advanced treatment technologies for recycling industrial wastewater in developing countries: Case study of Iranian industrial parks. *J. Clean. Prod.* **2018**, *170*, 1136–1150. [CrossRef]

31. Moud, H.I.; Shojaei, A.; Flood, I.; Zhang, X.; Hatami, M. Qualitative and Quantitative Risk Analysis of Unmanned Aerial Vehicle Flights over Construction Job Sites. In Proceedings of the Eighth International Conference on Advanced Communications and Computation (INFOCOMP 2018), Barcelona, Spain, 22–26 July 2018; pp. 22–26.

32. Hatami, M.; Shafieardekani, M. The Effect of Industrialization on Land Use Changes; Evidence from Intermediate Cities of Iran. *Int. J. Curr. Life Sci.* **2014**, *4*, 11899–11902.

33. Rincón, D.; Khan, U.; Armenakis, C. Flood Risk Mapping Using GIS and Multi-Criteria Analysis: A Greater Toronto Area Case Study. *Geosciences* **2018**, *8*, 275. [CrossRef]

34. Anane, M.; Bouziri, L.; Limam, A.; Jellali, S. Ranking suitable sites for irrigation with reclaimed water in the Nabeul-Hammamet region (Tunisia) using GIS and AHP-multicriteria decision analysis. *Resour. Conserv. Recycl.* **2012**, *65*, 36–46. [CrossRef]

35. Kim, Y.; Chung, E.-S.; Jun, S.-M.; Kim, S.U. Prioritizing the best sites for treated wastewater instream use in an urban watershed using fuzzy TOPSIS. *Resour. Conserv. Recycl.* **2013**, *73*, 23–32. [CrossRef]

36. Şengül, Ü.; Eren, M.; Shiraz, S.E.; Gezder, V.; Şengül, A.B. Fuzzy TOPSIS method for ranking renewable energy supply systems in Turkey. *Renew. Energy* **2015**, *75*, 617–625. [CrossRef]

37. Zyoud, S.H.; Kaufmann, L.G.; Shaheen, H.; Samhan, S.; Fuchs-Hanusch, D. A framework for water loss management in developing countries under fuzzy environment: Integration of Fuzzy AHP with Fuzzy TOPSIS. *Expert Syst. Appl.* **2016**, *61*, 86–105. [CrossRef]

38. Lahijanian, B.; Zarandi, M.F.; Farahani, F.V. Proposing a model for operating room scheduling based on fuzzy surgical duration. In Proceedings of the 2016 Annual Conference of the North American Fuzzy Information Processing Society (NAFIPS), El Paso, TX, USA, 31 October–4 November 2016; pp. 1–5.

39. Lahijanian, B.; Zarandi, M.F.; Farahani, F.V. Double coverage ambulance location modeling using fuzzy traveling time. In Proceedings of the 2016 Annual Conference of the North American Fuzzy Information Processing Society (NAFIPS), El Paso, TX, USA, 31 October–4 November 2016; pp. 1–6.

40. Adibfar, A.; Costin, A. Next Generation of Transportation Infrastructure Management: Fusion of Intelligent Transportation Systems (ITS) and Bridge Information Modeling (BrIM). In *Advances in Informatics and Computing in Civil and Construction Engineering*; Springer: Berlin/Heidelberg, Germany, 2019; pp. 43–50. [CrossRef]

41. Khorrami, M.; Alizadeh, B. Weighting parameters for selecting the appropriate urban public transportation system, using Analytical Hierarchy Process (AHP) method. In Proceedings of the 2nd International Congress On Structure, Architecture & Urban Development, Tabriz, Iran, 16–18 December 2014.

42. Riding, K.A.; Peterman, R.J.; Guthrie, S.; Brueseke, M.; Mosavi, H.; Daily, K.; Risovi-Hendrickson, W. Environmental and Track Factors That Contribute to Abrasion Damage. In Proceedings of the 2018 Joint Rail Conference, Pittsburgh, PA, USA, 18–20 April 2018; p. V001T001A009.

43. Costin, A.; Adibfar, A.; Hu, H.; Chen, S.S. Building Information Modeling (BIM) for transportation infrastructure—Literature review, applications, challenges, and recommendations. *Autom. Constr.* **2018**, *94*, 257–281. [CrossRef]

44. Choudhary, D.; Shankar, R. An STEEP-fuzzy AHP-TOPSIS framework for evaluation and selection of thermal power plant location: A case study from India. *Energy* **2012**, *42*, 510–521. [CrossRef]

45. Yasser, M.; Jahangir, K.; Mohmmad, A. Earth dam site selection using the analytic hierarchy process (AHP): A case study in the west of Iran. *Arab. J. Geosci.* **2013**, *6*, 3417–3426. [CrossRef]

46. Özcan, E.C.; Ünlüsoy, S.; Eren, T. A combined goal programming—AHP approach supported with TOPSIS for maintenance strategy selection in hydroelectric power plants. *Renew. Sustain. Energy Rev.* **2017**, *78*, 1410–1423. [CrossRef]

47. Esavi, V.; Karami, J.; Alimohammadi, A.; Niknezhad, S. Comparison the AHP and FUZZY-AHP Decision Making Methods in Underground Dam Site Selection in Taleghan Basin. *Geosciences* **2012**, *22*, 27–35. (In Persian)

48. Önüt, S.; Soner, S. Transshipment site selection using the AHP and TOPSIS approaches under fuzzy environment. *Waste Manag.* **2008**, *28*, 1552–1559. [CrossRef]

49. Mulliner, E.; Malys, N.; Maliene, V. Comparative analysis of MCDM methods for the assessment of sustainable housing affordability. *Omega* **2016**, *59*, 146–156. [CrossRef]

50. Balioti, V.; Tzimopoulos, C.; Evangelides, C. Multi-Criteria Decision Making Using TOPSIS Method Under Fuzzy Environment. Application in Spillway Selection. *Multidiscip. Digit. Publ. Inst. Proc.* **2018**, *2*, 11. [CrossRef]

51. Tzimopoulos, C.; Balioti, V.; Evangelides, C. Fuzzy multi-criteria decision making method for dam selection. In Proceedings of the 13th International Conference on Environmental Science and Technology, Athens, Greece, 5–7 September 2013.

52. Rezaei, P.; Rezaie, K.; Nazari-Shirkouhi, S.; Tajabadi, M.R.J. Application of fuzzy multi-criteria decision making analysis for evaluating and selecting the best location for construction of underground dam. *Acta Polytech. Hung.* **2013**, *10*, 187–205.

53. Jamali, I.A.; Mörtberg, U.; Olofsson, B.; Shafique, M. A spatial multi-criteria analysis approach for locating suitable sites for construction of subsurface dams in Northern Pakistan. *Water Resour. Manag.* **2014**, *28*, 5157–5174. [CrossRef]

54. Barkhordari, J. The pre-selection of suitable sites for small underground dams in arid areas using GIS (A case study in Yazd_Ardakan watershed). *Int. Geoinformat. Res. Dev. J.* **2015**, *6*, 18–27.

55. Minatour, Y.; Khazaie, J.; Ataei, M.; Javadi, A. An integrated decision support system for dam site selection. *Sci. Iran. Trans. A Civ. Eng.* **2015**, *22*, 319.

56. Choo, T.H.; Ahn, S.H.; Yang, D.U.; Yun, G.S. A Study on the Estimating Dam Suitable Site based on Geographic Information using AHP. In Proceedings of the 6th International Conference on Developments in Engineering and Technology (ICDET-2017), Bangkok, Thailand, 6–7 February 2017.

57. Njiru, F.M. Site Selection for an Earth Dam in Mbeere North, Embu County—Kenya. *J. Geosci. Environ. Prot.* **2018**, *6*, 113. [CrossRef]

58. Biranvandzadeh, M.; Sorkhkamal, K. Assessment of Development Level of Sistan and Baluchistan Province Compared to other Iran's Provinces. *Int. J. Archit. Urban Dev.* **2015**, *5*, 69–76.

59. Kottek, M.; Grieser, J.; Beck, C.; Rudolf, B.; Rubel, F. World map of the Köppen-Geiger climate classification updated. *Meteorol. Z.* **2006**, *15*, 259–263. [CrossRef]

60. Johnston, K.; Ver Hoef, J.M.; Krivoruchko, K.; Lucas, N. *Using ArcGIS Geostatistical Analyst*; Esri: Redlands, CA, USA, 2001; Volume 380.

61. Willmott, C.J.; Matsuura, K.; Robeson, S.M. Ambiguities inherent in sums-of-squares-based error statistics. *Atmos. Environ.* **2009**, *43*, 749–752. [CrossRef]

62. Roszkowska, E. Multi-criteria decision making models by applying the TOPSIS method to crisp and interval data. *Mult. Criteria Decis. Mak. Univ. Econ. Katow.* **2011**, *6*, 200–230.

63. Hyde, K.; Maier, H.; Colby, C. A distance-based uncertainty analysis approach to multi-criteria decision analysis for water resource decision making. *J. Environ. Manag.* **2005**, *77*, 278–290. [CrossRef]

64. Tzeng, G.-H.; Chen, T.-Y.; Wang, J.-C. A weight-assessing method with habitual domains. *Eur. J. Oper. Res.* **1998**, *110*, 342–367. [CrossRef]

65. Malczewski, J. *GIS and Multicriteria Decision Analysis*; John Wiley & Sons: Hoboken, NJ, USA, 1999.

66. Yue, Z. A method for group decision-making based on determining weights of decision makers using TOPSIS. *Appl. Math. Model.* **2011**, *35*, 1926–1936. [CrossRef]

67. Krohling, R.A.; Campanharo, V.C. Fuzzy TOPSIS for group decision making: A case study for accidents with oil spill in the sea. *Expert Syst. Appl.* **2011**, *38*, 4190–4197. [CrossRef]

68. Saaty, T.L.; Tran, L.T. On the invalidity of fuzzifying numerical judgments in the Analytic Hierarchy Process. *Math. Comput. Model.* **2007**, *46*, 962–975. [CrossRef]

69. Saaty, T.L. Decision-making with the AHP: Why is the principal eigenvector necessary. *Eur. J. Oper. Res.* **2003**, *145*, 85–91. [CrossRef]

70. Gavade, R.K. Multi-Criteria Decision Making: An overview of different selection problems and methods. *Int. J. Comput. Sci. Inf. Technol.* **2014**, *5*, 5643–5646.

71. Zardari, N.H.; Ahmed, K.; Shirazi, S.M.; Yusop, Z.B. Literature review. In *Weighting Methods and Their Effects on Multi-Criteria Decision Making Model Outcomes in Water Resources Management*; Springer: Berlin/Heidelberg, Germany, 2015; pp. 7–67. [CrossRef]

72. Tscheikner-Gratl, F.; Egger, P.; Rauch, W.; Kleidorfer, M. Comparison of multi-criteria decision support methods for integrated rehabilitation prioritization. *Water* **2017**, *9*, 68. [CrossRef]

73. Caterino, N.; Iervolino, I.; Manfredi, G.; Cosenza, E. A comparative analysis of decision making methods for the seismic retrofit of RC buildings. In Proceedings of the 14th world Conference on Earthquake Engineering, Beijing, China, 12–17 October 2008; pp. 12–17.

74. Ballı, S.; Korukoğlu, S. Operating system selection using fuzzy AHP and TOPSIS methods. *Math. Comput. Appl.* **2009**, *14*, 119–130. [CrossRef]

75. Jia, P.; Govindan, K.; Choi, T.-M.; Rajendran, S. Supplier selection problems in fashion business operations with sustainability considerations. *Sustainability* **2015**, *7*, 1603–1619. [CrossRef]

76. Dinmohammadi, A.; Shafiee, M. Determination of the most suitable technology transfer strategy for wind turbines using an integrated AHP-TOPSIS decision model. *Energies* **2017**, *10*, 642. [CrossRef]

77. Neisani Samani, Z.; Karimi, M.; Alesheikh, A.A. A Novel Approach to Site Selection: Collaborative Multi-Criteria Decision Making through Geo-Social Network (Case Study: Public Parking). *ISPRS Int. J. Geo-Inf.* **2018**, *7*, 82. [CrossRef]

Susceptibility Mapping on Urban Landslides using Deep Learning Approaches in Mt. Umyeon

Sunmin Lee [1,2]🆔, Won-Kyung Baek [1]🆔, Hyung-Sup Jung [1,*]🆔 and Saro Lee [3,4,*]🆔

[1] Department of Geoinformatics, University of Seoul, 163 Seoulsiripdaero, Dongdaemun-gu, Seoul 02504, Korea; smilee@kei.re.kr (S.L.); bekwkz@uos.ac.kr (W.-K.B.)

[2] Center for Environmental Assessment Monitoring, Environmental Assessment Group, Korea Environment Institute (KEI), Sejong-si 30147, Korea

[3] Geoscience Platform Division, Korea Institute of Geoscience and Mineral Resources (KIGAM), 124 Gwahak-ro, Yuseong-gu, Daejeon 34132, Korea

[4] Department of Geophysical Exploration, Korea University of Science and Technology, 217 Gajeong-ro, Yuseong-gu, Daejeon 34113, Korea

[*] Correspondence: hsjung@uos.ac.kr (H.-S.J.); leesaro@kigam.re.kr (S.L.)

Abstract: In recent years, the incidence of localized heavy rainfall has increased as abnormal weather events occur more frequently. In densely populated urban areas, this type of heavy rain can cause extreme landslide damage, so that it is necessary to estimate and analyze the susceptibility of future landslides. In this regard, deep learning (DL) methodologies have been used to identify areas prone to landslides recently. Therefore, in this study, DL methodologies, including a deep neural network (DNN), kernel-based DNN, and convolutional neural network (CNN) were used to identify areas where landslides could occur. As a detailed step for this purpose, landslide occurrence was first determined as landslide inventory through aerial photographs with comparative analysis using field survey data; a training set was built for model training through oversampling based on the landslide inventory. A total of 17 landslide influencing variables that influence the frequency of landslides by topography and geomorphology, as well as soil and forest variables, were selected to establish a landslide inventory. Then models were built using DNN, kernel-based DNN, and CNN models, and the susceptibility of landslides in the study area was determined. Model performance was evaluated through the average precision (AP) score and root mean square error (RMSE) for each of the three models. Finally, DNN, kernel-based DNN, and CNN models showed performances of 99.45%, 99.44%, and 99.41%, and RMSE values of 0.1694, 0.1806, and 0.1747, respectively. As a result, all three models showed similar performance, indicating excellent predictive ability of the models developed in this study. The information of landslides occurring in urban areas, which cause a great damage even with a small number of occurrences, can provide a basis for reference to the government and local authorities for urban landslide management.

Keywords: susceptibility mapping; urban landslides; debris flow; deep learning

---

1. Introduction

Natural disasters are becoming more frequent around the world due to extreme climates, and an increase in the probability of extreme rain is also associated with the occurrence of damage. In Korea, the current rainfall and rainfall intensity have gradually increased compared to the past, and the rainfall is concentrated in a small area over a short time. A lot of water-related disaster damage has occurred in this localized heavy rain, but the damage is also decreasing due to continuous river maintenance and the installation of facilities to prevent flooding. On the other hand, the number of casualties,

134 people, caused by landslides and slope failures has not decreased since the 2000s [1]. Recently, the incidence of localized rainfall increased 2.4 times, from an average of 5.1 in the 1970s to 12.3 in the 2000s [2]. Due to the increase in localized heavy rain, landslides and geomorphological hazards, such as debris flows, are rapidly increasing [3].

Landslides in Korea occur intensively in June to September, when heavy rains and typhoons occur, costing many lives and significant property damage. The area of landslide damage since the 2000s has been more than three times that of the 1980s [4]. In particular, serious damage to human lives has occurred in the years of torrential rains and typhoons, which is a serious social issue [5]. This kind of phenomenon is occurring not only in Korea, but also in other parts of the world, raising concerns about the occurrence of landslides. Damage increases due to a combination of natural and artificial causes; the increase in localized heavy rain is an example of a natural cause, and an indiscriminate development is a representative artificial cause. Development projects in mountainous regions have increased due to the tendency to prefer a nature-friendly environment. The risk of landslides is increasing due to the increasing number of cases where residential facilities, such as apartments, are located near steep slopes.

When torrential rainfall occurs, the soil begins to be saturated, the pore pressure increases, and the shear strength decreases. In addition, when rainfall infiltrates the bedrock, the adhesive strength weakens. The possibility of landslides increases when the soil is not able to support the increased weight of the saturated soil. Artificial steep slopes have the same characteristics; the expected damage range is a horizontal distance equal to twice the height of the slope. The steep slope in an urban area needs essential management for reducing the risk from landslides. In particular, debris flow is a phenomenon in which debris, such as rock fragments, soil, and mud move in groups with high speed. The debris, clay and rocks increase its volume and flow with strong erosion power, so that debris flow is the most dangerous type of landslide; research on debris flow is being conducted in various aspects [6–8]. It could lead to a large-scale disaster, with a loss of roads and damage to houses and people, especially when it occurs in urban areas.

In order to prevent or minimize damage from debris flow, it is important to install disaster prevention facilities by predicting movement paths and sedimentary areas before the event [9]. A debris barrier is a fundamental way to protect people and property from debris flow, but it is practically impossible to install it in all dangerous paths. Also, there are the difficulties related to modeling debris flows, due to the uncertainties in defining debris flow rheology behavior, such as the transition from a quasi-static regime to liquid regime [10]. The effects of grain concentration could lead to a dilatant or pseudoplastic fluid behavior, depending on grain volume concentration [11]; finally, the presence of coarse sediment fraction could affects both yield stress and the transition between solid-like to fluid-like behavior [12]. Therefore, it is more efficient to install facilities by predicting the extent of damage to the debris flow [13,14]. Landslide susceptibility analysis not only analyzes the cause of landslide occurrence, but also estimates the susceptibility of future occurrence, so it can be effectively used to prevent landslides in advance [15]. Accordingly, research on the susceptibility of landslides has been actively conducted by researchers.

Since a geographic information system (GIS) provides integrated analysis of various spatial data, it is useful to effectively analyze complex interrelationships between landslides and landslide-related variables. For this reason, there have been many attempts to explain the spatial pattern with various models based on GIS, in order to analyze the susceptibility of landslide occurrences in recent years. Most of the information for landslide prevention can come from modeling efforts using predictive models. Traditionally, to evaluate the susceptibility of landslide occurrence, simple, statistics-based frequency ratio (FR) [16–18] and logistic regression (LR) [19–21] were proposed and used as a representative prediction model. In recent years, various attempts have been made to produce machine learning (ML) models, such as support vector machines [22,23], artificial neural networks (ANNs) [24–26], ensembles [27,28], or hybrid models [29,30] including neuro-fuzzy [31,32] or metaheuristic ones [33,34]. Several spatial modeling studies were conducted on areas where landslides could occur in various regions by using these models with GIS [35,36].

In recent years, deep learning (DL), one of the ML technologies, has been receiving a lot of attention since it shows reliable results similar to or better than the existing ML methods [37]. In particular, it has been shown that model architectures, such as deep neural networks (DNNs), which are widespread, can significantly improve the quality of results [38]. In the area of landslide research, DL algorithms have mostly been used for landslide detection [39–42]. Additionally, among various DL methods, convolutional neural networks (CNNs) can identify expressions with extreme variability through convolution and pooling layers [43]. As a result, CNNs can play an important role in susceptibility analysis, as they can remove tedious functional engineering steps and extract useful information directly from the original data.

Although various techniques have been used, there is no academic consensus on which technique is most effective. Therefore, it is important to improve the accuracy of landslide susceptibility analysis by conducting comparative studies using various techniques, or by integrating multiple models [44,45]. Therefore, in this study, we tried to build this effort by using and comparing DL methodologies to map the susceptibility of landslides in the Mt. Umyeon area of Seoul, Korea. To this end, DNN, kernel-based DNN, and CNN models were applied among the DL models. This study proposes the application of DNNs for estimating landslide susceptibility, and compares their predictive ability with other models of ML techniques.

2. Study Area and Data

2.1. Study Area

As shown in Figure 1, the study area is Mt. Umyeon, located in Seocho-gu, Seoul. Seoul is the largest city in Korea (about 605 km^2), located in the center of the Korean Peninsula, and is one of the most populous cities in the world. This city, with about 10 million citizens, has a complex transportation and social infrastructure system. This density allows natural disasters, including landslides, to cause more recurrent disasters and serious damage [46]. The study area, Mt. Umyeon, is also a shallow mountain, with a height of 321.6 m, located in the center of a complex in a large city.

The slope near the summit is about 30 degrees, and most areas are less than 15 degrees. There is a gneissic rock outcrop in the area from weathering and geomorphological process. The biotite gneiss is weak to weathering, which could increase the susceptibility of debris flow in the area. The soil composition of this study area is mainly the brown forest soil developed on the parent rocks of biotite gneiss and granite gneiss. In some valley areas of the study area, brown forest soils with deep soil depth and high organic content are distributed, and the soil texture is silt and silt loam of sand, which drains well and easily causes severe weathering of the surface.

In July 2011, Typhoons Muifa and Kompasu generated 1092 mm of rainfall in Seoul for a month, and this resulted in simultaneous debris flows in the Mt. Umyeon area of Seoul. Specifically, 588 mm of heavy rain, which is about 40% of the annual average rainfall of 1451 mm, was concentrated in Mt. Umyeon for 3 days from July 26 to 28; debris flows caused 16 casualties and dozens of houses were damaged [47]. Debris flows that occurred simultaneously around Mt. Umyeon poured into streets and residential areas in downtown Seoul, damaging to apartments, vehicles, and pedestrians under the mountain, causing damage to life and property [48,49]. As shown in Figure 1, the debris flow of Mt. Umyeon occurred in the middle of the downtown area, different from the damage that occurred in the existing mountainous areas; the ripple effect was much greater than that in the past. This can be seen as a representative example, showing that even urban areas are not safe zones when a landslide, especially debris flow, occurs due to heavy rains [50,51].

Figure 1. The study area of Mt. Umyeon.

2.2. Spatial Datasets

In applying the data-driven methodology based on the established landslide (of debris flow in this study area) locations, accurate data is essential to build an appropriate database. In this study, data on the location of landslides was acquired based on aerial photographs. Additionally, 17 variables were selected as landslide-related variables; the variables were selected by referring to previous studies on various aspects of landslides, as follows: nine topographical variables (slope, aspect, curvature, topographic wetness index (TWI), stream power index (SPI), slope length factor (SLF), standardized height, valley depth, and downslope distance gradient (DDG)), four forest variables (timber type, timber diameter, timber density, and timber age) and four soil variables (soil depth, soil drainage, soil topography, and soil texture). The variables were selected as correlating between the landslides, and the variables have been investigated through the frequency ratio model in previous research [46,52,53].

Landslides in the Mt. Umyeon area received a lot of attention because damage occurred in areas where human activity was high, including major roads and residential areas in Seoul. Accordingly, a large-scale investigation into the cause and damage was conducted, so that a large amount of data, such as aerial photographs and field survey data, have been accumulated. Aerial photographs provide accurate status on the condition of the surface, with effectively captured complex urban structures. Also, despite the high cost of aerial photography, continuous monitoring using images is cost-effective for dense urban areas. The data on the location of landslides were selected by comparing the aerial photographs of Mt. Umyeon in 2010 before the occurrence and immediately after the occurrence in 2011 (Figure 2). Starting with the loss of the top slope, a debris flow occurred in which rocks and waste wood contained in the weathering and debris layers were lost together [54]. Accordingly, the landslide occurrence point was extracted based on the landslide occurrence starting point in this study, as in previous studies [27,31,55]. The traces of landslides in Mt. Umyeon remain for several years after the occurrence of debris flow, so it was easy to obtain through aerial photographic analysis. In addition, overlay and comparative analysis of 1:5000 digital topographic maps were performed, and 103 starting points of the landslide from landslide inventory were finally determined by comparing it with field survey data [56] conducted by the Seoul Metropolitan Government.

Figure 2. Landslide locations from aerial photographs after landslides in 2011: (**a**) Mt.Umyeon area after the landslides in 2011, (**b**) debris flows in northern part of Mt. Umyeon and (**c**) damage of residential areas including apartments.

Slopes have an important effect on the occurrence of landslides, as they affect water flow and soil characteristics [46,57]. Slope is a main landslide influencing factor that has the greatest impact on landslides, in that gentle slopes tend to have a lower frequency of landslides [55,58]. Therefore, the input layer for landslide susceptibility analysis was constructed by focusing on topographic and geomorphlogic variables among the variables that influence landslides (Figure 3). The topographic variables were calculated based on the acquisition of a digital topographic map (1:5000) generated by the National Geographic Information Institute (NGII); the slope was calculated using ArcGIS software (ESRI, Redlands, LA, United States) after geometric correction (Figure 3a), and the following additional topographic and geomorphologic variables were calculated using SAGA (Automated Geoscientific Analyses) GIS [59].

Slope aspect was classified into flat and north, northeast, east, southeast, south, southwest, west, and northwest, with 25 degree intervals (Figure 3b). The curvature variable represents the morphological characteristics of the research area by expressing the upwardly convex surface as a positive value, and the upwardly concave surface as a negative value. Curvature, a morphological characteristic in terms of affecting the erosion process and landslide evolution, is commonly used in landslide analysis, because it can account for local surface relief in the study area [58,60] (Figure 3c). TWI was used to express the influence of terrain on the presence of a runoff-generated saturated area [61,62], with higher values increasing the probability of landslide occurrence (Figure 3d) [63]. The erosive power of surface flow, measured by the SPI index, is one of the variables that influence landslides [64]. Higher SPI values indicate an increased likelihood of erosion that affects local stability [61], and allows a general determination of where soil conservation is needed to reduce the effects of surface runoff erosion (Figure 3e) [61]. SLF values refer to the ratio of soil loss to slope length, and were calculated through the revised universal soil loss equation (Figure 3f) [65]. The standardized height is calculated by multiplying the absolute height by the height of the study area normalized between 0 and 1 (Figure 3g). The standardized height yields parameters suitable for soil property prediction by simplifying complex terrain conditions using linear regression [66]. The standardized height represents a value suitable for predicting soil properties, by simplifying complex terrain conditions using linear regression [66]. Valley depth means the vertical distance at the basic level of the channel network [59]; valley depth has been used as a landslide-influencing variable in several studies (Figure 3h) [46,67]. The DDG is determined through quantitative hydrological gradient estimation and the downslope distance calculation when water loses a determined amount of energy due to precipitation (Figure 3i) [68].

Figure 3. Topographical variables influencing landslides used in this study: (**a**) slope gradient, (**b**) slope aspect, (**c**) curvature, (**d**) topographic wetness index (TWI), (**e**) stream power index (SPI), (**f**) slope length factor (SLF), (**g**) standardized height, (**h**) valley depth, and (**i**) downslope distance gradient (DDG).

The vegetation provides root strength to help stabilize the mountain slope. The effect of rainfall on the slope is likely to be mitigated by the degree of vegetation distribution. The attributes selected by several studies as landslide-influencing variables, including timber diameter, type, density, and age, were extracted from a digital vegetation map with a ratio of 1:25,000, provided by the Korea Forest Service (Figure 4a–d). As reported in numerous additional studies, landslides are strongly influenced by the geological structure and soil type covering the area [69,70]. Geological cover is one of the most used variables in the assessment of landslide susceptibility [58]; however, in this study, it was excluded, because it was mostly the same due to the size of the study area. Physical and organizational soil cover properties influence surface runoff and penetration processes [71]. Soil properties affect the flow of water, including surface and groundwater, in the way that landslides occur when the amount of water in the soil exceeds the drainage capacity and the cohesion drops sharply [72]. Therefore, forest and soil characteristics were considered as landslide-influencing variables in this study (Figure 4) [46]. Each variable was pre-processed into raster format with a grid size of 10 m and is illustrated in Table 1 with the source and scale.

Figure 4. Landslide-influencing variables related to vegetation and soil: (**a**) timber type, (**b**) timber diameter, (**c**) timber density, (**d**) timber age, (**e**) soil depth, (**f**) soil drainage, (**g**) soil topography, and (**h**) soil texture.

Table 1. Landslide influential variables used in this study.

Original Data	Variables	Data Type	Scale
Aerial photograph	Landslide location	Point	1:1000
Topographical map [a]	Slope gradient [°] Slope aspect Curvature Topographic wetness index (TWI) Stream power index (SPI) Slope length factor (SLF) Standardized height (m) Valley depth (m) Downslope distance gradient (DDG) (rad)	Grid	1:1000
Forest map [b]	Timber diameter Timber type Timber density Timber age	Polygon	1:25,000
Soil map [c]	Soil depth Soil drainage Soil topography Soil texture	Polygon	1:25,000

[a] The digital topographic map by the National Geographic Information Institute; [b] the forest map produced by the Korea Forest Service; [c] the soil map produced by Rural Development Administration.

3. Methodology

3.1. Deep Neural Network and Kernel-Based Deep Neural Network

ANNs are one of the most popular ML methods for prediction and classification across various fields. ANNs are composed of an input layer, a hidden layer, and an output layer; the input layer has as many input variables as the number of input variables in the model, and the output layer represents the prediction result. All layers between the input and output layers are referred to as hidden layers, and the number of hidden layers is determined through an empirical process [73]. The backpropagation algorithm is known to be the commonly used, effective method for training weight determination. Through the backpropagation algorithm, training is performed repeatedly, until a predefined error value is reached and finally is applied to predict for all the data by using the trained model [74].

DL methodology is an ML method based on this ANN, and it performs prediction and classification in the same way. Like the term "deep layer", learning is performed by transforming data levels into different levels in DL methodology [75]. Recent ML studies have proposed various DL architectures (e.g., autoencoders, Restricted Boltzmann Machine (RBM), CNN, and Recurrent Neural Network (RNN)) [76]. In this study, a DNN and kernel-based DNN were applied and compared; the DNN can be interpreted as a deep version of an ANN. As the number of layers increases, the network becomes more complex, and the most appropriate layer for data can be created.

Specifically, in the case of a DNN using a stochastic gradient descent (SGD) in this study, the architecture was constructed using 100, 100, 50, 50, 25, 25, 10, 10 sequential hidden nodes, as shown in Figure 5 and Table 2. The rectified linear unit (ReLU) was selected as the activation function, and the nonlinear properties of the network were improved through the ReLU activation function [77]. The loss function was designated as the mean squared error (MSE), and the max epochs was set to 300 (Table 1). The batch size and learning rate were set to 150.00 and 0.01, respectively. In the process of calculating the weight for each node during training, the influence of the node becomes too high when the weight of a certain node is high. For example, only the neuron that has the most influence among the input neurons is reflected, and the result may appear as if it was created only by that layer. Therefore, the weight decay was used, which is a concept that prevents the weight of all nodes from becoming larger than necessary. By increasing the weight decay, we kept the overall weight low and used the conditions for all input neurons. Increasing the weight decay keeps the overall weight low and uses the conditions for all input neurons. The weight decay of DNN was set to 0.015 in this study. For a comparison of the results, the DNN and kernel-based DNN were set to the same parameters. The image kernel size was set to only 7×7 in the case of kernel-based DNN, as shown in the structure from Figure 5b.

Table 2. Model parameters used in this study.

Model Parameters	DNN	KDNN	CNN
Input data dimension	-	7×7	7×7
Filter depth in convolutional layers	-	-	[48, 96, 96, 192, 192]
Filter size in convolutional layers	-	-	[3, 3, 3, 3, 1]
Number of nodes in fully connected layer	[100, 100, 50, 50, 25, 25, 10, 10]		[96, 48, 24]
Activation function	The rectified linear unit (ReLU)		
Optimizer	Stochastic gradient descent (SGD)		
Loss function	Mean squared error (MSE)		
Max epochs	300.00		
Batch size	150.00		
Learning rate	0.01		
Weight decay	0.015	0.015	0.030

Figure 5. Model architecture proposed in this study: (**a**) Deep Neural Network, (**b**) kernel-based Deep Neural Network, and (**c**) Convolutional Neural Network.

3.2. Convolutional Neural Network

Recently, CNN applications for classification and prediction in various fields are increasing [78]. As mentioned above, CNNs are one of the well-known DL methodologies that show strong performance in image processing fields based on an ANN [79]. Important features of raw data can be automatically extracted through CNN [80]. Researchers have developed various structures depending on the type or purpose of the data; the basic architecture consists of an input layer, multiple hidden layers, and an output layer. The difference between CNNs and simple neural networks is that the hidden layer is composed of one or more convolutional and pooling layers, which share weights with multiple layers, pooling, and local connections [40]. The convolution layer, composed of several convolution kernels, repeatedly extracts features of the original data. The pooling layer prevents overfitting by reducing the dimension of the feature map through downsampling after the convolutional layer. In addition, it has an advantage of fast processing speed by greatly reducing the number of parameters by interpreting the input data as an image.

The CNN architecture in this study also consists of a convolution layer that learns convolution, and a pooling layer that improves computation performance by controlling overfitting and reducing the number of structures due to convolution through stable transformation [81]. In particular, the sizes of the convolutional kernel and the pooling kernel determine the scale of the convolution and pooling operations, respectively [82]. The activation function converts a linear relationship into a nonlinear relationship [83], and the optimizer iteratively updates the parameters at various learning rates. The loss function updates the weight by measuring the error between the predicted and actual values. In this study, the hyperparameter setting, a key step in building CNN architecture, is as follows in Table 1.

In this study, ReLU was selected as an activation function to improve the nonlinear property of the network; hereby, landslide susceptibility with a value between 0 and 1 is finally calculated. Stochastic gradient descent (SGD) was used as the optimizer, and MSE was selected as the loss function. The max epochs, batch size, and learning rate were set to 300.00, 150.00, and 0.01, respectively. Like DNN, the weight decay was set to 0.03, so that the weights of all input neurons were reflected, and stable loss and accuracy could be calculated. Image kernel size was set to 7×7 in the same way as kernel-based DNN. The feature map extracted through the convolution and pooling layers was reconstructed by the connection layer, and a classification result was derived through a nonlinear activation function. Figure 5 shows the prediction process using the CNN classifier used in this study.

3.3. Susceptibility Modeling and Mapping

For landslide susceptibility modeling and mapping using DNN, Kernel-based DNN and CNN models, a landslide training dataset was first constructed based on landslide occurrence locations selected from the research area. Since the study area is a small mountain located in the downtown area, the oversampling step was performed, considering that the number of landslide locations was only 103. The same number of samples where a non-landslide would occur was extracted as the landslide location samples, by random sampling with a slope of less than 5 degrees. After that, the sample was randomly divided into two sections: 80% was allocated to the training set, and the remaining 20% was extracted separately to use for performance evaluation [84–86]. The training set and the test set were adjusted through oversampling, which randomly adjusts the number of data by extracting the same data several times from imbalanced data. Finally, a total 20,000 of training (16,000) and test (4000) datasets were extracted and used for modeling.

The influential landslide variables were resampled in a grid format with 5 m spatial resolution. The value of each variable was extracted by overlaying 17 landslide influential variables on the training dataset, consisting of landslide and non-landslide samples. Finally, after training the model, susceptibility modeling was performed by applying it to the entire study area. Then, the results of landslide susceptibility were visually mapped with values from 0 to 1 from the three models.

3.4. Performance Evaluation of the Models

Model performance evaluation was performed to measure the predictive ability of the landslide model and whether it was suitable for the region. In this study, in order to evaluate the error of the prediction result, root mean square errors (RMSEs) were calculated for the prediction result of each model and the actual landslide information [85,87]. RMSEs were calculated using the 20% landslide locations that were stored for testing in the previous step.

Additionally, the accuracy of the model was calculated by calculating the area under the curve (AUC) of the precision–recall (PR) curve [88]. The PR curve is a curve with recall on the x-axis and precision on the y-axis, and is a line connecting recall and precision values for each data while varying the threshold value based on the susceptibility value of the model result. The AUC value created from the landslide modeling result depends on the recall and precision index of the x-axis and y axis of the curve, respectively [88]. The AUC of the PR curve can be expressed as an average precision (AP) score, and it is the area of the graph when the PR curve is drawn. The AUC value approaches 1 when both recall and precision converge to the highest level, showing the model ability to efficiently predict the susceptibility of a landslide: the higher the value of the AUC of the PR, the higher the predictive ability of the model.

4. Results and Discussion

4.1. Susceptibility Maps

Figure 6 shows the three landslide susceptibility maps, derived from the model proposed in this study. Each map represents the susceptibility of landslide occurrence in the study area, derived through the DNN, kernel-based DNN, and CNN models, as described in the previous step. Due to the small study area, most areas can be classified into the same class when the map is graded. Therefore, the research area is represented by the susceptibility of occurrence from 0 to 1, and continuous susceptibility can be confirmed. The three models generally showed similar spatial distribution.

Figure 6. Landslide susceptibility maps generated using deep learning models: Using a (**a**) deep neural network (DNN), (**b**) kernel-based DNN, NS (**c**) convolutional neural network (CNN) model.

The proportions with susceptibility of less than 0.2 were 38.55%, 39.04%, and 38.30% in DNN, kernel-based DNN, and CNN models, respectively. On the other hand, the ratios of 0.8 or higher with a high susceptibility of landslide occurrence were 39.59%, 33.13%, and 34.47%, respectively. Relatedly, the DNN model shows a large number of regions with higher susceptibility, and in the case of kernel-based DNN, the resolution is slightly lowered. This is because the smoothing effect appeared as a result of using the kernel. Areas with high susceptibility of landslides were identified mainly along the north and south sides, where large debris flows occurred.

Since the slopes of Mt. Umyeon consists of gneiss, the forces remaining in the weathered gneiss layer after a torrential downpour reduce the amount of surface runoff and absorb a large amount of moisture into the weathered layer. Moisture in the saturated weathering layer reduces the resistance to bedrock, facilitating the downslope movement by gravity. As the resistance was sharply decreased, the physical erosion power was strengthened by the increase of the surface water flow, which erodes and mobilizes the mud and rocks. In kernel-based DNN and CNN, the effect on not only a single pixel but also neighboring pixels is considered at the same time, thus reflecting this phenomenon more realistically. In Figure 6, the probability of landslide is low in the mountain ridge in all models, and the valley slope forms a steep slope by lateral erosion.

4.2. Evaluation Results

The modeling process using the training dataset defined the relationship between landslide influencing variables with the location of the past landslides and future landslides, and the results were evaluated through AP scores with PR curves and statistical indicators of RMSE. The AP score curve showed that the performance of the DNN, kernel-based DNN, and CNN models was 99.45%, 99.44%, and 99.41%, respectively (Figure 7d–f). All showed over 99%, indicating that each model has very good accuracy in predicting landslides. RMSE was calculated based on the test dataset for the predictive model (Figure 7a–c). There were highest (0.1806) and lowest (0.1694) RMSEs for the DNN and kernel-based DNN models, respectively, and the CNN model had an RMSE of 0.1747. A low RMSE indicates good predictive power of the developed model. All three models show similar values, but the kernel-based DNN and CNN models were slightly better than the DNN model. This is obviously because the models are based on the kernel approach. However, the benefits of kernel processing were not significant in this study, because (1) the spatial resolutions of the forest and soil maps are low, and (2) the topographic relationships between adjacent pixels are given by the input data, such as TWI, SPI, SLF, etc. We can find one more difference between the DNN and the kernel-based models in Figure 6a–c. The kernel-based models shown in Figure 6b,c are smoothed due to the kernel calculation.

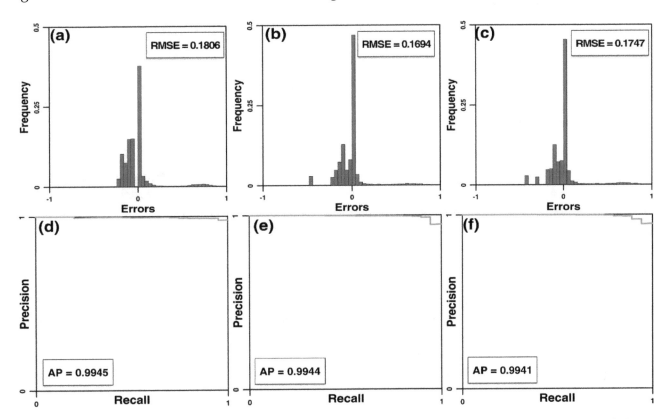

Figure 7. Model performance evaluation results: (**a–c**) root mean square errors (RMSEs) and (**d–f**) average precision (AP) scores of the DNN, kernel-based DNN, and CNN models.

The AP scores were almost identical in all models, indicating that the model used in this study was reasonably accurate in mapping the susceptibility of landslides in the study area. The DNN showed the highest accuracy among three models, but the difference compared to the kernel-based DNN was very small, and the RMSE was relatively high. As a result of using an additional kernel, the RMSE value decreased by about 6.2%, and such inconsistency did not occur. Therefore, it can be interpreted that the kernel-based DNN model exhibits better performance for predicting future landslides than the DNN model. In addition, the weight decay was increased to adjust the total weight, in order to prevent the weight of a specific node from being higher than necessary, thereby preventing the result of being weighted by a specific variable. By preventing overestimation, it was possible to prevent the results from being greatly affected by the slightly adjusted training data. Through this, the results of improving the precision of the test data, as well as the precision of the training data, were confirmed. The three results compared in this study all showed similar patterns due to the limited input neurons, and it is difficult to specify which model is better. However, reliable results were obtained in all of the DNN, kernel-based DNN, and CNN models, and can be applied not only to landslides, but also to modeling the susceptibility of natural phenomena and various regions, such as flood and forest fire mapping.

5. Conclusions

In the present study, a methodological approach was introduced to develop a DNN model, a kernel-based DNN that combines a kernel with the DNN model, and a CNN model, all of which were trained based on a SGD. The location data for training the susceptibility of a landslide was constructed by analyzing aerial photographs and field survey results. Topographic variables were extracted from aerial photographs and digital topographic maps, and soil and forest influencing variables were also extracted. A landslide inventory was constructed using the training dataset built through oversampling for the landslide area and non-landslide area, with the landslide influencing variables; modeling was performed for three models based on the inventory. Finally, the resulting map was evaluated using a test set (20%) of landslide data that was not used for training. The training and prediction functions of the developed DL model were evaluated and compared with the results of RMSE estimation and AP score analysis. The experimental results showed that the optimized DNN model had the highest predictive power (99.45%), followed by the kernel-based DNN (99.44%) and CNN model (99.41%) methods. The recall and precision of all models showed sufficiently high results, and the performance of all models was more than 99% satisfactory.

The expected landslide susceptibility was high near the top of the mountain and was similar to the existing landslide pattern. The susceptibility of landslides decreases according to the slope of the susceptibility map of this study, which is a logical result of the occurrence of debris flow due to the shape of the topography. Nevertheless, there are some limitations to the implementation of neural network (NN)-based DL models. In selecting variables, it is necessary to apply a different methodology in advance, in order to understand the fundamental relationship between variables and landslides, just as variables were selected based on previous studies in this study. Also, the model architecture is less intuitive due to the black box of the NN model. That means selecting the right hyperparameters is a task that requires experiential expertise.

Compared to water-related disasters, landslides were not previously considered a serious problem in urban areas. Most cities developed rapidly without taking into account the damage caused by natural disasters in the process of urbanization. However, due to the recent increase in local heavy rain, extreme precipitation and population pressure in any part of the city can have serious consequences. Therefore, it is important to establish basic scientific data to prevent landslides in urban areas such as Seoul. Regarding the applicability of the developed methodology, the results of this study are expected to support decision-making to identify spatial problems related to urban landslides and effective risk reduction policies.

Author Contributions: S.L. (Sunmin Lee), W.-K.B., and S.L. (Saro Lee) collected data. W.-K.B. and H.-S.J. processed input data. S.L. (Sunmin Lee) and H.-S.J. managed the paperwork and interpreted the results. H.-S.J. suggested the idea and prepared the data. All of the authors contributed to the writing of each part. All authors have read and agreed to the published version of the manuscript.

Funding: This work was supported by the National Research Foundation of Korea, funded by the Korean government under grant NRF-2018M1A3A3A02066008. This research was supported by the Basic Research Project of the Korea Institute of Geoscience and Mineral Resources (KIGAM) and Project of Environmental Business Big Data Platform and Center Construction funded by the Ministry of Science and ICT. This research was supported by a grant (20010017) of R&D Program of Responding Technology to Climate Disaster such as Heat Wave funded by Ministry of Interior and Safety (MOIS, Korea) which was conducted by the Korea Environment Institute (KEI).

References

1. Landslide Information System. Available online: http://sansatai.forest.go.kr/ (accessed on 26 September 2020).
2. Kang, S.; Lee, S.; Nikhil, N.; Park, J.-Y. Analysis of Differences in Geomorphological Characteristics on Initiation of Landslides and Debris Flows. *J. Korean Soc. Hazard Mitig.* **2015**, *15*, 249–258. [CrossRef]
3. Lee, J. Management system for landslides hazard area using GIS. *J. Korea Soc. For. Eng. Technol.* **2005**, *3*, 245–255.
4. Lee, H.-G.; Kim, G.-H. Landslide Risk Assessment in Inje Using Logistic Regression Model. *J. Korean Soc. Surv. Geodesy Photogramm. Cartogr.* **2012**, *30*, 313–321.
5. Kim, G. *A Basic Study on the Development of the Guidelines on Setting Debris Flow Hazards*; Research Institute for Gangwon: Chuncheon-s, Korea, 2011; p. 170.
6. Vagnon, F.; Ferrero, A.M.; Pirulli, M.; Segalini, A. Theoretical and experimental study for the optimization of flexible barriers to restrain debris flows. *Geam. Environ. Min. Geo. Eng.* **2015**, *149*, 29–35.
7. Vagnon, F.; Ferrero, A.M.; Alejano, L.R. Reliability-based design for debris flow barriers. *Landslides* **2019**, *17*, 49–59. [CrossRef]
8. Vagnon, F. Design of active debris flow mitigation measures: A comprehensive analysis of existing impact models. *Landslides* **2020**, *17*, 313–333. [CrossRef]
9. Miyamoto, K. Two dimensional numerical simulation of landslide mass movement. *J. Eros. Control Eng.* **2002**, *55*, 5–13.
10. Pellegrino, A.M.; Schippa, L. Macro viscous regime of natural dense granular mixtures. *Int. J. Geomate* **2013**, *4*, 482–489. [CrossRef]
11. Schippa, L. Modeling the effect of sediment concentration on the flow-like behavior of natural debris flow. *Int. J. Sediment Res.* **2020**, *35*, 315–327. [CrossRef]
12. Pellegrino, A.M.; Schippa, L. A laboratory experience on the effect of grains concentration and coarse sediment on the rheology of natural debris-flows. *Environ. Earth Sci.* **2018**, *77*, 749. [CrossRef]
13. Kim, P. Numerical modeling for the detection and movement of debris flow using detailed soil maps and GIS. *J. Korean Soci. Civ. Eng.* **2017**. [CrossRef]
14. VanDine, D. *Debris Flow Control Structures for Forest Engineering*; Working paper; Ministry of Forests: Victoria, BC, Canada, 1996.
15. Lee, S.; Lee, M.-J.; Won, J.-S. Study on Landslide using GIS and Remote Sensing at the Kangneung Area (II)-Landslide Susceptibility Mapping and Cross-Validation using the Probability Technique. *Econ. Environ. Geol.* **2004**, *37*, 521–532.
16. Kose, D.D.; Turk, T. GIS-based fully automatic landslide susceptibility analysis by weight-of-evidence and frequency ratio methods. *Phys. Geogr.* **2019**, *40*, 481–501. [CrossRef]
17. Yan, F.; Zhang, Q.; Ye, S.; Ren, B. A novel hybrid approach for landslide susceptibility mapping integrating analytical hierarchy process and normalized frequency ratio methods with the cloud model. *Geomorphology* **2019**, *327*, 170–187. [CrossRef]
18. Zhang, Y.-X.; Lan, H.-X.; Li, L.-P.; Wu, Y.-M.; Chen, J.-H.; Tian, N.-M. Optimizing the frequency ratio method for landslide susceptibility assessment: A case study of the Caiyuan Basin in the southeast mountainous area of China. *J. Mt. Sci.* **2020**, *17*, 340–357. [CrossRef]

19. Feby, B.; Achu, A.; Jimnisha, K.; Ayisha, V.; Reghunath, R. Landslide susceptibility modelling using integrated evidential belief function based logistic regression method: A study from Southern Western Ghats, India. *Remote Sens. Appl. Soc. Environ.* **2020**, *20*, 100411. [CrossRef]

20. Nhu, V.-H.; Shirzadi, A.; Shahabi, H.; Singh, S.K.; Al-Ansari, N.; Clague, J.J.; Jaafari, A.; Chen, W.; Miraki, S.; Dou, J. Shallow Landslide Susceptibility Mapping: A Comparison between Logistic Model Tree, Logistic Regression, Naïve Bayes Tree, Artificial Neural Network, and Support Vector Machine Algorithms. *Int. J. Environ. Res. Public Health* **2020**, *17*, 2749. [CrossRef]

21. Soma, A.S.; Kubota, T.; Mizuno, H. Optimization of causative factors using logistic regression and artificial neural network models for landslide susceptibility assessment in Ujung Loe Watershed, South Sulawesi Indonesia. *J. Mt. Sci.* **2019**, *16*, 383–401. [CrossRef]

22. Abedini, M.; Ghasemian, B.; Shirzadi, A.; Bui, D.T. A comparative study of support vector machine and logistic model tree classifiers for shallow landslide susceptibility modeling. *Environ. Earth Sci.* **2019**, *78*, 560.

23. Saha, A.; Saha, S. Comparing the efficiency of weight of evidence, support vector machine and their ensemble approaches in landslide susceptibility modelling: A study on Kurseong region of Darjeeling Himalaya, India. *Remote Sens. Appl. Soc. Environ.* **2020**, *19*, 100323.

24. Moayedi, H.; Mehrabi, M.; Mosallanezhad, M.; Rashid, A.S.A.; Pradhan, B. Modification of landslide susceptibility mapping using optimized PSO-ANN technique. *Eng. Comput.* **2019**, *35*, 967–984. [CrossRef]

25. Sevgen, E.; Kocaman, S.; Nefeslioglu, H.A.; Gokceoglu, C. A novel performance assessment approach using photogrammetric techniques for landslide susceptibility mapping with logistic regression, ANN and random forest. *Sensors* **2019**, *19*, 3940. [CrossRef] [PubMed]

26. Yu, C.; Chen, J. Landslide Susceptibility Mapping Using the Slope Unit for Southeastern Helong City, Jilin Province, China: A Comparison of ANN and SVM. *Symmetry* **2020**, *12*, 1047. [CrossRef]

27. Arabameri, A.; Pradhan, B.; Rezaei, K.; Lee, S.; Sohrabi, M. An ensemble model for landslide susceptibility mapping in a forested area. *Geocarto Int.* **2019**, *35*, 1680–1705. [CrossRef]

28. Sachdeva, S.; Bhatia, T.; Verma, A.K. A novel voting ensemble model for spatial prediction of landslides using GIS. *Int. J. Remote Sens.* **2020**, *41*, 929–952. [CrossRef]

29. Thai Pham, B.; Shirzadi, A.; Shahabi, H.; Omidvar, E.; Singh, S.K.; Sahana, M.; Talebpour Asl, D.; Bin Ahmad, B.; Kim Quoc, N.; Lee, S. Landslide susceptibility assessment by novel hybrid machine learning algorithms. *Sustainability* **2019**, *11*, 4386. [CrossRef]

30. Zhang, T.-Y.; Han, L.; Zhang, H.; Zhao, Y.-H.; Li, X.-A.; Zhao, L. GIS-based landslide susceptibility mapping using hybrid integration approaches of fractal dimension with index of entropy and support vector machine. *J. Mt. Sci.* **2019**, *16*, 1275–1288. [CrossRef]

31. Jaafari, A.; Panahi, M.; Pham, B.T.; Shahabi, H.; Bui, D.T.; Rezaie, F.; Lee, S. Meta optimization of an adaptive neuro-fuzzy inference system with grey wolf optimizer and biogeography-based optimization algorithms for spatial prediction of landslide susceptibility. *Catena* **2019**, *175*, 430–445. [CrossRef]

32. Mehrabi, M.; Pradhan, B.; Moayedi, H.; Alamri, A. Optimizing an Adaptive Neuro-Fuzzy Inference System for Spatial Prediction of Landslide Susceptibility Using Four State-of-the-art Metaheuristic Techniques. *Sensors* **2020**, *20*, 1723. [CrossRef]

33. Moayedi, H.; Mehrabi, M.; Kalantar, B.; Mu'Azu, M.A.; Rashid, A.S.A.; Foong, L.K.; Nguyen, H. Novel hybrids of adaptive neuro-fuzzy inference system (ANFIS) with several metaheuristic algorithms for spatial susceptibility assessment of seismic-induced landslide. *Geomat. Nat. Hazards Risk* **2019**, *10*, 1879–1911. [CrossRef]

34. Panahi, M.; Gayen, A.; Pourghasemi, H.R.; Rezaie, F.; Lee, S. Spatial prediction of landslide susceptibility using hybrid support vector regression (SVR) and the adaptive neuro-fuzzy inference system (ANFIS) with various metaheuristic algorithms. *Sci. Total. Environ.* **2020**, *741*, 139937. [CrossRef] [PubMed]

35. Arabameri, A.; Pradhan, B.; Rezaei, K.; Sohrabi, M.; Kalantari, Z. GIS-based landslide susceptibility mapping using numerical risk factor bivariate model and its ensemble with linear multivariate regression and boosted regression tree algorithms. *J. Mt. Sci.* **2019**, *16*, 595–618. [CrossRef]

36. Lee, S. Current and future status of GIS-based landslide susceptibility mapping: A literature review. *Korean J. Remote Sens.* **2019**, *35*, 179–193.

37. Schmidhuber, J. Deep Learning: Our Miraculous Year 1990–1991. *arXiv* **2020**, arXiv:2005.05744.

38. Qi, C.; Fourie, A.; Ma, G.; Tang, X. A hybrid method for improved stability prediction in construction projects: A case study of stope hangingwall stability. *Appl. Soft Comput.* **2018**, *71*, 649–658. [CrossRef]

39. Can, A.; Dagdelenler, G.; Ercanoglu, M.; Sonmez, H. Landslide susceptibility mapping at Ovacık-Karabük (Turkey) using different artificial neural network models: Comparison of training algorithms. *Bull. Eng. Geol. Environ.* **2019**, *78*, 89–102. [CrossRef]

40. Ghorbanzadeh, O.; Blaschke, T.; Gholamnia, K.; Meena, S.R.; Tiede, D.; Aryal, J. Evaluation of different machine learning methods and deep-learning convolutional neural networks for landslide detection. *Remote Sens.* **2019**, *11*, 196. [CrossRef]

41. Ghorbanzadeh, O.; Meena, S.R.; Blaschke, T.; Aryal, J. UAV-based slope failure detection using deep-learning convolutional neural networks. *Remote Sens.* **2019**, *11*, 2046. [CrossRef]

42. Mohan, A.; Singh, A.K.; Kumar, B.; Dwivedi, R. Review on remote sensing methods for landslide detection using machine and deep learning. *Trans. Emerg. Telecommun. Technol.* **2020**, e3998. [CrossRef]

43. Wang, Y.; Fang, Z.; Hong, H. Comparison of convolutional neural networks for landslide susceptibility mapping in Yanshan County, China. *Sci. Total Environ.* **2019**, *666*, 975–993. [CrossRef]

44. Aghdam, I.N.; Varzandeh, M.H.M.; Pradhan, B. Landslide susceptibility mapping using an ensemble statistical index (Wi) and adaptive neuro-fuzzy inference system (ANFIS) model at Alborz Mountains (Iran). *Environ. Earth Sci.* **2016**, *75*, 1–20. [CrossRef]

45. Li, Y.; Chen, W. Landslide susceptibility evaluation using hybrid integration of evidential belief function and machine learning techniques. *Water* **2020**, *12*, 113. [CrossRef]

46. Lee, S.; Lee, M.-J.; Jung, H.-S. Data mining approaches for landslide susceptibility mapping in Umyeonsan, Seoul, South Korea. *Appl. Sci.* **2017**, *7*, 683. [CrossRef]

47. Yune, C.-Y.; Chae, Y.-K.; Paik, J.; Kim, G.; Lee, S.-W.; Seo, H.-S. Debris flow in metropolitan area—2011 Seoul debris flow. *J. Mt. Sci.* **2013**, *10*, 199–206. [CrossRef]

48. Jun, K.; Oh, C.; Jun, B. Study on analyzing characteristics which causes a debris flow–focusing on the relation with slope and river. *J. Saf. Crisis Manag.* **2011**, *7*, 223–232.

49. Pradhan, A.M.S.; Kim, Y.-T. Spatial data analysis and application of evidential belief functions to shallow landslide susceptibility mapping at Mt. Umyeon, Seoul, Korea. *Bull. Eng. Geol. Environ.* **2017**, *76*, 1263–1279. [CrossRef]

50. Kim, S.; Paik, J.; Kim, K.S. Run-out modeling of debris flows in Mt. Umyeon using FLO-2D. *J. Korean Soc. Civ. Eng.* **2013**, *33*, 965–974. [CrossRef]

51. Ko, S.M.; Lee, S.W.; Yune, C.-Y.; Kim, G. Topographic analysis of landslides in Umyeonsan. *J. Korean Soc. Surv. Geod. Photogramm. Cartogr.* **2014**, *32*, 55–62. [CrossRef]

52. Lee, M.-J.; Park, I.; Lee, S. Forecasting and validation of landslide susceptibility using an integration of frequency ratio and neuro-fuzzy models: A case study of Seorak mountain area in Korea. *Environ. Earth Sci.* **2015**, *74*, 413–429. [CrossRef]

53. Park, S.-J.; Lee, C.-W.; Lee, S.; Lee, M.-J. Landslide susceptibility mapping and comparison using decision tree models: A Case Study of Jumunjin Area, Korea. *Remote Sens.* **2018**, *10*, 1545. [CrossRef]

54. KIM, J.-H.; OH, J.-S. A Study on the Geological Characteristics and Slope Landform Changes of Mt. Umyeon. *Korean J. Nat. Conserv.* **2013**, *11*, 1–14. [CrossRef]

55. Dehnavi, A.; Aghdam, I.N.; Pradhan, B.; Varzandeh, M.H.M. A new hybrid model using step-wise weight assessment ratio analysis (SWARA) technique and adaptive neuro-fuzzy inference system (ANFIS) for regional landslide hazard assessment in Iran. *Catena* **2015**, *135*, 122–148. [CrossRef]

56. Seoul City. Supplemental investigation report of causes of Umyeonsan landslides. *Seoul: Seoul City* **2014**.

57. Capitani, M.; Ribolini, A.; Bini, M. The slope aspect: A predisposing factor for landsliding? *Comptes Rendus Geosci.* **2013**, *345*, 427–438. [CrossRef]

58. Pourghasemi, H.R.; Yansari, Z.T.; Panagos, P.; Pradhan, B. Analysis and evaluation of landslide susceptibility: A review on articles published during 2005–2016 (periods of 2005–2012 and 2013–2016). *Arab. J. Geosci.* **2018**, *11*, 193. [CrossRef]

59. Conrad, O.; Bechtel, B.; Bock, M.; Dietrich, H.; Fischer, E.; Gerlitz, L.; Wehberg, J.; Wichmann, V.; Böhner, J. System for automated geoscientific analyses (SAGA) v. 2.1. 4. *Geosci. Model Dev. Discuss.* **2015**, *8*. [CrossRef]

60. Oh, H.-J.; Pradhan, B. Application of a neuro-fuzzy model to landslide-susceptibility mapping for shallow landslides in a tropical hilly area. *Comput. Geosci.* **2011**, *37*, 1264–1276. [CrossRef]

61. Moore, I.D.; Grayson, R.; Ladson, A. Digital terrain modelling: A review of hydrological, geomorphological, and biological applications. *Hydrol. Process.* **1991**, *5*, 3–30. [CrossRef]

62. Wilson, J.P.; Gallant, J.C. *Terrain Analysis: Principles and Applications*; John Wiley & Sons: Hoboken, NJ, USA, 2000.

63. Conoscenti, C.; Di Maggio, C.; Rotigliano, E. GIS analysis to assess landslide susceptibility in a fluvial basin of NW Sicily (Italy). *Geomorphology* **2008**, *94*, 325–339. [CrossRef]

64. Nefeslioglu, H.A.; Gokceoglu, C.; Sonmez, H. An assessment on the use of logistic regression and artificial neural networks with different sampling strategies for the preparation of landslide susceptibility maps. *Eng. Geol.* **2008**, *97*, 171–191. [CrossRef]

65. Wischmeier, W.H.; Smith, D.D. *Predicting Rainfall Erosion Losses: A Guide to Conservation Planning*; United States Department of Agriculture, Science and Education Administration: Beltsville, MD, USA, 1978.

66. Böhner, J.; Selige, T. Spatial prediction of soil attributes using terrain analysis and climate regionalisation. In *SAGA—Analyses and Modelling Applications*; Goltze: Göttingen, Germany, 2006.

67. Nguyen, V.V.; Pham, B.T.; Vu, B.T.; Prakash, I.; Jha, S.; Shahabi, H.; Shirzadi, A.; Ba, D.N.; Kumar, R.; Chatterjee, J.M. Hybrid machine learning approaches for landslide susceptibility modeling. *Forests* **2019**, *10*, 157. [CrossRef]

68. Hjerdt, K.; McDonnell, J.; Seibert, J.; Rodhe, A. A new topographic index to quantify downslope controls on local drainage. *Water Resour. Res.* **2004**, *40*. [CrossRef]

69. Ohlmacher, G.C.; Davis, J.C. Using multiple logistic regression and GIS technology to predict landslide hazard in northeast Kansas, USA. *Eng. Geol.* **2003**, *69*, 331–343. [CrossRef]

70. Van Westen, C.; Van Asch, T.W.; Soeters, R. Landslide hazard and risk zonation—why is it still so difficult? *Bull. Eng. Geol. Environ.* **2006**, *65*, 167–184. [CrossRef]

71. Mogaji, K.A.; San Lim, H.; Abdullah, K. Modeling of groundwater recharge using a multiple linear regression (MLR) recharge model developed from geophysical parameters: A case of groundwater resources management. *Environ. Earth Sci.* **2015**, *73*, 1217–1230. [CrossRef]

72. JONES, H.G. Estimation of an effective soil water potential at the root surface of transpiring plants. *Plant Cell Environ.* **1983**, *6*, 671–674.

73. Du, S.S.; Wang, Y.; Zhai, X.; Balakrishnan, S.; Salakhutdinov, R.R.; Singh, A. How many samples are needed to estimate a convolutional neural network? In Proceedings of the Advances in Neural Information Processing Systems 31 (NeurIPS 2018), Montreal, QC, Canada, 2–8 December 2018; pp. 373–383.

74. Liu, W.; Wang, Z.; Liu, X.; Zeng, N.; Liu, Y.; Alsaadi, F.E. A survey of deep neural network architectures and their applications. *Neurocomputing* **2017**, *234*, 11–26. [CrossRef]

75. Patterson, J.; Gibson, A. *Deep Learning: A practitioner's Approach*; O'Reilly Media Inc.: Boston, MA, USA, 2017.

76. Alom, M.Z.; Taha, T.M.; Yakopcic, C.; Westberg, S.; Sidike, P.; Nasrin, M.S.; Hasan, M.; Van Essen, B.C.; Awwal, A.A.; Asari, V.K. A state-of-the-art survey on deep learning theory and architectures. *Electronics* **2019**, *8*, 292. [CrossRef]

77. Agarap, A.F. Deep learning using rectified linear units (relu). *arXiv* **2018**, arXiv:1803.08375.

78. Zuo, R.; Xiong, Y.; Wang, J.; Carranza, E.J.M. Deep learning and its application in geochemical mapping. *Earth-Sci. Rev.* **2019**, *192*, 1–14. [CrossRef]

79. LeCun, Y.; Bengio, Y.; Hinton, G. Deep learning. *Nature* **2015**, *521*, 436–444. [CrossRef] [PubMed]

80. Lawrence, S.; Giles, C.L.; Tsoi, A.C.; Back, A.D. Face recognition: A convolutional neural-network approach. *IEEE Trans. Neural Netw.* **1997**, *8*, 98–113. [CrossRef]

81. Nair, V.; Hinton, G.E. Rectified linear units improve restricted boltzmann machines. In Proceedings of the 27th International Conference on Machine Learning, Haifa, Israel, 21–24 June 2010.

82. Golik, P.; Tüske, Z.; Schlüter, R.; Ney, H. Convolutional neural networks for acoustic modeling of raw time signal in LVCSR. In Proceedings of the 16th Annual Conference of the International Speech Communication Association, Dresden, Germany, 6–10 September 2015.

83. Audebert, N.; Le Saux, B.; Lefèvre, S. Deep learning for classification of hyperspectral data: A comparative review. *IEEE Geosci. Remote Sens. Mag.* **2019**, *7*, 159–173. [CrossRef]

84. Pradhan, B.; Lee, S. Landslide susceptibility assessment and factor effect analysis: Backpropagation artificial neural networks and their comparison with frequency ratio and bivariate logistic regression modelling. *Environ. Model. Softw.* **2010**, *25*, 747–759. [CrossRef]

85. Tien Bui, D.; Shahabi, H.; Shirzadi, A.; Chapi, K.; Alizadeh, M.; Chen, W.; Mohammadi, A.; Ahmad, B.B.; Panahi, M.; Hong, H. Landslide detection and susceptibility mapping by airsar data using support vector machine and index of entropy models in cameron highlands, malaysia. *Remote Sens.* **2018**, *10*, 1527. [CrossRef]

86. Mersha, T.; Meten, M. GIS-based landslide susceptibility mapping and assessment using bivariate statistical methods in Simada area, northwestern Ethiopia. *Geoenviron. Disasters* **2020**, *7*, 1–22. [CrossRef]

87. Pham, B.T.; Prakash, I.; Singh, S.K.; Shirzadi, A.; Shahabi, H.; Bui, D.T. Landslide susceptibility modeling using Reduced Error Pruning Trees and different ensemble techniques: Hybrid machine learning approaches. *Catena* **2019**, *175*, 203–218. [CrossRef]

88. Davis, J.; Goadrich, M. The relationship between Precision-Recall and ROC curves. In Proceedings of the 23rd International Conference on Machine Learning, Pittsburgh, PA, USA, 25–29 June 2006; pp. 233–240.

3

Opportunities and Challenges of Geospatial Analysis for Promoting Urban Livability in the Era of Big Data and Machine Learning

Anna Kovacs-Györi [1,*], **Alina Ristea** [2], **Clemens Havas** [3], **Michael Mehaffy** [4], **Hartwig H. Hochmair** [5], **Bernd Resch** [3,6], **Levente Juhasz** [7], **Arthur Lehner** [3], **Laxmi Ramasubramanian** [8] **and Thomas Blaschke** [3]

[1] IDA Lab, University of Salzburg, 5020 Salzburg, Austria
[2] Boston Area Research Initiative, School of Public Policy and Urban Affairs, Northeastern University, Boston, MA 02115, USA; a.ristea@northeastern.edu
[3] Department of Geoinformatics—Z_GIS, University of Salzburg, 5020 Salzburg, Austria; clemensrudolf.havas@sbg.ac.at (C.H.); bernd.resch@sbg.ac.at (B.R.); arthur.lehner@stud.sbg.ac.at (A.L.); thomas.blaschke@sbg.ac.at (T.B.)
[4] Centre for the Future of Places, KTH Royal Institute of Technology, 114 28 Stockholm, Sweden; michael.mehaffy@gmail.com
[5] Fort Lauderdale Research and Education Center, University of Florida, Davie, FL 33314, USA; hhhochmair@ufl.edu
[6] Center for Geographic Analysis, Harvard University, Cambridge, MA 02138, USA
[7] GIS Center, Florida International University, Miami, FL 33199, USA; ljuhasz@fiu.edu
[8] Department of Urban & Regional Planning, San José State University, San Jose, CA 95192-0185, USA; laxmi.ramasubramanian@sjsu.edu
*   Correspondence: anna.gyori@sbg.ac.at

Abstract: Urban systems involve a multitude of closely intertwined components, which are more measurable than before due to new sensors, data collection, and spatio-temporal analysis methods. Turning these data into knowledge to facilitate planning efforts in addressing current challenges of urban complex systems requires advanced interdisciplinary analysis methods, such as urban informatics or urban data science. Yet, by applying a purely data-driven approach, it is too easy to get lost in the 'forest' of data, and to miss the 'trees' of successful, livable cities that are the ultimate aim of urban planning. This paper assesses how geospatial data, and urban analysis, using a mixed methods approach, can help to better understand urban dynamics and human behavior, and how it can assist planning efforts to improve livability. Based on reviewing state-of-the-art research the paper goes one step further and also addresses the potential as well as limitations of new data sources in urban analytics to get a better overview of the whole 'forest' of these new data sources and analysis methods. The main discussion revolves around the reliability of using big data from social media platforms or sensors, and how information can be extracted from massive amounts of data through novel analysis methods, such as machine learning, for better-informed decision making aiming at urban livability improvement.

Keywords: spatial data science; livability; urban planning; big data; urban assessment; spatio-temporal analysis

1. Introduction

International landmark agreements such as the Sustainable Development Goals (SDG) or the New Urban Agenda (NUA) have focused attention on the improvement of the quality of urban life in the past few years [1,2]. These agreements are addressing several natural and social challenges in cities incaluding phenomena such as resilience to natural disasters, climate change, air pollution, access to clean water, public health, or equity in the access of public spaces. By tackling these issues, the overall goal is to provide better and safer circumstances for the growing urban population worldwide. Nowadays, also thanks to these international agreements, the existence and severity of these problems is getting more obvious and acknowledged, along with the importance of raising awareness amongst citizens and decision-makers. However—especially from the urban planners' point of view, by whom the majority of the local actions should be initiated—the NUA and SDG targets represent only the starting point of the long and challenging endeavor of actually solving the mentioned problems [3–6]. Although researchers and decision-makers are aware of what goals should be achieved and why, the actions should be adapted to the specific local circumstances and shorter timeframes, which is highly challenging mainly due to the complex nature of cities. In this paper, we intend to illustrate the opportunities and limitations of Geographic Information Systems (GIS)-based urban analysis techniques and novel data sources such as social media or sensor data to address these challenges of urban livability improvement, which includes big data analysis and machine learning approaches, among others.

The last decade has seen an increasing integration of GIS in everyday life and an almost ubiquitous production of big data across various contexts. These coupled technological and societal developments have resulted in the emergence of new fields, such as urban data science, and urban informatics [7–9]. These new intellectual domains enrich the study of urban systems and the practice of urban and regional planning or policy by making use of new datasets and novel methodological approaches [10,11]. Various disciplines are involved in shaping these interdisciplinary fields, including computer science, geography, geographic information science (GIScience), as well as urban studies. Such a revolutionary take on the investigation of cities brings along several advantages and can also benefit practical domains like urban planning.

However, understanding cities and their processes thoroughly, which is a key to improve urban livability, requires exploration beyond the merely technical and data-driven approaches to analysis techniques offered traditionally by methodological disciplines. This is what urban informatics and urban data science aim to achieve as a field "at the intersection of people, place and technology with a focus on cities" [7] (p. 1). Doing so will result in a better understanding of the theoretical foundations underlying human behavior in urban environments. This, in turn, is necessary to conceptualize and assess livability and to provide information for urban planners about the processes at different spatial (and often temporal) scales.

Although machine learning, urban data science and in general the use of big data in urban analytics have already received attention in the literature [10,12–14], the goal of this paper is to discuss some synergies and challenges of the interdisciplinary approach on GIS-based urban analysis in improving the livability of cities, by reviewing the key findings and open questions emerging from the state-of-the-art literature. The current work addresses three interrelated research questions concerning new data sources and novel analysis techniques in the context of cities and urban planning.

- RQ1: How can we assure the reliability of results based on social media data and Volunteered Geographic Information (VGI) for urban livability analysis and planning?
- RQ2: How can geospatial analysis aid urban livability assessment and planning that relies on machine learning methods?
- RQ3: How can relevant information be identified in urban big data to facilitate urban livability improvement?

Before investigating these questions in detail, Section 2 reviews urban theories related to cities and livability, as well as highlights use cases to demonstrate the potential of spatio-temporal analysis for urban planning using Twitter data. Section 3 provides a summary of methodological and data-related aspects of RQ1 and RQ2, whereas Section 4 describes a potential approach to extract information ('signal') from the 'noise' of vast amounts of data focusing on the usage patterns and livability of cities instead of purely data-driven solutions, which is closely related to RQ3. Finally, Section 5 summarizes relevant findings concerning the potential and pitfalls of urban analysis and presents open questions for future research. We hope that this work will put forward the dialogue between methodological and academic research and benefits urban planning practitioners as well. Also, we hope to trigger a more intense debate about the values and challenges implied by making more extensive use of data in urban planning along with livability assessment and improvement.

2. Urban Theories and Assessment in the Light of Livability and BIG Data

In the 1930s, Lewis Mumford posed a simple question, "What is a City?" [15]. The simple answer, of course, is that cities are surprisingly complex. We argue that it is this complexity that has confounded legions of architects, planners, engineers, data scientists, and city managers who seek to impose a spatial (and social) order to make cities more 'manageable'. Historically, social scientists have proposed theoretical frameworks that help to explain and predict the morphological structure of cities beginning in the early part of the 20th century such as Park and Burgess [16] to contemporary theorists like David Harvey [17] and Sassen [18]. For these social scientists, the city's complexity is shaped by demographic, social, political, social, and technological changes. Collectively, these scientists take a process-oriented approach to city transformations. Yet, for the past 200 years or so, modern city development has also been shaped by positive visions of the future; many of which were products of individual imaginations [19,20] specifically intended to address the challenges of urban living. All these historic visions have addressed the challenges of density (some visions have advocated for more density than less) and spatial scale (large megalopolises versus small and intimate), alongside considerations of efficiency, access, and mobility. It is important to note that these are the same concerns that drive the conversations about 'smart cities' and the use of technology and data in urban planning.

The intention to assess cities and their livability has a long history, and the ubiquity of digital technology and big data opened up new horizons also in this regard [21]. The emergence of new aspects and methodologies such as big data [22], VGI [23,24], or citizen-generated data, coupled with social media analysis [25,26], and even remote sensing or earth observation [27] might offer relevant alternatives to traditional methodologies and concepts mentioned above. These approaches provide ways to extract useful information from data, also at finer spatial and/or temporal scales than national statistics, and if needed, even for individuals or various social groups. To a certain extent, it is now at least technologically possible to analyze how people interact with the urban environment. However, quite often this analysis focuses on only quantitative approaches [26,28–32] and neglects the subjective needs and perception of the residents. Therefore, it is the socio-spatial theories of urban planning that offer the most promise when trying to utilize advantages provided by big data to explain, predict, and manage future cities.

2.1. Urban Morphology Assessment

Urban planners, particularly scholar-practitioners have long championed bottom-up theory development, in other words, the development of explanatory frameworks about urban living through inductive reasoning. For instance, Kevin Lynch [33] proposed that urban environments that have clearly "imageable" (memorable) characteristics are more likely to be recognized, i.e., legible. Lynch put forward the notion that when cities or urban environments are memorable, that they create feelings of safety and security, and that they encourage movement and participation in the life of the city. Lynch proposed that humans created environmental images of their built environment by creating mental maps. These maps were easier to create for some cities/urban environments than others. Lynch also proposed a vocabulary to understand and describe urban environments—paths, edges, nodes, landmarks, and districts. Lynch used observational and survey data from everyday participants in three cities—Boston, Jersey City, and Los Angeles—to arrive at his seminal conclusions. The power of urban analytics nowadays provides the opportunity to better understand how individuals create and store their mental maps and how these maps are deployed under different circumstances—for example, in times when they are under time constraints, or in other stressful situations. In addition, it facilitates to understand how people move spatially and temporally through urban environments, Lynch's theoretical frameworks can help identify and replicate successful urban morphologies—some street patterns work well and hold up to the test of time, others do not. Big data approaches help rapid prototyping and design of new environments in ways previously not feasible [34].

2.2. Urban Livability Assessment

In an environment of rapid transformation and locational competition on a global scale, research on the livability of cities has gained much attention during the last years. Research investigating the quality of life in cities has already started very early to deal with the spatial relationships of and within cities due to the complexity and diversity of livability standards, taking into account the global aspect of such standards and indices [35–37]. One declared important objective of such research is the objectification of economic, social, and structural-spatial factors of influence. In this context, Onnom et al. [38] discuss several indices that allow for measuring the quality, the development and the livability of cities. Others investigate the relationship between the livability of an urban area and its (sustainable) development [37,39].

Based on the research of Kovacs-Györi et al. [40] we argue that livability can serve as a useful conceptual and analytical framework to assess and improve the quality of urban life, also considering the personal aspects of the residents by focusing on the personal-environment relationship instead of merely statistical indices with no spatial relevance. Figure 1 shows the key elements and their relationship in livability to clarify what is meant conceptually when using this term throughout the article [40]. The lower part represents the urban environment (i.e., actual state), which consists, on the one hand of the built and natural environment along with the infrastructure, described as urban form. It also includes what this environment can provide for the citizens as urban functions, reflecting how the people use the urban space in the light of the functionality. This component of livability is more tangible, and spatial data for further analysis are available or acquirable, respectively. The upper part of the figure corresponds to the individual preferences (i.e., expectations) in livability and the person–environment relationship, consisting of human needs that the urban functions (or even urban form) need to satisfy and the personal values that define and influence the person's preferences. Due to the high level of subjectivity, individual preferences bring an inherent limitation to livability assessment, as means or data to extract these preferences are often not available in an intersubjective way. Based on this framework, urban livability describes how well the environment can satisfy citizen needs and expectations, i.e., to what degree the actual state matches the goal state.

However, as mentioned above, in the context of urban planning analysis, the difficulty exists in two main circumstances.

First, it is often challenging to derive recommendations for actions to urban planners, because they are often limited in the actions they can take. In private space, planners can only be active in supervision or demand for specifications in a restricted framework. In public space, they have more freedom of action, yet they are bound by various guidelines and social necessities and must take into account the private space. The translation of identified indicators and parameters into planning-relevant actions must therefore take place in such a way that it facilitates the identification of space for maneuver. Such methodologies target the simplification of the identification of the effects of measures and facilitate scenario formation when there are alternative courses of action.

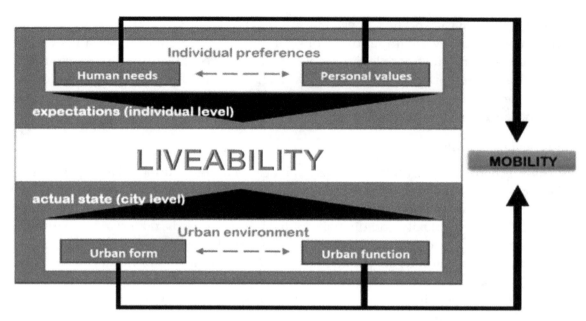

Figure 1. Conceptual framework for livability [40].

Furthermore, the mere accessibility depicted by minutes or meters cannot reflect the purpose or the quality of the satisfaction of needs at the target destination nor the actual quality on the way itself; the crucial question why a way or a means of transport has not been chosen; as such, this question cannot be directly mapped or only indirectly depicted. In times of resource scarcity and decreasing urban budgets, providing answers to such questions is essential for efficient urban planning. In this way, and as highlighted in Figure 1, mobility has a special role within livability assessment both in terms of conceptual and practical aspects. Collecting data on mobility is quite well-established and also various analysis techniques are available depending on the purpose of the analysis. Besides direct information about the paths or routes of residents, different approaches exist to provide information about mobility and its qualitative aspects. Pigliautile et al. [41] present wearable sensing techniques and automatized data-processing for urban microclimate mapping. Such mapping could provide information about the comfort of pedestrians and for other outside traffic participants that move within the urban area. A single trajectory can tell a lot in terms of subjective preferences and the satisfaction of needs that are both cornerstones of livability assessment as described above. This is because the destination usually reveals the need to be satisfied whereas the transportation mode and the route taken can provide insights into personal characteristics or routing preferences. Overall, we think that mobility analysis where livability aspects are considered (even only at a conceptual level) can combine the strong theoretical knowledge from urban planning with the new technological advancement from GIS and big data resulting in new approaches to improve urban quality of life.

2.3. Illustrations of GIS-Based Social Media Data Analysis to Support Urban Planning

In this section, we provide practical examples from previous research, to illustrate the potential of geospatial analysis for urban planning purposes using social media data. These examples provide only an insight into how geospatial analysis of Twitter data can facilitate urban planning but we do not provide an extensive review of the state-of-the art approaches in this field. However, Section 3 reviews more in depth the potential of VGI in general for these purposes.

In Sections 2.3.1 and 2.3.2, it is demonstrated how the analysis of Twitter data can support urban planners by providing new insights about urban phenomena and challenges. Concept-wise, morphology (2.1) and livability (2.2) assessments are good examples of methodological solutions from the GIS domain to provide new insights for urban theories that have a long history. This often happens by applying mixed methods to grasp the complexity of urban phenomena. Mixed methods research originated in the social sciences and it is a third category of approaching an analysis, by combining the first two categories, namely qualitative and quantitative. The main idea of this methodology of research is that such complex integration permits developing better synergies in using the data than by following separated collection and analysis [42]. More than two decades ago, urban researchers discussed the possible integration of mixed methods strategies in planning, with examples from New York City [43]. However, with the growth of data and technologies, the in-depth mixed methods got into development about a decade ago. Although these approaches are often difficult to evaluate, due to the data-rich environment they can still be informative, as it is illustrated throughout this section.

2.3.1. Towards Citizen-Contributed Urban Planning Using Twitter Data—Case Study for Planned Large Events

In a 2018 study, Kovacs-Györi et al. [44] investigated how social media data can be utilized to investigate planned large events in cities, using the Olympic Games in London, 2012. Similar studies have been performed also for the Rio Olympics [45], transportation during planned (and unplanned) events [46], or to detect events based on Twitter data [47–49].

Using over 12 million tweets, the authors in [44] identified potential residents and visitors of the Olympic Games in 2012 and analyzed their tweeting behavior in terms of spatial and temporal patterns, as well as the sentiments of their tweets. The analysis of the spatio-temporal and sentiment patterns revealed that it was possible to identify (spatial) hotspots for both groups during the Olympic Games, and these patterns were significantly different. The real potential of this result is to provide information for planners and decision-makers even during an event with a quite high spatial and especially temporal resolution, which is often a limitation with traditional surveys or other existing authoritative methods. However, it is important to emphasize that the usefulness of social media data (or VGI in general) highly depends on the purpose of the analysis. As different social media and VGI platforms are not used by each demographic groups equally, researchers must consider how this inherent selection bias might influence their results. In the case of a planned event or disasters the high spatial and temporal resolution of the data and their near-real time availability bring clearly more advantage and information even if it is not fully representative, compared to other scenarios, where representativeness might be essential such as the availability of specific urban services. Section 3 describes further considerations and analysis techniques regarding the reliability of different VGI data sources.

Another method demonstrated in the paper [44] was topic modeling to extract topics from the tweets of the previously identified groups of visitors and residents. This method can also be interesting for urban planners and other stakeholders to get indirect feedback regarding different aspects (e.g., transportation) from the two groups during an event. In a near-real-time scenario, it can be informative, especially in bigger amounts. The paper includes further assessment potentials both on micro and macro scale, usually by suggesting the combination of different data sources to dig even deeper into urban phenomena.

2.3.2. Classifying Parks and Their Visitors in London Based on Twitter Data Analysis

In another study, Kovacs-Györi et al. [50] investigated how social media data analysis can help the planners monitoring urban park visits and the perception of the visitors. Twitter data is often used for this purpose [51], for example to investigate seasonal variation in physical activities [52], or determine the emotional responses of people to urban green spaces [53]. Similar to the above-presented case study for the Olympic Games and other large events, the high spatial and temporal resolution of the tweets can provide new insights compared to other methods, such as in-situ assessments (surveys, counting, etc.). The study included the analysis of the distance between park tweets and the activity center of the user (geometric average of the coordinates where the given user tweeted from), followed by a sentiment analysis and emotion extraction of the tweets. By performing spatio-temporal analysis and clustering it was possible to identify different types of parks based on the time of the visit (during the day, week, and year) and also to identify parks where people tended to tweet more positively or experiencing specific emotions. The research demonstrated, also in this case, the potential of using social media and GIS analysis for urban planning in this regard, as well as pointed out some of the limitations (most of all the lack of representativeness of the dataset). Overall, this method can serve as an input for more in-depth analysis performed by planners or other stakeholders, by revealing new insights or providing information about parks in greater amount and for a longer time than it is possible with traditional techniques, such as questionnaires.

3. Valuable Data Source Types and Methodologies for Analyzing Complex Urban Systems and Their Quality

Urban informatics can tap into a rapidly growing number of data sources providing a large variety of data sets of different spatial and temporal scales, completeness, and reliability in different combinations [54]. For example, metro card user data or social media check-ins allows analyzing and modeling people's individual movement, property appraisal data allow for an estimation of parcel values and determining factors using hedonic models, whereas traffic cameras or inductive loops provide information about traffic congestion at certain road segments. On the temporal scale, weather station data from fixed sensors provide both hourly information but can also be used for analyzing long term trends in climate change. Similarly, floating car data provide continuous information about speed, however at varying locations depending on the cars' locations. Other geo-data, such as household traffic surveys or on-board transit surveys, are collected less frequent at a smaller sample size which limits suitability for local analysis tasks. The different channels provide a large pool of data, which renders machine learning algorithms and artificial intelligence potentially useful for the identification of spatio-temporal activity patterns of various phenomena in urban environments at all spatial and temporal levels. Artificial intelligence, including machine learning and natural language processing as methods for urban data science, have, for example, been used for the early detection of disease outbreaks, such as COVID-19, Zika, or H1N1 Influenza, based on big data from health-wearable devices [55], or for urban flood risk mapping in urban environments [56].

3.1. User-Generated Data Sources

Quantitative urban research has traditionally relied on data from censuses, surveys, and specialized sensor systems. Low response rates of surveys and the high cost associated with surveys and replacing sensor systems, have, however, led to increasing interest in alternative ways of supplementing the urban data infrastructure [28]. The emergence of Web 2.0 has led to a plethora of VGI and social media data as part of big data that has been used for a wide range of urban analyses also addressing societal questions. This includes exploring urban dynamics from open and user-generated content posted on location-based social networks Foursquare, Google Places, Twitter, and Airbnb [57], assessing the effect of natural disasters on urban mobility patterns using tweets [58], or for monitoring the development of the COVID-19 epidemic through WeChat keyword analysis [59]. User-generated data sources provide different types of geo-data, including points of interest, event patterns, travel trajectories, traffic information, sentiments, or social activities. Such data form the building blocks for geo-applications and analysis tools in the context of urban planning. Consequently, knowledge about the quality of crowdsourced data is important to determine their fitness for purpose [60]. Whereas the abundance and multi-dimensionality of crowd-sourced big data lend itself to machine learning techniques in the spatial and temporal domain, machine learning has also been used to assess and improve the data quality of VGI, such as OpenStreetMap [61], and to remove automated bots in social media data, such as tweets [62].

The International Organization for Standardization Technical Committee (ISO/TC) 211 developed a set of international standards that define the measures of geographic information quality and provide a structure for describing digital geographic data. The ISO metadata standard 19,115 [63] defines the schema required for describing geographic information and services by means of metadata. It distinguishes between five quantitative measures (completeness, consistency, positional accuracy, temporal accuracy, and thematic accuracy) and three qualitative quality indicators (purpose, usage, and lineage). At the emergence of crowdsourced data comparison to authoritative data (called extrinsic methods) was a common approach to determine data quality [64,65]. With the absence of available comparison data, additional qualitative indicators as a proxy for traditional ISO-defined data quality indicators [66] were developed, partially mitigating the need for reference data but rather relying on contribution history. Hence, data quality assessment began to utilize intrinsic methods recently as well [67]. These methods rely exclusively on the analyzed data source itself. Examples include the analysis of historical metadata as a means of inferring the inherent quality of the data [68], or the measurement of text metrics (e.g., number of words, structure, spelling) to determine the quality of crowd-sourced text content [69].

Alternatively, to determine their data quality, crowd-sourced data can also be compared to other non-authoritative data [70–72], to the surrounding environment [73], or data contributed by other volunteers [74]. Furthermore, a volunteer's trustworthiness, credibility, experience, recognition, or reputation can be used as a proxy for the quality of contributed data [75,76]. Other measures to determine the quality of user-generated content include its localness [77,78], or its vagueness [79]. Vagueness can, for example, be caused by low-resolution imagery, or by capturing only a snapshot of a temporal phenomenon. Data quality of crowdsourced datasets vary between world regions [80], and the spatial distribution of such datasets is uneven [81]. Since urban environments are centers of human activities (jobs, social life, leisure, high-density housing), urban areas provide a rich source of geo-data, which are potentially suited for spatial and spatio-temporal analytics in different research domains including human mobility, health and environment, social interaction, food and nutrition, gentrification, land-use, and urban functionality. These geo-data are of different lineage and type, and therefore may be best described with different quality measures or indicators. A detailed overview of existing methods to assess the quality of map-based VGI (e.g., positional accuracy, completeness), image-based VGI (e.g., thematic accuracy of tags), and text-based VGI (e.g., credibility) is provided in the work of Senaratne et al. [82]. Table 1 summarizes commonly used VGI quality assessment methods (extrinsic vs. intrinsic) together with the type of used reference datasets and quality measures. An x indicates the primary use of these measures, whereas (x) indicates less common use for a given method.

Table 1. Different methods and measures for VGI quality assessment

Assessment Method	Extrinsic		Intrinsic
Comparison Reference Dataset	Authoritative	Non-Authoritative	None
Measures			
Quantitative measures (completeness, consistency, positional/temporal/thematic accuracy)	x	x	(x)
Qualitative indicators (purpose, usage, lineage)			x
Proxy quality indicators (trustworthiness, credibility, text content quality, vagueness, local knowledge, experience, recognition, reputation)	(x)	(x)	x

3.2. Improving Data Quality and Reliability in Urban Analysis by Combining Data from Several Crowd-Sourcing Platforms

A user's credibility has traditionally been derived from contribution patterns to individual citizen science projects, mapping platforms, or social media apps—such as Ushahidi [76], Flickr [83], or Twitter [84]—which also led to the development of methods for detecting data vandalism [85]. However, lately, individual users began to become active in multiple crowdsourcing platforms [86], and to use multiple social media services simultaneously. Therefore, we propose to extend the assessment of a user's credibility and trustworthiness through consideration of activities across multiple platforms. One obvious challenge in that aspect is to identify which users contribute to multiple platforms. This is not a trivial task and largely unaddressed in the GIScience literature. An online survey—conducted by Levente Juhász [87]—was performed to test this hypothesis on a small sample. Participants were asked about their geospatial application usage and requested to share their online activity in 10 geospatial applications. Participants were recruited through university mailing lists, discussion forums and social media channels, and the study mainly focused on Europe and the United States. Figure 2 illustrates the results where 32 out of 53 responding users indicated that they use more than one geolocated social media or crowdsourcing platform. Considering the identification of individuals who are using multiple social media platforms, a common approach is to try to match user names between different platforms. More computationally complex and resource-intensive methods could apply concepts of space-time geography on user contribution patterns to identify contributions to different platforms from the same user [88]. The basic idea behind this would be that space-time prisms, which show an individual's reachable positions based on a person's capability constraints (e.g., travel speed) or authority constraints (e.g., rules and regulations), determine whether locations extracted from different data sources could have been visited by the same individual user or not. With millions of users contributing to different platforms, matching up contributions from different platforms and identifying them to be from the same user, if so, will require the use of advanced space-time indexing, data mining approaches, and machine learning techniques [89]. There are also ethical challenges associated with linking individuals and their activity between multiple online platforms, which stem from the right to privacy. Therefore, combining data from multiple platforms should also consider locational privacy. This is best addressed by an approach that combines legal, ethical, technological, and education aspects of the issue [90,91].

Figure 2. Social media and crowdsourcing platforms used by a set of 53 users.

Although the small sample size of the study mentioned above does not allow to draw generalized conclusions about the percentage of social media users who use multiple platforms, we assume that data from multiple data sources (e.g., Twitter and Instagram) can be combined to provide a more complete spatio-temporal representation of an individual's activities compared to information extracted from a single platform (e.g., Twitter). An illustration of this is given in Figure 3, where maps a–d show locations and activities extracted from Web 2.0 platforms over the course of a day for one user in a region in southern Austria. Figure 3e overlays all geocoded activities of the same individual during the same day and therefore provides a more complete spatial representation of the user's activities than one single platform. It also provides information about transportation between the cities (i.e., Strava cycling activity). Although this example suggests that the approach leads to increased credibility and completeness, this step of data fusion from different sources needs to be repeated for a larger, more representative, and diverse user base to be able to draw conclusions about its usefulness and applicability for a more general population. Challenges, such as feature matching, which are inherent to data fusion from different sources [92], need to be addressed within this process.

Figure 3. Spatial representation of user activities extracted from single platforms for a day (**a–d**), and data sources combined together providing a more complete picture of the user's activity locations (**e**).

Exploring the relationship between social media activity spaces of a user on different platforms and quantifying their similarity is a novel research direction [93]. Understanding the limitations of using data from only one platform and the potential benefits of combining user data on the individual level is relevant, since urban analytics already utilizes social media and crowdsourced datasets to extract valuable spatio-temporal information to understand human mobility and citizen opinions on certain topics [40,44].

A new trend in contribution patterns of user-generated content is also to cross-reference information between platforms, i.e., to obtain information from one data source and insert it into another, which can be referred to as "cross-tagging" [86]. An example is to use the hotel name found on a Mapillary image, which is then attached to an OSM feature as a name attribute value during the OSM editing procedure, based on which the Mapillary photo ID is linked to the map feature. Automated systems can then utilize this link to display an image of the linked map feature. In addition, contributors can also look up information from various data sources and platforms and use that information in other platforms, without explicitly documenting this process in the edited data. This process can be referred to as "cross-viewing" [94]. It is yet an open question if and how cross-tagging and cross-viewing affect data quality of the connected data sources. On the one side, it can be argued that cross-tagging and cross-viewing add more eyes on the data, which should improve data quality, based on Linus' Law [95]. However, if the original data is not cross-checked, its potentially inferior data quality may propagate across platforms and negatively affect the data quality of the targeted data source, similar to how correct (or incorrect) news can propagate within the social media landscape [96] or how errors propagate in measurement-based GIS into the indirect values of the end product [97]. Joint use of datasets from different sources can provide better completeness, e.g., in feature change detection over time, which has been applied in urban environments [98]. However, different demographic groups do use crowd-sourcing tools, such as bicycle tracking apps [99] or social media apps [100] differently, which is important to consider in efforts of data fusion from different apps and in conclusions drawn from this. As far as cross-tagging goes, we propose to add another quality measure of user-generated content, namely the abundance of cross-linked data sources in the data set of interest.

In conclusion, it can be stated that the evolvement of novel data sources from user-generated content, and their combination, necessitate a review and adaptation of quality metrics to appropriately capture and describe their validity and usefulness.

3.3. Potential of Machine Learning Algorithms in Geospatial Urban Analysis and Assessment

In a recent transformational development, the GIScience community has faced a transition from a data-scarce to a data-rich environment [101]. This advancement also reshapes the scientific discipline of geoinformatics through the more acknowledged investigation of data-driven approaches. Therefore, the development of new methods for data acquisition, storage, and analysis become inevitable, including unsupervised machine learning algorithms or semi-supervised learning systems. This also applies to the particular idea of assessing walkability, where researchers have used machine learning algorithms, for instance, on Google Street View imagery to generate three measures on visual enclosure [102]. The proposed methodology addresses the dimension of urban design that has usually been analyzed through urban characteristics like street network density or block size. The authors adopt an approach that uses Artificial Neural Networks (ANN) and Support Vector Machines (SVM) to identify sky areas for measuring the proportion of visible sky from using Google Street View imagery, from which a measure for visual enclosure is derived.

Also, Resch et al. [25] developed a methodology that extracts emotion information from social media for improving urban planning processes. The authors present a semi-supervised machine learning approach that is based on similarities between social media posts in geographic space, temporal space, and semantic space. They conclude that social media are a valuable and reliable data source for urban planning related issues, but that traditional GIS methods are oftentimes not appropriate for analyzing social media data. These and other recent machine learning-based approaches show that the dichotomy between purely hypothesis-driven approaches (mostly developed by domain experts) and data-driven approaches (oftentimes pursued by computer scientists and statisticians) are currently closing up. In other words, urban planners tend to have a better understanding of the technical capabilities of cutting-edge technology, while computer science-oriented researchers become more and more familiar with the concepts developed in the urban planning community. This makes the field of urban analysis a highly interdisciplinary one by bringing together experts from a variety of scientific fields.

With respect to the analysis of human-generated data like social media to better understand urban contexts, the high degree of uncertainty—including textual ambiguities, positional and temporal inaccuracies, and semantic irregularities—has sparked the development of new machine learning-based analysis methods [12]. Various machine learning methods have recently been used and developed for analyzing urban spaces, including self-organizing maps to map recreational beneficiaries [103], to analyze urban traffic conditions [104], and to mine latent urban activity patterns [105], or methods of natural language processing (NLP) to extract urban perceptions [106]. These research efforts allow for generating new, unseen insights into urban processes through analyzing a new set of data sources that previously posed severe challenges to scientific research due to their unstructured, unstandardized, and voluminous nature.

Concluding, these new algorithms seem to be a promising avenue for several reasons: First, they reduce or eliminate the need for a priori knowledge with respect to linguistic structures, geospatial correlations, or semantic meaning. Second, it is possible to incorporate the geospatial dimension into the analysis process, in many cases in a simple fashion without having to modify the original method. This step is essential because many machine learning algorithms have originally not been designed for handling and analyzing geospatial data. Third, machine learning methods have the power to deal with large amounts of data in that data-driven approaches can be applied to mine latent, unanticipated patterns in human-generated data [12].

4. Potential Methodological Problems for Urban Planning and Livability

4.1. Finding the Signal in the Noise

For urban planners, the era of large-scale geodata sets, and other forms of big data, is creating remarkable opportunities, but also raising a number of important problems. One of the most challenging problems is the sheer volume of the data available, which is often noisy in character. That is, it has inherent imprecision, ambiguity and inaccuracy, often obscuring both the full dynamics of urban challenges, and the path to their resolution. Researchers have an opportunity to help find strategies to clarify the relevant information—the 'signal'—amongst an increasing deluge of information noise.

To be sure, many specific benefits of urban big data, and its related smart cities movement, have been described enthusiastically and in detail by proponents [107,108]. Among the benefits touted, information from sensors can be used to ease traffic jams, optimize energy demand, and give users real-time updates on transit and other systems. New mapping technologies can identify, and help to correct, urban problems that were previously much more difficult to identify and manage. Social media and crowdsourcing can target problems large and small, down to which potholes to fix. In this and other respects, it is claimed, the new technologies can even make governments more responsive to citizen needs [109].

There is, of course, a darker view of big data and smart cities. Both are criticized as fads, as technology marketing schemes, or as tools of covert manipulation and control, threatening to abridge political and personal rights (notably including the right to privacy). At best, they over-simplify the complex challenges of the city, with dubious results. For example, Hollands [110] asserts that with smart cities, researchers and decision-makers know too little about "what the label ideologically reveals as well as hides". Söderström, Paasche, and Klauser [111] attack smart cities as "corporate storytelling" that can "obfuscate more urgent needs" such as "technology-poor affordable housing or sewage systems (that) are arguably more urgent in many of the world's cities". Kitchin [112] argues that "data-informed urbanism" is being replaced by "data-driven urbanism", and its efforts too often "fail to recognize that cities are complex, multifaceted, contingent, relational systems, full of contestation and wicked problems that are not easily captured or steered, and that urban issues are often best solved through political/social solutions and citizen-centred deliberative democracy, rather than technocratic forms of governance".

However, the vast streams of data are already here, and so are the challenges. The more pointed question is, therefore, how will planners, decision-makers and other stakeholders manage this already existing (and growing) technology in order to make progress? In what ways might it be engaged as a helpful resource to deal with the most pressing human challenges, also addressed by SDG and the NUA? To avoid "data-driven urbanism" planners and researchers can start by asking, what is it that they wish the data in a "data-informed urbanism" to reveal to them? They must be clear in the first instance what they are counting, and why.

The focus of this paper, the goal of livability, points in an important and helpful direction. If livability is defined as the ability to endure, survive, inhabit, and experience quality of life in an environment, from that we can develop objective criteria to measure (for example) access to resources, economic opportunities, income, health, recreation, and freedom from dangers, threats, obstacles, pollution, and so on. It is also possible to develop subjective and intersubjective measurements of satisfaction and well-being, which help to indicate problem sources. From there, researchers and planners can identify both existing and new datasets that give us a picture of the current situation, and the specific ways that we might target improvements, most likely through an iterative process of action, measurement, and refinement.

Here is where another crucial point arises: Is it only the experts (researchers and planners), working as external agents who can contribute to the achievement of goals also defined by them? A number of researchers have demonstrated that a city is a partly self-organizing system, emerging from myriad incremental actions by distributed agents, i.e., the citizens [113,114]. In part, those actions create adaptations aimed at achieving a more livable environment with a higher quality of life. The data must therefore inform not only the experts working externally upon the city, but the actors themselves working within (and as part of) the city.

4.2. Distinguishing Information about the City, and Information within the City

To consider further opportunities and challenges, it might be important to draw a basic distinction between the information about a city, and the information within a city. The former is the customary focus of technologists and officials, who make measurements and draw conclusions for their own externally applied actions. The latter includes citizens, together with the structures of their environments. This addition is key: the interaction between information with the environment itself, and the people within it, generates new information and new outcomes, often quite apart from any external data process. In biology, this self-organizing phenomenon is known as stigmergy [115]. Applied to cities, it creates the capacity for urban self-organization [116]. This concept is also explored in actor network theory (ANT), developed in sociology [117]. Edelenbos et al. [118] argue that ANT points to an alternative working framework to the 'data determinism' that tends to dominate smart city discourse, and "accords specifically well with an understanding of cities as complex, adaptive, self-organizing systems".

A question then arises concerning the relation between the information about a city, and the information within a city, including stigmergic information. These two forms of information do not exist in isolation, but rather, they occur together and in characteristic interactions. For example, planners and policymakers create infrastructure and regulatory frameworks, which then go on to produce stigmergic patterns by users, which can then be studied and acted upon externally, and so on. As a particular example, planners might create street networks with traffic signals, which then produce the stigmergic patterns of traffic as it self-organizes, which is then modified according to feedback from smart city sensors, and so on. A more distributed example is the stigmergic response of users to smartphone traffic congestion data.

All of these different forms of information and stages of process require clarity of understanding, lest the tasks of geodata application become commingled and confused. We, therefore, close by enumerating several critical ones.

First is the distinction between centralized information and distributed information—and especially, distributed information that is interacted with, acted upon, often created by, distributed agents who are acting at a range of scales. Particularly important are the distinct and often overlapping scales of democratic governance, in what the political scientist Elinor Ostrom referred to as "polycentric" governance [119]. Those who engage geodata in the pursuit of urban goals (including livability) must surely honor the need of these democratic institutions and individuals to be empowered by this distributed information, and not merely to be controlled by it.

Second is the distinction between three different kinds of information, which operate in distinct modes:

- Descriptive information about the city, used to guide actions usually by centralized agents. Examples of descriptive information include maps, measurement datasets, and user survey data.

- Prescriptive information in the form of rules that prescribe actions. They generally produce static configurations, e.g., "all cars stop at the line when a light is red", and so on. Examples include zoning codes, traffic laws, and other regulations.
- Generative information. This is information within the city, operating iteratively between distributed agents and/or their environments (actor-networks). This kind of information is capable of generating emergent structures through self-organization, and more particularly, through the dynamic of stigmergy [116]. Examples include built-up environmental patterns (like the modest pathway changes in Figure 4), pop-up structures, neighborhood-scale cooperative projects, and other 'bottom-up' iterative changes by residents.

Figure 4. A modest example of stigmergy, where pedestrians build up pathway patterns over time that depart from what planners anticipate. Some planners revise their permanent structures based on these information patterns. Image: George Redgrave via Flickr.

It is important to note that descriptive, prescriptive, and generative information can all exist independently or in combination, and in either centralized or distributed forms—or indeed in combinations. There are of course cases in which distributed information would operate in tandem with centralized information, combining descriptive, prescriptive and/or generative kinds—say, in a Smartphone app. The key conclusion is that a geodata strategy that overlooks distributive and generative aspects is not necessarily wrong, but it is incomplete—perhaps dangerously so.

To be sure, the new era of geodata applied to urban livability offers exciting prospects. There is understandable excitement at the usefulness (and yes, the potential economic value) of vast new streams of information and new kinds of analysis about the city, with undeniably great capability and promise. However, as this discussion suggests, no less important is to consider how information moves within the city, within its distributed actors, and within its structures. The new technologies must surely work with both kinds of information, empowering residents and actors at all scales to identify their own signals to pursue within the noise, providing appropriate tools and strategies to do so—some new, and some no doubt yet to be developed.

5. Conclusions

Big data, coupled with machine learning algorithms and geospatial analysis bring without a doubt new insights into urban planning and livability improvement through the acquisition and extraction of vast amount of information about different urban systems. There are numerous disciplines contributing to this effort, therefore understanding the complexity and interrelationship of these urban systems clearly requires an interdisciplinary approach. In this paper, we highlighted some of the potential synergies of applying geospatial data sources and methodologies for urban issues based on reviewing the literature and discussing relevant issues and open questions, especially regarding the application of VGI and various geospatial analytics techniques for livability assessment and improvement. We want to acknowledge that this interdisciplinary take can be expected to strengthen the collaboration between urban planners, urban scientists, and researchers from the GIS and other domain rather than substituting traditional urban theories and urban analysis.

Through our research questions, we addressed some key aspects of urban analysis and livability assessment in the era of big data, such as the reliability of these new data sources, the potential and need of applying machine learning algorithms for big data related to urban problems, and the extraction of truly meaningful information from the vast amount of data in the case of cities and planning. The issue of data ethics and geoprivacy are gaining more and more relevance due to the advanced usage, especially when different data sources are linked to analyze users [120]. Future research should address these aspects as well more in depth; however, it was beyond the scope of the current article.

This paper although investigated just a subset of potential approaches or limitations—the field of studying cities might be as (or probably even more) complex than cities themselves. Our goal was to demonstrate the strength of these new data types and methodologies in livability assessment and improvement as key elements in urban planning.

Also, we want to emphasize the importance of collaborative work, and diversity integration. One example can be the relationship between academics and stakeholders such as policymakers, governmental bodies, non-profit organizations, urban planning businesses. Framing a research problem and a possible solution can make the difference between being listened or not by a large crowd or a group of decision-making experts. Overall, we cannot emphasize enough the relevance of livability, as an overarching concept in the case of urban analysis. Cities are for and about people, the ultimate goal of any urban analysis or planning action is (or should be) to enhance the quality of life in cities. This requires the consideration of qualitative aspects (such as the perception or satisfaction of the dwellers) as well, beyond the merely data-driven and technology-focused analysis. However, the fact should not be neglected, that this qualitative approach has inherent limitations, which can only be partially conquered by having big data or machine learning methods. Further research should consider and address these qualitative aspects by deepening the cooperation in making cities more livable between traditional urban research domains and the newer, more technology-based fields such as urban data science and GIS.

Author Contributions: Conceptualization, Anna Kovacs-Györi, Alina Ristea, Clemens Havas, and Michael Mehaffy; Writing—original draft preparation, Anna Kovacs-Györi, Alina Ristea, Clemens Havas, Michael Mehaffy, Hartwig H. Hochmair, Bernd Resch, Levente Juhasz, Arthur Lehner, Laxmi Ramasubramanian, and Thomas Blaschke; Writing—review and editing, Anna Kovacs-Györi, Hartwig H. Hochmair, and Levente Juhasz. All authors have read and agreed to the published version of the manuscript.

Acknowledgments: Open Access Funding by the Austrian Science Fund (FWF).

References

1. United Nations General Assembly. Transforming our World: The 2030 Agenda for Sustainable Development. 2015. Available online: https://sustainabledevelopment.un.org/post2015/transformingourworld/publication (accessed on 1 October 2020).
2. United Nations General Assembly. *Habitat III New Urban Agenda: Quito Declaration on Sustainable Cities and Human Settlements for All*; United Nations General Assembly: Quito, Ecuador, 2016.
3. Caprotti, F.; Cowley, R.; Datta, A.; Broto, V.C.; Gao, E.; Georgeson, L.; Herrick, C.; Odendaal, N.; Joss, S. The New Urban Agenda: Key opportunities and challenges for policy and practice. *Urban Res. Pract.* **2017**, *10*, 367–378. [CrossRef]
4. Kabisch, S.; Finnveden, G.; Kratochvil, P.; Sendi, R.; Smagacz-Poziemska, M.; Matos, R.S.; Bylund, J. New Urban Transitions towards Sustainability: Addressing SDG challenges (Research and Implementation

Tasks and Topics from the Perspective of the Scientific Advisory Board (SAB) of the Joint Programming Initiative (JPI) Urban Europe). *Sustainability* **2019**, *11*, 2242. [CrossRef]

5. Mehaffy, M.W.; Haas, T.; Elmlund, P. Public Space in the New Urban Agenda: Research into Implementation. *Urban Plan.* **2019**, *4*, 134–137. [CrossRef]

6. Elmlund, P.; Haas, T.; Mehaffy, M.W. Public Space in the New Urban Agenda. The Challenge of Implementation. *J. Public Space* **2018**, *3*, 165–170. [CrossRef]

7. Foth, M.; Choi, J.H.; Satchell, C. Urban informatics. In Proceedings of the ACM 2011 Conference on Computer Supported Cooperative Work—CSCW '11, Hangzhou, China, 19–23 March 2011.

8. Kang, W.; Oshan, T.; Wolf, L.J.; Boeing, G.; Frias-Martinez, V.; Gao, S.; Poorthuis, A.; Xu, W. A roundtable discussion: Defining urban data science. *Environ. Plan. B Urban Anal. City Sci.* **2019**, *46*, 1756–1768. [CrossRef]

9. Law, T.; Legewie, J. Urban Data Science. In *Emerging Trends in the Social and Behavioral Sciences*; Wiley: Hoboken, NJ, USA, 2018; pp. 1–12.

10. Batty, M. Big Data and the City. *Built Environ.* **2016**, *42*, 321–337. [CrossRef]

11. Singleton, A.D.; Spielman, S.; Folch, D. *Urban Analytics*; Sage: Thousand Oaks, CA, USA, 2017; ISBN 9781473958630.

12. Resch, B.; Szell, M. Human-Centric Data Science for Urban Studies. *ISPRS Int. J. Geo-Inf.* **2019**, *8*, 584. [CrossRef]

13. Balduini, M.; Brambilla, M.; Della Valle, E.; Marazzi, C.; Arabghalizi, T.; Rahdari, B.; Vescovi, M. Models and Practices in Urban Data Science at Scale. *Big Data Res.* **2019**, *17*, 66–84. [CrossRef]

14. Nosratabadi, S.; Mosavi, A.; Keivani, R.; Ardabili, S.F.; Aram, F. State of the Art Survey of Deep Learning and Machine Learning Models for Smart Cities and Urban Sustainability. In *Green Technology for Smart City and Society*; Springer Science and Business Media LLC: Berlin/Heidelberg, Germany, 2020; pp. 228–238.

15. Mumford, L. What's a City? *Archit. Rec.* **1937**, *37*, 235–264.

16. Park, R.E.; Burgess, E.W. *The City: Suggestions for Investigation of Human Behavior in the Urban Environment*; Chicago Press: Chicago, IL, USA, 1925.

17. Harvey, D. *Justice, Nature and the Geography of Difference*; Wiley: Hoboken, NJ, USA, 1997; ISBN 9781557866813.

18. Sassen, S. *The Global City: New York, London, Tokyo*; Book Collections on Project MUSE; Princeton University Press: Princeton, NJ, USA, 2001; ISBN 9780691070636.

19. Howard, E. *Garden Cities of Tomorrow*; Swan Sonnenschein & Co.: London, UK, 1902; ISBN 0571061893 9780571061891.

20. le Corbusier; Etchells, F. *The City of To-Morrow and Its Planning*; Rodker, J., Ed.; Courier Corporation: Chelmsford, MA, USA, 1929.

21. Batty, M. *The New Science of Cities*; MIT Press: Cambridge, MA, USA, 2013; ISBN 9780262019521.22.

22. Batty, M. Big data, smart cities and city planning. *Dialog Hum. Geogr.* **2013**, *3*, 274–279. [CrossRef]

23. Jokar Arsanjani, J.; Helbich, M.; Bakillah, M.; Hagenauer, J.; Zipf, A. Toward mapping land-use patterns from volunteered geographic information. *Int. J. Geogr. Inf. Sci.* **2013**, *27*, 2264–2278. [CrossRef]

24. Elwood, S.; Goodchild, M.F.; Sui, D.Z. Researching Volunteered Geographic Information: Spatial Data, Geographic Research, and New Social Practice. *Ann. Assoc. Am. Geogr.* **2012**, *102*, 571–590. [CrossRef]

25. Resch, B.; Summa, A.; Zeile, P.; Strube, M. Citizen-Centric Urban Planning through Extracting Emotion Information from Twitter in an Interdisciplinary Space-Time-Linguistics Algorithm. *Urban Plan.* **2016**, *1*, 114–127. [CrossRef]

26. Crooks, A.; Pfoser, D.; Jenkins, A.; Croitoru, A.; Stefanidis, A.; Smith, D.; Karagiorgou, S.; Efentakis, A.; Lamprianidis, G. Crowdsourcing urban form and function. *Int. J. Geogr. Inf. Sci.* **2015**, *29*, 720–741. [CrossRef]

27. Netzband, M.; Stefanov, W.L.; Redman, C. (Eds.) *Applied Remote Sensing for Urban Planning, Governance and Sustainability*; Springer: Berlin/Heidelberg, Germany, 2007.

28. Thakuriah, P.; Tilahun, N.; Zellner, M. *Introduction to Seeing Cities Through Big Data: Research, Methods and Applications in Urban Informatics*; Springer Science and Business Media LLC: Berlin/Heidelberg, Germany, 2016; pp. 1–9.

29. Wu, L.; Zhi, Y.; Sui, Z.; Liu, Y. Intra-Urban Human Mobility and Activity Transition: Evidence from Social Media Check-In Data. *PLoS ONE* **2014**, *9*, e97010. [CrossRef] [PubMed]

30. Sagl, G.; Resch, B.; Hawelka, B.; Beinat, E. *From Social Sensor Data to Collective Human Behaviour Patterns—Analysing and Visualising Spatio-Temporal Dynamics in Urban Environments*; Jekel, T., Car, A.,

Strobl, J., Griesebner, G., Eds.; GI_Forum 2012 Geovizualisation; Herbert Wichmann Verlag, VDE VERLAG GMBH: Berlin/Offenbach, Germany, 2012; pp. 54–63. ISBN 978-3-87907-521-8.

31. Jacobs-Crisioni, C.; Rietveld, P.; Koomen, E.; Tranos, E. Evaluating the Impact of Land-Use Density and Mix on Spatiotemporal Urban Activity Patterns: An Exploratory Study Using Mobile Phone Data. *Environ. Plan. A Econ. Space* **2014**, *46*, 2769–2785. [CrossRef]

32. Calabrese, F.; Colonna, M.; Lovisolo, P.; Parata, D.; Ratti, C. Real-Time Urban Monitoring Using Cell Phones: A Case Study in Rome. *IEEE Trans. Intell. Transp. Syst.* **2010**, *12*, 141–151. [CrossRef]

33. Lynch, K. *The Image of the City*; The MIT Press: Cambridge, MA, USA, 1960; ISBN 978-0-262-62001-7.

34. Boeing, G. Spatial information and the legibility of urban form: Big data in urban morphology. *Int. J. Inf. Manag.* **2019**, 102013. [CrossRef]

35. Giap, T.K.; Thye, W.W.; Aw, G. A new approach to measuring the liveability of cities: The Global Liveable Cities Index. *World Rev. Sci. Technol. Sustain. Dev.* **2014**, *11*, 176. [CrossRef]

36. Tan, K.G.; Woo, W.T.; Tan, K.Y.; Low, L.; Aw, G.E.L. *Ranking the Liveability of the World's Major Cities: The Global Liveable Cities Index (GLCI)*; World Scientific: Singapore, 2012; ISBN 981-4417-32-7.

37. Newton, P. Liveable and Sustainable? Socio-Technical Challenges for Twenty-First-Century Cities. *J. Urban Technol.* **2012**, *19*, 81–102. [CrossRef]

38. Onnom, W.; Tripathi, N.K.; Nitivattananon, V.; Ninsawat, S. Development of a Liveable City Index (LCI) Using Multi Criteria Geospatial Modelling for Medium Class Cities in Developing Countries. *Sustainability* **2018**, *10*, 520. [CrossRef]

39. Ley, A.; Newton, P.; Kallidaikurichi, S.; Yuen, B. Cretaing and Sustaining Liveable Cities. In *Developing Living Cities*; World Scientific Publisher Co Pte Lt.: Singapore, 2010; pp. 191–229.

40. Kovacs-Gyori, A.; Cabrera-Barona, P.; Resch, B.; Mehaffy, M.W.; Blaschke, T. Assessing and Representing Livability through the Analysis of Residential Preference. *Sustainability* **2019**, *11*, 4934. [CrossRef]

41. Pigliautile, I.; D'Eramo, S.; Pisello, A.L. Intra-urban microclimate mapping for citizens' wellbeing: Novel wearable sensing techniques and automatized data-processing. *J. Clean. Prod.* **2021**, *279*, 123748. [CrossRef]

42. Shorten, A.; Smith, J. Mixed methods research: Expanding the evidence base. *Evid. Based Nurs.* **2017**, *20*, 74–75. [CrossRef] [PubMed]

43. Gaber, J.; Gaber, S.L. Utilizing Mixed-Method Research Designs in Planning: The Case of 14th Street, New York City. *J. Plan. Educ. Res.* **1997**, *17*, 95–103. [CrossRef]

44. Kovacs-Gyori, A.; Ristea, A.; Havas, C.; Resch, B.; Cabrera-Barona, P. #London2012: Towards Citizen-Contributed Urban Planning through Sentiment Analysis of Twitter Data. *Urban Plan.* **2018**, *3*, 75–99. [CrossRef]

45. Condeço-Melhorado, A.; Mohino, I.; Moya-Gómez, B.; García-Palomares, J.C. The Rio Olympic Games: A Look into City Dynamics through the Lens of Twitter Data. *Sustainability* **2020**, *12*, 7003. [CrossRef]

46. Zhang, Z.; Ni, M.; He, Q.; Gao, J. Mining Transportation Information from Social Media for Planned and Unplanned Events, Buffalo, NY, USA, 2016. Available online: https://rosap.ntl.bts.gov/view/dot/30838 (accessed on 2 October 2020).

47. Lee, R.; Sumiya, K. Measuring geographical regularities of crowd behaviors for Twitter-based geo-social event detection. In *Proceedings of the 2nd ACM SIGSPATIAL International Workshop on Location Based Social Networks—LBSN '10, New York, NY, USA, 2 November 2010*; Association for Computing Machinery (ACM): New York, NY, USA, 2010; pp. 1–10.

48. Li, R.; Lei, K.H.; Khadiwala, R.; Chang, K.C.-C. TEDAS: A Twitter-based Event Detection and Analysis System. In *Proceedings of the 2012 IEEE 28th International Conference on Data Engineering, Arlington, VA, USA, 1–5 April 2012*; Institute of Electrical and Electronics Engineers (IEEE): Piscatway, NJ, USA, 2012; pp. 1273–1276.

49. Weng, J.; Yao, Y.; Leonardi, E.; Lee, F. Event Detection in Twitter. *Development* **2011**, *2*, 179–182. [CrossRef]

50. Kovacs-Gyori, A.; Ristea, A.; Kolcsar, R.; Resch, B.; Crivellari, A.; Blaschke, T. Beyond Spatial Proximity—Classifying Parks and Their Visitors in London Based on Spatiotemporal and Sentiment Analysis of Twitter Data. *ISPRS Int. J. Geo-Inf.* **2018**, *7*, 378. [CrossRef]

51. Roberts, H.V. Using Twitter data in urban green space research: A case study and critical evaluation. *Appl. Geogr.* **2017**, *81*, 13–20. [CrossRef]

52. Roberts, H.; Sadler, J.P.; Chapman, L. Using Twitter to investigate seasonal variation in physical activity in urban green space. *Geo: Geogr. Environ.* **2017**, *4*, e00041. [CrossRef]

53. Roberts, H.; Sadler, J.; Chapman, L. The value of Twitter data for determining the emotional responses of people to urban green spaces: A case study and critical evaluation. *Urban Stud.* **2019**, *56*, 818–835. [CrossRef]

54. Krishnamurthy, R.; Smith, K.L.; DeSouza, K.C. Urban Informatics: Critical Data and Technology Considerations. In *Springer Geography*; Springer Science and Business Media LLC: Berlin/Heidelberg, Germany, 2017; pp. 163–188.

55. Allam, Z.; Dey, G.; Jone, D.S. Artificial Intelligence (AI) Provided Early Detection of the Coronavirus (COVID-19) in China and Will Influence Future Urban Health Policy Internationally. *AI* **2020**, *1*, 156–165. [CrossRef]

56. Darabi, H.; Choubin, B.; Rahmati, O.; Haghighi, A.T.; Pradhan, B.; Kløve, B. Urban flood risk mapping using the GARP and QUEST models: A comparative study of machine learning techniques. *J. Hydrol.* **2019**, *569*, 142–154. [CrossRef]

57. Martí, P.; García-Mayor, C.; Serrano-Estrada, L. Taking the urban tourist activity pulse through digital footprints. *Curr. Issues Tour.* **2020**, 1–20. [CrossRef]

58. Han, S.Y.; Tsou, M.-H.; Knaap, E.; Rey, S.; Cao, G. How Do Cities Flow in an Emergency? Tracing Human Mobility Patterns during a Natural Disaster with Big Data and Geospatial Data Science. *Urban Sci.* **2019**, *3*, 51. [CrossRef]

59. Lu, Y.; Zhang, L. Social media WeChat infers the development trend of COVID-19. *J. Infect.* **2020**, *81*, e82–e83. [CrossRef]

60. Hochmair, H.H.; Juhász, L.; Cvetojevic, S. Data Quality of Points of Interest in Selected Mapping and Social Media Platforms. *Lect. Notes Geoinforma. Cartography* **2017**, 293–313. [CrossRef]

61. Vargas-Muñoz, J.E.; Tuia, D.; Falcão, A.X. Deploying machine learning to assist digital humanitarians: Making image annotation in OpenStreetMap more efficient. *Int. J. Geogr. Inf. Sci.* **2020**, 1–21. [CrossRef]

62. Yang, K.; Varol, O.; Davis, C.A.; Ferrara, E.; Flammini, A.; Menczer, F. Arming the public with artificial intelligence to counter social bots. *Hum. Behav. Emerg. Technol.* **2019**, *1*, 48–61. [CrossRef]

63. ISO 19115-1:2014 Geographic Information—Metadata—Part 1: Fundamentals. Available online: https://www.iso.org/standard/53798.html?browse=tc (accessed on 28 November 2020).

64. Haklay, M. How Good is Volunteered Geographical Information? A Comparative Study of OpenStreetMap and Ordnance Survey Datasets. *Environ. Plan. B Plan. Des.* **2010**, *37*, 682–703. [CrossRef]

65. Zielstra, D.; Hochmair, H.H. Comparative Study of Pedestrian Accessibility to Transit Stations Using Free and Proprietary Network Data. *Transp. Res. Rec. J. Transp. Res. Board* **2011**, *2217*, 145–152. [CrossRef]

66. Antoniou, V.; Skopeliti, A. Measures and indicators of VGI quality: An overview. *ISPRS Ann. Photogramm. Remote Sens. Spat. Inf. Sci.* **2015**, *1*, 345–351. [CrossRef]

67. Degrossi, L.C.; de Albuquerque, J.P.; Rocha, R.D.S.; Zipf, A. A taxonomy of quality assessment methods for volunteered and crowdsourced geographic information. *Trans. GIS* **2018**, *22*, 542–560. [CrossRef]

68. Barron, C.; Neis, P.; Zipf, A. A Comprehensive Framework for Intrinsic OpenStreetMap Quality Analysis. *Trans. GIS* **2014**, *18*, 877–895. [CrossRef]

69. Bordogna, G.; Carrara, P.; Criscuolo, L.; Pepe, M.; Rampini, A. A linguistic decision-making approach to assess the quality of volunteer geographic information for citizen science. *Inf. Sci.* **2014**, *258*, 312–327. [CrossRef]

70. Ballatore, A.; de Sabbata, S. Charting the Geographies of Crowdsourced Information in Greater London. *Lect. Notes Geoinform. Cartogr.* **2018**, 149–168. [CrossRef]

71. Alivand, M.; Hochmair, H.H. Spatiotemporal analysis of photo contribution patterns to Panoramio and Flickr. *Cartogr. Geogr. Inf. Sci.* **2016**, *44*, 170–184. [CrossRef]

72. Juhász, L.; Hochmair, H.H. User Contribution Patterns and Completeness Evaluation of Mapillary, a Crowdsourced Street Level Photo Service. *Trans. GIS* **2016**, *20*, 925–947. [CrossRef]

73. Zielstra, D.; Hochmair, H.H. Positional accuracy analysis of Flickr and Panoramio images for selected world regions. *J. Spat. Sci.* **2013**, *58*, 251–273. [CrossRef]

74. Foody, G.M.; See, L.; Fritz, S.; van der Velde, M.; Perger, C.; Schill, C.; Boyd, D.S. Assessing the Accuracy of Volunteered Geographic Information arising from Multiple Contributors to an Internet Based Collaborative Project. *Trans. GIS* **2013**, *17*, 847–860. [CrossRef]

75. Flanagin, A.J.; Metzger, M.J. The credibility of volunteered geographic information. *GeoJournal* **2008**, *72*, 137–148. [CrossRef]

76. Hung, K.-C.; Kalantari, M.; Rajabifard, A. Methods for assessing the credibility of volunteered geographic information in flood response: A case study in Brisbane, Australia. *Appl. Geogr.* **2016**, *68*, 37–47. [CrossRef]

77. Hecht, B.J.; Gergle, D. On the "localness" of user-generated content. In *Proceedings of the 2010 ACM Conference on Computer Supported Cooperative Work—CSCW '10, Savannah, GA, USA, 6–10 February 2010*; Association for Computing Machinery (ACM): New York, NY, USA, 2010; pp. 229–232.

78. Johnson, I.L.; Sengupta, S.; Schöning, J.; Hecht, B. The Geography and Importance of Localness in Geotagged Social Media. In *Proceedings of the 2016 CHI Conference on Human Factors in Computing Systems, San Jose, CA, USA, 7–12 May 2016*; Association for Computing Machinery (ACM): New York, NY, USA, 2016; pp. 515–526.

79. de Longueville, B.; Ostländer, N.; Keskitalo, C. Addressing vagueness in Volunteered Geographic Information (VGI)-A case study. *Int. J. Spat. Data Infrastruct. Res.* **2010**, *5*, 463–1725.

80. Rice, M.T.; Paez, F.I.; Mulhollen, A.P.; Shore, B.M.; Caldwell, D.R. *Crowdsourced Geospatial Data: A Report on the Emerging Phenomena of Crowdsourced and User-Generated Geospatial Data*; Defense Technical Information Center (DTIC): Fort Belvoir, VA, USA, 2012.

81. Graham, M.; Zook, M. Augmented Realities and Uneven Geographies: Exploring the Geolinguistic Contours of the Web. *Environ. Plan. A Econ. Space* **2013**, *45*, 77–99. [CrossRef]

82. Senaratne, H.; Mobasheri, A.; Ali, A.L.; Capineri, C.; Haklay, M. A review of volunteered geographic information quality assessment methods. *Int. J. Geogr. Inf. Sci.* **2017**, *31*, 139–167. [CrossRef]

83. Senaratne, H.; Bröring, A.; Schreck, T. Using Reverse Viewshed Analysis to Assess the Location Correctness of Visually Generated VGI. *Trans. GIS* **2013**, *17*, 369–386. [CrossRef]

84. Zhao, B.; Sui, D.Z. True lies in geospatial big data: Detecting location spoofing in social media. *Ann. GIS* **2017**, *23*, 1–14. [CrossRef]

85. Neis, P.; Goetz, M.; Zipf, A. Towards Automatic Vandalism Detection in OpenStreetMap. *ISPRS Int. J. Geo-Inf.* **2012**, *1*, 315–332. [CrossRef]

86. Juhász, L.; Hochmair, H.H. Cross-Linkage Between Mapillary Street Level Photos and OSM Edits. *Lect. Notes Geoinf. Cartogr.* **2016**, *64*, 141–156. [CrossRef]

87. Juhász, L. Geo-Social Activity Research. Available online: https://research.jlevente.com/tool/ (accessed on 28 November 2020), (unpublished work).

88. Miller, H.J. A Measurement Theory for Time Geography. *Geogr. Anal.* **2005**, *37*, 17–45. [CrossRef]

89. Huang, P.-C.; Lee, S.-S.; Kuo, Y.-H.; Lee, K.-R. A flexible sequence alignment approach on pattern mining and matching for human activity recognition. *Expert Syst. Appl.* **2010**, *37*, 298–306. [CrossRef]

90. Keßler, C.; McKenzie, G. A geoprivacy manifesto. *Trans. GIS* **2017**, *22*, 3–19. [CrossRef]

91. McKenzie, G.; Keßler, C.; Andris, C. Geospatial Privacy and Security. *J. Spat. Inf. Sci.* **2019**. [CrossRef]

92. McKenzie, G.; Janowicz, K.; Adams, B. A weighted multi-attribute method for matching user-generated Points of Interest. *Cartogr. Geogr. Inf. Sci.* **2014**, *41*, 125–137. [CrossRef]

93. Juhász, L.; Hochmair, H.H. Cross-checking user activities in multiple geo-social media networks. In Proceedings of the 21st AGILE Conference on Geo-Information Science, Lund, Sweden, 12–15 June 2018.

94. Juhász, L.; Hochmair, H.H. How do volunteer mappers use crowdsourced Mapillary street level images to enrich OpenStreetMap? In Proceedings of the 20th AGILE Conference on Geo-Information Science, Wageningen, The Netherlands, 9–12 May 2017.

95. Haklay, M.; Basiouka, S.; Antoniou, V.; Ather, A. How Many Volunteers Does it Take to Map an Area Well? The Validity of Linus' Law to Volunteered Geographic Information. *Cartogr. J.* **2010**, *47*, 315–322. [CrossRef]

96. Pei, S.; Muchnik, L.; Andrade, J.S., Jr.; Zheng, Z.; Makse, H.A. Searching for superspreaders of information in real-world social media. *Sci. Rep.* **2015**, *4*, srep05547. [CrossRef]

97. Leung, Y.; Ma, J.-H.; Goodchild, M.F. A general framework for error analysis in measurement-based GIS Part 1: The basic measurement-error model and related concepts. *J. Geogr. Syst.* **2004**, *6*, 325–354. [CrossRef]

98. Ahmouda, A.; Hochmair, H.H. Using Volunteered Geographic Information to measure name changes of artificial geographical features as a result of political changes: A Libya case study. *GeoJournal* **2017**, *83*, 237–255. [CrossRef]

99. Griffin, G.P.; Jiao, J. Where does bicycling for health happen? Analysing volunteered geographic information through place and plexus. *J. Transp. Health* **2015**, *2*, 238–247. [CrossRef]

100. Smith, A.; Andersen, M. Social Media Use in 2018. *Pew Res. Cent.* **2018**. Available online: https://www.pewinternet.org/wp-content/uploads/sites/9/2018/02/PI_2018.03.01_Social-Media_FINAL.pdf (accessed on 28 November 2020).

101. Miller, H.J.; Goodchild, M.F. Data-driven geography. *GeoJournal* **2014**, *80*, 449–461. [CrossRef]

102. Yin, L.; Wang, Z. Measuring visual enclosure for street walkability: Using machine learning algorithms and Google Street View imagery. *Appl. Geogr.* **2016**, *76*, 147–153. [CrossRef]

103. Gosal, A.; Geijzendorffer, I.R.; Václavík, T.; Poulin, B.; Václavík, T. Using social media, machine learning and natural language processing to map multiple recreational beneficiaries. *Ecosyst. Serv.* **2019**, *38*, 100958. [CrossRef]

104. Steiger, E.; Resch, B.; de Albuquerque, J.P.; Zipf, A. Mining and correlating traffic events from human sensor observations with official transport data using self-organizing maps. *Transp. Res. Part C Emerg. Technol.* **2016**, *73*, 91–104. [CrossRef]

105. Steiger, E.; Resch, B.; Zipf, A. Exploration of spatiotemporal and semantic clusters of Twitter data using unsupervised neural networks. *Int. J. Geogr. Inf. Sci.* **2016**, *30*, 1694–1716. [CrossRef]

106. Liu, Y.; Yuan, Y.; Zhang, F. Mining urban perceptions from social media data. *J. Spat. Inf. Sci.* **2020**. [CrossRef]

107. Nam, T.; Pardo, T.A. Conceptualizing smart city with dimensions of technology, people, and institutions. In *Proceedings of the 12th Annual International Digital Government Research Conference on Digital Government Innovation in Challenging Times—dg.o '11, College Park, MD, USA, 12–15 June 2011*; Association for Computing Machinery (ACM): New York, NY, USA, 2011; pp. 282–291.

108. Albino, V.; Berardi, U.; Dangelico, R.M. Smart Cities: Definitions, Dimensions, Performance, and Initiatives. *J. Urban Technol.* **2015**, *22*, 3–21. [CrossRef]

109. Pereira, G.V.; Parycek, P.; Falco, E.; Kleinhans, R. Smart governance in the context of smart cities: A literature review. *Inf. Polity* **2018**, *23*, 143–162. [CrossRef]

110. Hollands, R.G. Will the real smart city please stand up? *City* **2008**, *12*, 303–320. [CrossRef]

111. Söderström, O.; Paasche, T.; Klauser, F. Smart cities as corporate storytelling. *City* **2014**, *18*, 307–320. [CrossRef]

112. Kitchin, R. The ethics of smart cities and urban science. *Philos. Trans. R. Soc. A Math. Phys. Eng. Sci.* **2016**, *374*, 20160115. [CrossRef] [PubMed]

113. Portugali, J.; Haken, H.; Benenson, I.; Omer, I.; Alfasi, N. *Self-Organization and the City*; Springer Series in Synergetics; Springer Berlin Heidelberg: Berlin/Heidelberg, Germany, 2012; ISBN 9783662040997.

114. Bettencourt, L.M.A. The Kind of Problem a City Is: New Perspectives on the Nature of Cities from Complex Systems Theory. *Decod. City* **2014**. [CrossRef]

115. Theraulaz, G.; Bonabeau, E. A Brief History of Stigmergy. *Artif. Life* **1999**, *5*, 97–116. [CrossRef] [PubMed]

116. Mehaffy, M.; Elmlund, P. Smart cities: Missing the stigmergy? In *Spatial Knowledge as a Tool for Strategic and Data-Based Regional Policy*; Regional Science Academy and Université de Lyon: Lyon, France, 2020.

117. Latour, B. On actor-network theory: A few clarifications. *Soz. Welt* **1996**, *4*, 369–381.

118. Edelenbos, J.; Hirzalla, F.; van Zoonen, L.; van Dalen, J.; Bouma, G.; Slob, A.; Woestenburg, A. *Governing the Complexity of Smart Data Cities: Setting a Research Agenda*; Springer Science and Business Media LLC: Berlin/Heidelberg, Germany, 2017; pp. 35–54.

119. Ostrom, E. Beyond Markets and States: Polycentric Governance of Complex Economic Systems. *Am. Econ. Rev.* **2010**, *100*, 641–672. [CrossRef]

120. Kounadi, O.; Resch, B. A Geoprivacy by Design Guideline for Research Campaigns That Use Participatory Sensing Data. *J. Empir. Res. Hum. Res. Ethic* **2018**, *13*, 203–222. [CrossRef]

4

Road Safety Risk Evaluation using GIS-Based Data Envelopment Analysis—Artificial Neural Networks Approach

Syyed Adnan Raheel Shah [1,2,*], **Tom Brijs** [1] , **Naveed Ahmad** [2], **Ali Pirdavani** [4] , **Yongjun Shen** [3] and **Muhammad Aamir Basheer** [1]

[1] Transportation Research Institute (IMOB), Hasselt University, Diepenbeek 3590, Belgium; tom.brijs@uhasselt.be (T.B.); muhammadaamir.basheer@student.uhasselt.be (M.A.B.)
[2] Taxila Institute of Transportation Engineering, Department of Civil Engineering, University of Engineering & Technology, Taxila 47050, Pakistan; n.ahmad@uettaxila.edu.pk
[3] School of Transportation, Southeast University, Nanjing 210096, China; yongjunshen@outlook.com
[4] Faculty of Engineering Technology, Hasselt University, Diepenbeek 3590, Belgium; ali.pirdavani@uhasselt.be
* Correspondence: syyed.adnanraheelshah@uhasselt.be or shahjee.8@gmail

Abstract: Identification of the most significant factors for evaluating road risk level is an important question in road safety research, predominantly for decision-making processes. However, model selection for this specific purpose is the most relevant focus in current research. In this paper, we proposed a new methodological approach for road safety risk evaluation, which is a two-stage framework consisting of data envelopment analysis (DEA) in combination with artificial neural networks (ANNs). In the first phase, the risk level of the road segments under study was calculated by applying DEA, and high-risk segments were identified. Then, the ANNs technique was adopted in the second phase, which appears to be a valuable analytical tool for risk prediction. The practical application of DEA-ANN approach within the Geographical Information System (GIS) environment will be an efficient approach for road safety risk analysis.

Keywords: road safety; risk evaluation; data envelopment analysis; artificial neural networks; crash data analysis

1. Introduction

Crash injury severity has always been a major concern in highway safety research. To model the relationship between crash occurrence along with severity outcomes, related traffic features, and contributing factors, a large number of advanced models have been proposed. Road safety research incorporates a broad exhibit of research territories, and the most successful of them is crash information investigation. There have been a lot of discussion about crash information-based safety analysis and other distinguishable activity attributes have been proposed, more regularly than crashes, as an option. In any case, investigation of crash information remains the most broadly received way to deal with the safety of a transportation system (e.g., expressways, arterials, crossing points, etc.). The traditional approach is to build up connections between crash recurrence, traffic flow attributes, and geometry of the roads [1]. On the one hand, the impact of the geometric design on the probability of a driver behavior has been very much archived in conventional safety studies. This course of research is useful in settling on choices in such things as installing cautioning signs on roadway areas, etc. On the other hand, Average Annual Daily Traffic (AADT) is a generally used indicator for measuring the traffic movement conditions, as it is recorded by most organizations around the nation/the world, is accessible to all roadway areas, and gives a measure of introduction to the specific roadway segment. Crash recurrence examination in view of AADT is a total or aggregate approach to take a glance at the crash information where the recurrence of crashes is computed, by amassing the crash information over particular eras (months or years) and areas (particular roadway segments) [2].

During road safety analysis of a road, a major target is to locate those segments which are dangerous, and then to identify the factors influencing its safety level. This study focuses on the concept that crashes can be decreased by better assessment of road hazard incremental elements, and by recognizable proof of hazardous segments at the initial stage, and after that, assessment of very dangerous sections with reference to the major contributing components is conducted in the second stage. In doing so, a combination of Data Envelopment Analysis (DEA) with Artificial Neural Networks (ANNs) is applied to evaluate the performance of roads with reference to safety conditions. The outcome is able to help decision makers/safety engineers to build a valuable system to analyze risk and significant attributes. Although it is new in the road safety research field, such an integrated mechanism has been popular in other sectors like banks, hospitals, schools, and corporations. Some researchers have used a combination of DEA-ANNs to evaluate performance (efficiency/risk) of rail transport, power suppliers, etc. [3–5]. DEA-ANNs was also used for efficiency classification by different researchers for banks and corporate companies [5–7]. For analysis regarding hospitals and large companies, screening of training data was also evaluated by using the DEA-ANNs technique. In addition, DEA-ANNs was also introduced for data processing [8–13], and will be more useful when applied within a Geographical Information System (GIS) environment. In this study, this integrated concept, which is popular in other sectors, is introduced to evaluate the safety performance (risk evaluation) of motorways. This evaluation of risk helps decision makers to decide on economical investment for risky segments, along with related factors, and consequently to reduce the cost of risk evaluation.

2. Literature Review

2.1. Risk and Road Safety Analysis

Usually, road safety performance is evaluated on the basis of 'Risk' which is associated with the number of crashes and casualties, known as the road safety outcome. In the field of road safety, the risk is defined as 'the road safety outcome to the amount of exposure' as shown in Equation (1):

$$Risk = \frac{Road\ Safety\ Outcome}{Exposure} \tag{1}$$

Exposure can be measured using different parameters; while comparing the performance of road segments, it can be measured as vehicle miles traveled, vehicle hours traveled, volume and number of trips, etc., however for countries it can be passenger kilometers travelled, population and number of registered vehicles, etc. [14,15]. Risk assessment is necessary for road safety performance analysis. Although risk can be analyzed on the basis of direct calculation using outcome by exposure, in the case of multiple outcomes and multiple input, it is difficult to deal with the calculation. Crashes are random events, and their outcome can also vary, as in one crash there may be zero fatalities, or fifty or more fatalities. Thus, a method that can deal with multiple outputs can be beneficial in calculating risk for road safety performance analysis of different units.

2.2. DEA for Road Safety Analysis

Road safety performance analysis of highways is an important task for the safety of travelers. To analyze the safety performance of certain attributes, a benchmarking mechanism has remained a basic procedure to be adopted by researchers [16–18]. With reference to the applied techniques for this purpose, DEA has been a popular technique with its theoretical basis in linear programming. Evolving the concept of DEA from research work in 1978, Charnes et al. [19] applied a linear program to estimate an empirical production technology frontier (bench marking) for the first time [19]. In the basic DEA model, the definition of the best practices relies on the assumption that inputs have to be minimized and outputs have to be maximized (such as in the economics field). However, to use DEA for road safety risk evaluation, the target becomes the output, i.e., the number of traffic crashes, to be as low as possible with respect to the level of exposure to risk. Therefore, the DEA frontier based Decision Making Units(DMUs) or the best-performing road segments are those with minimum output levels given the input levels, and other segments' risk is then measured relative to this frontier [20].

Mathematically, to use DEA for road safety evaluation, the model is shown as follows:

$$\min R_0 = \sum_{r=1}^{s} u_r y_{r0}$$
$$\text{s.t.} \quad \sum_{i=1}^{m} v_i x_{i0} = 1,$$
$$\sum_{i=1}^{m} v_i x_{ij} - \sum_{r=1}^{s} u_r y_{rj} \leq 0, \quad j = 1, \cdots, n$$
$$u_r, v_i \geq 0, \quad r = 1, \cdots, s, \quad i = 1, \cdots, m$$

(2)

where y_{rj} and x_{ij} are the rth output and ith input respectively of the jth DMU, u_r is the weight given to output r, and v_i is the weight given to input i.

In view of the model applications for road safety analysis, road safety condition was compared to 21 European countries [16] and an ideal trauma management record score was also calculated by using DEA [21]. Furthermore, using population, passenger-kilometers, and passenger cars as inputs, and the number of fatalities as output, DEA was used for the evaluation of risk level of countries [22]. Monitoring of yearly progress in road safety was also conducted by utilizing the DEA technique [23]. Adding to road safety determination on a national level for 27 Brazilian states, two fundamental indicators accessible in Brazil: death rate (fatalities per capita) and casualty rate (fatalities per vehicle and fatalities per vehicle kilometer traveled) were focused upon [24]. From the literature review on DEA application in the field of road safety, it was confirmed that DEA is one of the established techniques to evaluate the risk level of road safety.

2.3. ANNs for Road Safety Analysis

Artificial Neural Networks is a model instrument of nonlinear statistical data that can be used to model a complex relationship between input and output to seek patterns. ANNs has been often implemented in many fields of science for prediction [25]. In road safety research, ANNs was applied to investigate crashes with reference to driver, vehicle, roadway, and condition attributes [26]. After application of ANNs, the impact of factors like seatbelt usage, light condition, and driver's liquor utilization on driver's safety was evaluated [27]. ANNs was also applied to determine the relationship between crash severity and the model parameters including years, highway sections, section length (km), AADT, the degree of horizontal curvature, the degree of vertical curvature, heavy vehicles (percentage), and season summer (percentage). The results shown that degree of vertical curvature has strong impact on number of crashes [28]. By modeling AADT, SL (Posted speed limit), Gradient (Average segment gradient), and Curvature (Average segment curvature) against road crashes, it was concluded that ANN was superior to multivariate Poisson-lognormal models [29]. From the literature review, we can summarize that ANNs was previously used as a crash data analysis model, which was a useful technique to study road-related features, geometry, and other contributing factors to road safety.

2.4. DEA-ANNs Approach

The combination of DEA and ANNs has not been applied in the road safety field, but it is popular in other fields like banks and corporate sectors. From the previous studies it was concluded that DEA is powerful for efficiency calculation, but for prediction purposes ANNs is ahead, so a discussion started after [30] on combining these two techniques to obtain the best possible outputs, i.e., efficiency calculation for ranking and prioritizing and then efficiency prediction for factor analysis purpose. To validate this combination, efficiency prediction was performed for 50 companies [31], 19 power plants [32], 49 Indian business schools [5], 102 bank branches [7], and 45 countries [33]. Efficiency classification was also tested by studying 142 bank branches [34] and 23 supplier companies [35]. Following the similar pattern, the DEA-ANNs approach is selected in this study for road safety risk evaluation and analysis of factors affecting risk.

2.5. GIS for Road Safety Analysis

"While geometrical concept can be enriched by culture-specific devices like maps, or the terms of a natural language, underneath this variability lies a shared set of geometrical concepts. These concepts allow adults and children with no formal education, and minimal spatial language, to categorize geometrical forms and to use geometrical relationships to represent the surrounding spatial layout." (Elizabeth S. Spelke-Harvard). GIS has gained a reputation that provides a better visualization of a large data set for understanding and decision making processes. GIS-provided maps which helped in identifying the crash concentration areas, located along the major road in the main urban areas [36]. During the road safety analysis of motorway (M-25), GIS provided relevant data on road accidents, traffic and road characteristics for 70 segments [37]. High risky sections on the basis of potential crash cost for expressways of Shanghai with the application of GIS has been clearly mapped [38]. Zonal crash frequency has also been expressed through GIS, showing association with several social-economic, demographic, and transportation system factors [39]. Through spatial analysis of high risk areas, pedestrian crashes have also been mapped in Tehran [40]. In Belgium, through the use of GIS and point pattern techniques, mapping road-accident black zones has been conducted within urban agglomerations [41]. GIS has also been used to explore the spatial variations in relationship between Number of Crashes and other explanatory variables of 2200 Traffic Analysis Zones (TAZs) in the study area, Flanders, Belgium [42,43]. GIS was used for modelling crash data at a small-scale level in Belgium, which permitted the identification of several areas with exceptionally high crash data. It endorsed more effective reallocation of resources and more efficient road safety management in Belgium [44].

2.6. ANN-GIS Approach

ANNs has been introduced as a mapping tool to GIS to perform a predictive capability for joint operations [45]. Although GIS in combination with ANNs was popular in the fields of geoscience, irrigation, meteorology and Agriculture, it has been tested in the field of road safety by applying deep learning models using a Recurrent Neural Network (RNN) to predict the injury severity of traffic crashes for the North-South Expressway-Malaysia [46]. Previously this technique had been applied for sediment prediction in Gothenburg harbor [45], landslide susceptibility using the landslide occurrence factors produced with the help of a ANNs model [47], detection of flood hazards in the Blue Nile, White Nile, Main Nile, and River Atbara [48], macrobenthos habitat potential mapping regarding *Macrophthalmus dilatatus, Cerithideopsilla cingulata,* and *Armandia lanceolate* [49], learning the patterns of development in the region [50], tunneling performance prediction required in routine tunnel design works and performance in terms of stability as well as impact on surrounding environment [51], and deforestation maps production to determine the relationship between deforestation and various spatial variables such as the vicinity to roads and to expenditures, forest disintegration, elevation, slope, and soil type [52].

2.7. Research Gap

DEA is popular as an optimization tool with its theoretical background in linear programming. DEA is popular with reference to benchmarking mechanism for efficiency and risk evaluation [3,6]. Previously, DEA was popular with its multi stage properties, but it has shortcomings with respect to its prediction capabilities, which reduces its application. A powerful technique, ANN, has been joined with DEA to fill that gap. Finally, with the predictive potential of ANNs and the optimization capacity of DEA performing complementary features, a prominent modeling option is envisioned [3,6]. The performance of the DEA-ANNs technique in the field of road safety for decision making mechanisms for road safety performance analysis was evaluated. This is the first study for an application of the DEA-ANN approach within a GIS environment for road safety performance analysis, using a case study on Motorways. This will lead traffic engineers and decision makers to better visualize the risky sections and key factors for road safety condition improvement.

3. Materials and Methods

3.1. Basic Framework of the Analysis

Road authorities have to prioritize the sites which require safety treatment, due to budget limitations. So in this study, a two phase framework was proposed for road safety risk evaluation, as shown in Figure 1. In the first phase, the number of crashes and fatalities was evaluated against exposure variables, with the help of DEA to calculate the risk level of road segments. In the second phase, that risk was predicted and evaluated with the help of ANNs.

Figure 1. The Proposed Data Envelopment Analysis-Artificial Neural Networks (DEA-ANNs) Framework for Risk Evaluation.

3.2. Data Description and Selection of Variables

The study area selected for this study was two motorways in Belgium named E-313 and E-314 (Limburg Province Sections with a total length of 103 km). Each Motorway has segments, traffic-related characteristics, and road network segmentation derived from the FEATHERS model [53]. In this study, a segment with at least one crash was considered as a decision-making unit (DMU) to analyze the road safety condition. According to this criterion, 67 segments are selected for these two motorways. The crash data used in this study consisted of a geographically coded set of crash data that occurred between 2010 and 2012, which was provided by the Flemish Ministry of Mobility and Public Works, as shown in Table 1. The first and very critical step in conducting an analysis is the selection of inputs and outputs variables. For this purpose in the first stage (DEA), those variables which were the exposure variables and could not be directly affected by a traffic engineer/decision maker were selected to calculate risk, while in the second stage (ANN) those variables (i.e., Horz and Vert Curve design, speed, and flow) which could be altered or improved by directly changing certain parameters, were selected. So, the target while calculating risk was to reduce the number of crashes (NoC) and casualties (NoAP) with the increase of average volume to capacity on each segment (V/C), total daily vehicle miles travelled on each segment (VMT) and total daily vehicle hours travelled on each segment (VHT). A traffic engineer cannot directly change V/C, VMT, or VHT; however, the geometric design (Horz and Vert Curve), speed (speed limit) and flow (by controlling access) so practically, a selection of variables was targeted according to the feasibility of the problem's solution. To confirm the validity of the DEA model condition, an isotonicity test [54] was conducted. An isotonicity test comprises the intention of all inter-correlations between inputs and outputs for detecting whether increasing amounts of inputs lead to greater outputs. As positive correlations were established, the isotonicity test was accepted and the presence of the inputs and outputs was reasonable. However there are no diagnostic checks for improper model specification detection in DEA [55]. However, a general rule of thumb, the minimum number of DMUs is higher than three times the number of inputs plus outputs [56]. In our study with a total of three inputs and two outputs, so a set of 15 data points would be optimal; we have 67 data segments.

Table 1. Description Statistics of the Variables.

Stage	Variables	Description	Mean	SD	Min.	Max.
1st Stage DEA	NoC	No. of Crashes	9.58	13.12	1	74
	NoAP	No. of Affected Persons (Injured and Killed)	14.36	19.55	1	105
	V/C	Average Volume to Capacity on each segment	0.4405	0.1807	0.08	0.6435
	VMT	Total daily Vehicles Miles Travelled on each Segment	1828	1388	77	5186
	VHT	Total daily Vehicles hours Travelled on each Segment	1093	879	38	3616
2nd Stage ANN	Flow	Average annual daily traffic on each segment (vph)	968.1	449.6	31.5	1483.4
	Speed	Average Travel Speed for each segment (kph)	110.99	8.23	96.89	120
	Horz_Curve	0 = Tangent, 1 = Curve	–	–	0	1
	Vert_Curve	1 = Upward, 2 = Downward, 3 = Flat	–	–	1	3

3.3. Phase-I: Application of DEA for Risk Calculation and Ranking

As there were two major phases of modeling, we had decide on the variables for both phases. The initial target was to evaluate risk with reference to the variables that were basically exposure variables. In the basic DEA model, the definition of the best practice relied on the assumption that inputs had to be minimized, and outputs have to be maximized (such as in the economics field). However, to use DEA for road safety risk evaluation, the target became the output, i.e., the number of traffic crashes, to be as low as possible with respect to the level of exposure to risk.

There are two basic concepts in application of DEA, starting from efficiency as in Equation (3), and converting into calculation of risk as shown in Equation (4).

Efficiency: The basic concept of DEA-Efficiency calculation is as follows:

$$Efficiency = \frac{Weighted\ Sum\ of\ Output}{Weighted\ Sum\ of\ Input} = \frac{Maximize\ Output}{Minimize\ Input} \tag{3}$$

Risk: The basic concept of DEA-Risk calculation in connection between Equations (1) and (3):

$$Risk = \frac{Weighted\ Sum\ of\ Output}{Weighted\ Sum\ of\ Input} = \frac{Minimize\ Output}{Maximize\ Input} = \frac{Road\ Safety\ Outcome}{Exposure} \tag{4}$$

So the equation to calculate the Risk value through DEA is as follows:

$$Risk = \frac{U_2(NoC) + U_1\ (NoAP)}{V_1\ (V/C) + V_2\ (VMT) + V_3\ (VHT)} \tag{5}$$

where U_1 = weights for 1st output (NoC), U_2 = weights for 2nd output (NoAP); V_1 = Weights for 1st Input (V/C), V_2 = weights for 2nd Input (VMT), V_3 = weights for 3rd Input (VHT).

After calculation of Risk value, for ranking purposes, a cross-efficiency approach was one of the best methods to calculate A cross-risk value for ranking purposes. DEA has an attractive feature in that each DMU can have its own input and output weights, which leads to difficulty in making a comparison between DMUs. To compare DMUs, a Cross efficiency matrix (CEM) was developed as a DEA extension tool to assist in identifying the overall best or worst performer among all DMUs and rank them. Its basic idea is to apply DEA in a peer assessment instead of a self-assessment mode. Specifically, the CEM calculates the performance of a DMU with a concept by using not only its own optimal input and output weights, but also those of all other DMUs. Results can then be accumulated in a CEM as shown in Table 2. In the CEM, the element in the *ith* row and *jth* column signifies the risk scores of DMU *j* using the optimal weights of DMU *i*. The basic DEA risk is thus positioned in the principal diagonal. The average of each column of the CEM is calculated as a mean cross risk value for each DMU [20]. Since the same weighting process is applied for all the DMUs, their evaluations can then be made on a comparison basis, with a higher cross-risk score indicating a higher risky DMU.

Table 2. A Generalized Cross-Efficiency Matrix (CEM) [20].

Rating DMU	Rated DMU				
	1	2	3	. . . . . .	n
1	E_{11}	E_{12}	E_{13}	. . . . . .	E_{1n}
2	E_{21}	E_{22}	E_{23}	. . . . . .	E_{2n}
.	.	.	.	.	.
n	E_{n1}	E_{n2}	E_{n3}	. . . . . .	E_{nm}
Mean	$\overline{E}_1$	$\overline{E}_2$	$\overline{E}_3$	. . . . . .	$\overline{E}_n$

For those DMUS, which have illogical weights in the basic DEA model, a relatively low or higher risk value will be calculated. Therefore, for ranking purpose, this method serves a type of sensitivity analysis by applying a method of a different set of weights to each DMU, with a back channel mechanism of self-generated weights rather than an externally imposed [20]. So the target value, which is a value of 1 to be considered for the best DMU, can now be changed, and after application of CEM it can vary, but the selection of best DMU (with the lowest Risk) will be easier.So after applying model (1) for calculating risk R_0 in road safety field, the lowest level has been considered as the frontier of safety. As explained above, for ranking purposes, a cross risk procedure [20] has been adopted to obtain the best ranking, as shown in Table 3.

The major advantage of using DEA here is that it can handle multiple inputs and multiple outputs. Moreover, DEA has some benefits as it does not require an assumption of a functional form relating inputs to outputs; DMUs considered in DEA are directly compared against a peer or combination of peers; Inputs and outputs used in DEA can have different measurement units. In this study, number of crashes (NoC) and number of affected persons-injured or killed (NoAP) are considered as two outputs, while exposure variables—average volume to capacity on each segment (V/C), total daily vehicle miles travelled on each segment (VMT), and total daily vehicle hours travelled on each segment (VHT)were considered as three inputs. Although the segment length also varied, it was not included here because it was already been involved in the backup calculation of VMT.

Table 3. DEA-Based Risk Evaluation and Ranking Segments.

DMUs Road Seg.	Input 1 V/C	Input 2 VMT	Input 3 VHT	Output 1 NoC	Output 2 NoAP	CE-RISK VALUE	RANK
1	0.368518	3039.221	1541.607	74	105	91.06902	1
29	0.139052	109.169	54.58458	6	8	71.72984	2
19	0.603085	183.7303	118.162	12	20	69.92395	3
2	0.384021	2494.327	1268.376	49	76	65.10151	4
34	0.07999	82.51051	41.25526	3	6	62.90294	5
5	0.277711	2190.904	1096.683	38	50	62.28254	6
25	0.139052	76.73981	38.36996	3	6	58.10739	7
26	0.236548	202.3093	101.2386	9	11	57.15026	8
3	0.360649	2683.937	1361.267	40	61	53.2604	9
21	0.53409	594.7792	336.9631	13	24	35.40002	10
-	-	-	-	-	-	-	-
53	0.631117	4275.086	3046.694	2	3	1.47492	64
67	0.592324	1093.968	734.8267	1	1	1.312419	65
49	0.498964	3214.268	1780.697	1	2	1.08319	66
66	0.574944	1714.219	1098.003	1	1	1.068806	67

Based on Model (1), the range of risk value began at 1 and proceeded to a higher value, so a segment with a value of 1 was considered safest, while the road segment with the highest value was considered the most dangerous. Moreover, the cross risk method [20] was used to make all the DMUs comparable. Table 3 presents the results. As the ranking was done on a priority basis to evaluate the safety condition of that segment, the risk value of 91.07 was the highest value in the table and was ranked first (i.e., the most risky segment). The top 10 riskiest segments are shown in Table 3 to explain an idea of risky segment selection for improvement.

Furthermore, risk value was normalized by applying natural log, and with the help of GIS, a complete spatial map of both motorways is shown in Figure 2. A straight line demonstration provides an insight in locating the most riskiest segment on a motorway or highway.

Figure 2. Risk based Straight Line Map for Motorways (E-313&r314)

3.4. Phase-II: Application of ANNs Model for Risk Prediction and Evaluation

In the second phase, the dependent variable is the risk value generated by the DEA model, was transformed by applying natural log to have data normalized. For independent variables, speed could be controlled by controlling the speed limit; flow was directly related to the number of vehicles, and could be controlled by controlling access; horizontal curve could be removed or altered as per infrastructural changes, and the same was the case for vertical curve as a geometric design feature. So for the application of ANN, data was distributed on the basis of a K-Fold mechanism with five folds (i.e., distribution is as 53 segments-DMUs for Training and 14 segments-DMUs for validation).

ANNs, unlike other modeling platforms, requires some form of model validation to aid in the model-building process and to help prevent overfitting of the model. The basic idea behind validation (or cross-validation) is to hold a subset of the data out of the model-building process. This process forms two partitions of the data, a training set and a validation set (note that a third set, or test set, can also be used). The model is built using the training set, while the k-fold validation set is then used to assess how well the model performs, and to aid in model selection. The most mainstream decision for the quantity of concealed layers is used. A solitary concealed layer is typically adequate to catch even extremely complex connections between the indicators. The quantity of links in the shrouded layers likewise decides the level of multifaceted nature of the connection between the indicators that the system catches [57].

From one viewpoint, utilizing an excessive couple of links is not adequate to catch complex connections (e.g., review the unique instances of a straight relationship in direct and calculated relapse, in the extraordinary instance of zero links or no shrouded layer). Then again, an excessive number of links may prompt overfitting. A dependable guideline is, to begin with (number of indicators) links and reduce or increment gradually while checking for overfitting. Another approach is to start with the default neural model, with one layer and four nodes, and then run a much more complex model with two layers and several nodes, and different activation functions. If the fit statistics do not improve substantially with a more complex model, then a simpler model may suffice [57]. We applied a simpler model to check the performance of the ANNs model in our case of risk evaluation, as shown in Figure 3. After running the model as shown in Figure 3, we displayed the model structure. We saw input variables mapping to each of the activation functions in the hidden layer, and nodes in the hidden layer mapping to the output layer. The background mechanism in each of the nodes in the hidden layer designated that the Gaussian activation function was used. Model results for both the training and validation sets are shown in Table 4. The response variable (risk) for this model was continuous. Like other techniques, it was necessary to follow the validation mechanism. With the validation mechanism and separation of the data into two sets, unbiased results were provided.

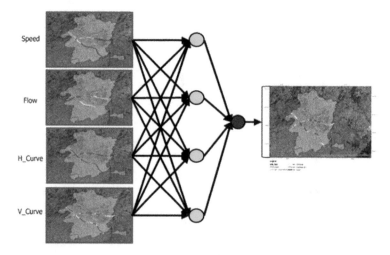

Figure 3. Risk Prediction GIS Map based on the ANNs Model.

In this study, as discussed earlier, original data were distributed into two parts. Out of 67 segments, 53 as the major data set were used for the training of the ANNs setup, while the remaining 14 were used for validation after model building. The training set was the part that estimated model parameters. The validation set was the part that assessed or validated the predictive ability of the model. In addition, the most critical validation was applied in this study. Specifically, the K-Fold technique was adopted, which divided the original data into K subsets. In turn, each of the K sets is used to validate the model fit on to the rest of the data, fitting a total of K models. The model giving the best validation statistic was chosen as the final model. This method was best for small data sets because it made efficient use of limited amounts of data [57,58].The ANN based predicted value of risk was mapped in the GIS environment as shown in Figure 4, showing a red line as the riskiest segments, while dark green segments are the safest as there zero crashes on these segments.

Table 4. Parametric estimates of the ANNs Model.

Parameters	Estimates-Hidden Layer				
	Code	H1_1	H1_2	H1_3	H1_4
Flow		0.258908	−2.00717	0.868246	4.984629
Speed		−1.26756	3.435496	−1.83834	2.048267
Horz_Curve	0	2.150204	13.66468	−1.03056	1.968045
Vert_Curve	1	2.141838	−2.07175	1.313892	0.014534
	2	−3.21511	7.986301	−3.09312	−0.53461
Intercept		1.90514	−1.87443	0.481882	−4.76064
	Int	H1_1	H1_2	H1_3	H1_4
NLog_Risk	2.221	−2.34861	6.612427	−1.8978	2.041027
Cross Validation					
Sample Size	Training	53	Validation	14	
R² (Training)	0.788	R² (Validation)	0.775	RMSE	0.624

Legend
ANN_Risk
━━ 0 (Low)
━━ 0.01 - 1 (Low-medium)
━━ 1.01 - 2 (Medium)
━━ 2.01 - 3 (Medium-high)
━━ 3.01 - 5 (High)

Figure 4. Geographical information system (GIS)-based ANNs predicted risk spatial map

The values of R-square and Root Mean Square Error (RMSE) are the two basic validation indicators for testing the goodness-of-fit of the model. ANNs is a very flexible model and has a tendency to overfit data. When that happens, the model predicts the fitted data very well, but predicts future observations poorly. To mitigate overfitting, the neural platform applies a penalty on the model parameters and uses an independent dataset to assess the predictive power of the model. The applied technique to control overfitting is the squared method. This method is applied if it is considered that independent variables are contributing to the predictive ability of the model.

During the analysis, the graphical representation of data showed a better performance both in the case of training and validation data. The data distribution was adopted in five segments, having a distribution of 53 segments for training and 14 segments for validation. The plots showing the perfection of predictability were shown in Figure 5 for both training and validation data. The values of R-square were also almost similar for both major and training and validation data sets.

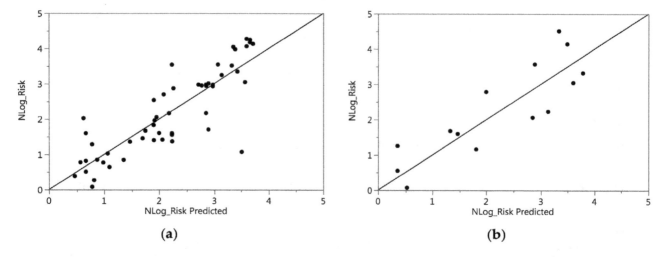

(a)   **(b)**

Figure 5. Actual By Predicted Plot (**a**) Training (**b**) Validation.

The contribution of factors associated with risk can be analyzed by the importance of the variables (i.e., Flow, Speed, Vertical and Horizontal Curve).The impact of variables is one of the necessary targets to analyze and improve the safety performance of the roads. Traffic safety engineers always search for the relationships between factors and safety performance indicators (i.e., Risk). The relationship can be observed in Table 5, which shows that speed and flow were two major factors which are having a high impact on risk.

Table 5. Factors Association with the Risk.

Factor	Main Effect	Total Effect	Comparison
Flow	0.224	0.908	
Speed	0.064	0.47	
Vert_Curve	0.072	0.426	
Horz_Curve	0.052	0.288	

A comprehensive analysis to overview the importance of factors provides traffic engineers to take a decision during road safety analysis and implementation procedure.

3.5. Model Selection Criteria

In order to assess the performance of the DEA-based Risk prediction models, a number of evaluation criteria were used to evaluate these models. These criteria were applied to measure how close the real values were to the values predicted using the developed models. They included Root

Mean Square Error (RMSE) and the correlation coefficient R or R^2. These are given in Equations (6) and (7) respectively [59].

$$RMSE = \sqrt{\frac{1}{n} \sum_{i=1}^{n} (y_i - \hat{y}_i)^2} \tag{6}$$

$$R = \frac{\sum_{i=1}^{n} (y_i - \bar{y})(\hat{y}_i - \bar{\hat{y}})}{\sqrt{\sum_{i=1}^{n} (y_i - \bar{y})^2} \sqrt{\sum_{i=1}^{n} (\hat{y}_i - \bar{\hat{y}})^2}} \tag{7}$$

where y is actual Risk values, $\hat{y}$ is the estimated Risk values using the proposed techniques, and n is the total number of observations of DMUs.

4. Results

In order to find the factors influencing the road safety risk, ANNs and multiple linear regression (MLR) were generated using the Road Traffic and Crash data obtained for European routes (E-313&E-314) of Limburg (Belgium). Although the basic target was to implement the ANNs model, regression analysis was also conducted to assess the performance of ANNs.

4.1. Performance of Model

The main objective of the methods (ANNs and MLR) was to fit an accurate model for risk prediction. The adequacy of such models are typically measured either by the coefficient of determination of the predictions against actual values (R^2) or by RMSE. Figure 6 shows the comparative diagram of prediction between ANNs and MLR.

The graph shown in Figure 6 suggests that ANNs is a better predictor than MLR. Moreover, if we considered the comparative assessment of model predicting capability, we see that R^2 from ANNs (0.788) is much higher than that from MLR (0.276). Another major tester of the capability of the model is the RMSE; a smaller value indicating better fit. It also indicated that ANNs (0.624) has performed better in comparison with MLR (1.0789), as shown in Table 6.

Figure 6. Comparative Analysis for Predicted Vs Actual Risk Values

Table 6. Comparative Analysis of ANNs Vs Multiple Linear Regression (MLR).

Model	R^2 Predicted	R^2 (K-Fold) Validation	RMSE
Sample Size	53	14	
ANN	0.788	0.774	0.624109
MLR	0.276	0.147	1.0789985

Note: RMSE = Root Mean Square Error.

4.2. Analysis of Factors

As far as a solution to the problem is concerned, we can also analyze data on a graphical basis were the relationship can serve as a better understanding of our problem. After successfully applying the DEA-ANNs model for the road safety risk evaluation, we focused on the contributing factors used in the risk prediction. Decision makers/traffic safety engineer aim for low-cost treatments for problematic/risky segments. Thus from graphical analysis of the contributing factors, we saw that the majority of the crashes were on the curved portions of the motorways. Decision makers usually avoid infrastructural changes because redesigning and reconstruction is a costly procedure, so if they focus on the low-cost treatments, they can focus on speed and flow control. Figure 7 presents the relationship between the risk and the different contributing factors. The red lines represent mean speed levels, while the green lines represent mean flow level. We can see from Figure 7 that the risk level could be reduced by controlling just these two factors.

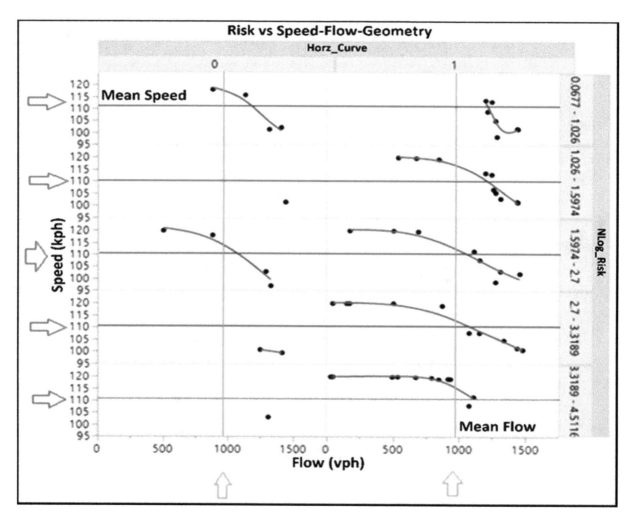

Figure 7. Contributing Factor based Risk Analysis.

4.2.1. Speed

In the case of motorways, a high speed limit is preferred to provide for free and easy maneuvering, but excessive speed is a very important factor having an impact on the number of crashes and injuries. In high-income countries, speed is one of the major factors (probably one third) of fatal and serious crashes [60–62]. We observed from the data that 35 out of 67 segments (52%) were above the mean speed limit 110 kph. So a reduction in speed limit could help in reducing risk level.

4.2.2. Flow

Flow is one of the major factors related to road safety, in parallel speed, the analysis showed that 39 out of 67 segments (58%) of the portion had above mean levels of traffic flow, i.e., 1000 vehicles. Traffic flow is one of the major contributing factor in road crashes [63,64]. "Based on the fluid mechanics theory of the traffic flow, the traffic flow parameters were specified, and the models of compressibility and viscosity of traffic flow were established respectively. Traffic control measures such as restricting the traffic flow at the upstream and downstream of the accident section should be carried out to control the crashes" [65]. So controlling the flow factor for the risky segments could assist in reducing the risk level of those segments.

4.2.3. Horizontal Curve

For the road safety analysis, the horizontal alignment designed cannot be ignored, especially the horizontal curve [66]. From previous research , the occurrence of accidents occurring on the curve is higher than the tangent (straight line), and it is necessary to design a horizontal curve [66]. In this analysis 80% of the risky level was along with the horizontal curves, so continuous marking of road marking signs for horizontal curves and straightening of curves can help in reducing crashes.

4.2.4. Vertical Curve

Research related to geometric characteristics showed that vertical curves had a significant effect on road crashes, and also while estimating speeds on highways [67]. Researchers also concluded that roads with vertical curves and higher speed limits tended to have more severe crashes [68]. Sometimes, a combination of horizontal and vertical curves is dangerous for road safety. The upward and downward gradient of the road contributed 76% in risk segment contribution, so a change in level could also help in reducing the risk level of road segments.

4.3. Safety Management and Financial Decision Making

Road safety management system and decision making is linked with econometrics i.e., funding and investments. Most countries need to enhance their understanding of spending on the significances of road safety, both by administration and organizations, and investment in road safety improvement. Road safety establishments need this knowledge to prepare financial and economic indication on the costs and usefulness of proposed solutions in order to win public and state support for funding road safety programs. There are prospects for targeted road safety funds that provide competitive revenues [69]. Road safety consultants and specialists develop business cases for this investment by applying such methods (i.e., the proposed DEA-ANNs method). A step change in funds invested in road safety management and in safer transport systems is compulsory to comprehend the success of motivated road safety targets in most of the world [69].

"Even though the implementation and maintenance costs of motorways vary significantly between the countries, in some cases also due to the different tendering systems, they are usually high, comparing to the implementation costs of other road safety road infrastructure related initiatives" [70]. During Cost-benefit analysis (CBA), the "cost-effectiveness of motorways also varies from case to case, especially due to the different implementation costs. In most cases, though, CBA results reveal relatively small ratios for new motorway development comparing to respective ratios regarding other road safety investments, mainly due to the very high implementation costs. However, even these ratios are considered as adequate to support the decisions for motorway development or the upgrade of existing rural network into motorways and apart from the strict financial criteria, the significant benefits for the road users can enhance the investment's effectiveness and should also be taken into account by the appropriate authorities" [70].

After safety analysis, we can target a different type of solutions: low-cost solutions, relatively costly solutions, and costly solutions. Speed limit change is considered as a low-cost solution because by changing sign boards for speed limit can help in the implementation of safety related alternatives. Consultants sometimes even recommend to installing permanent solution of electronic speed limit signs which help in controlling speed limit, some may be electronically related to the flow of the road and speed limit, to automatically change according to requirements. Flow limit is also another problem on the road, and can be solved by implementing the option of controlling access. The controlling flow option needs to have structures (e.g., toll installation) which lead to an investment higher than speed controlling signs. Flow can be controlled by applying tolls on those segments which are under high risk, which leads towards higher investments. Infrastructural change is one of the costly solutions during the safety solution process. Horizontal curve and vertical curve change can be backed by higher investments. Decision makers are always reluctant to change the structural pattern because a proper structural design change and construction is required to implement the decision. Sometimes, cost can increase by additional super elevation changes, in combination to horizontal and vertical changes.

4.4. Advantages and Limitations of Using the DEA-ANN Method

Since DEA offers some benefits to other approaches such "as "(1) DEA is able to handle multiple inputs and outputs (2) DEA does not require a functional form that relates inputs and outputs (3) DEA optimizes on each individual observation and compares them against the "best practice" observations. (3) DEA can handle inputs and outputs without knowing a price or knowing the weights and (4) DEA produces a single measure for every DMU that can be easily compared with other DMUs and also have some limitation as (1) DEA only calculates relative efficiency measures and (2) As a nonparametric technique statistical hypothesis test are quite difficult" [71,72]. "Neural networks offer a number of advantages, including requiring less formal statistical training, ability to implicitly detect complex nonlinear relationships between dependent and independent variables, ability to detect all possible interactions between predictor variables, and the availability of multiple training algorithms. Disadvantages include its "black box" nature, greater computational burden, proneness to overfitting, and the empirical nature of model development" [73]. However, overfitting can be controlled by the penalty method. Previously, DEA was popular with its multi-stage properties, but it has shortcomings with respect to its prediction capabilities, which reduces/limits its application. So, a powerful technique, ANNs, has been joined with DEA to fill that gap. Finally, the predictive potential of ANNs and the optimization capacity of DEA perform complementary features, thus envisioning a prominent modelling option [3,6].

5. Conclusions

This study focuses on road safety risk evaluation and connection between risk recurrence with respect to contributing factors. To enhance the estimation accuracy, a joint technique has been proposed and applied to achieve the risk evaluation, i.e., a benchmarking mechanism of DEA in combination with a prediction model of ANNs has been introduced to the road safety field. A crash dataset extracted from the Flemish Road safety department is stratified by two factors: the number of total crashes and number of affected persons, and is utilized to exhibit the proposed model formation of DEA and neural network performance. Notwithstanding the over-scattered crash information and the high relationship between the crash frequencies of the distinctive damage degrees, the outcomes demonstrate that comprehensive neural systems beat the multiple linear regression, shows in fitting and prescient execution. It demonstrates the neural system's prevalence over linear regression.

Risk has been calculated with the help of DEA for two motorways. Calculating risk has another advantage if segment length or volume of traffic is high or low; if we had just analyzed on the basis of number of crashes, it would not be a fair way to evaluated the most problematic segments out of a length of highway. Thus, using maximum information to evaluate an overall risk, DEA is a better option. In addition, we can rank them on the basis of risk value, and we can select our priorities on the basis that it could lead us to better decision making. So, for selecting problematic segments, it is a great achievement if we are able to indicate the most dangerous (risky) segments.

The predictability of risk values were checked with the assistance of ANNs: speed, flow, and horizontal and vertical curve. These are the most important factors which could be influenced by decision makers/transportation engineers. Selecting the contributing factors and changing the speed limit and flow limitation for risky segments could provide low-cost safety solutions. On the other hand, an infrastructural change like an amendment in the horizontal and vertical curve can cost much. However, this system can also help with designing better solutions as no one would prefer to change the structure of an entire highway (i.e., if a 100 km long highway), thus, selecting the most problematic section and solving the safety problem of only those segments would also provide low-cost decision making outcomes. Furthermore, combining ANN with GIS in a road safety analysis system can further encompass the functionality of the ANNs and, at the same time, increase the set of potential applications of GIS. The main advantage of using an ANNs system within a GIS environment for road safety and crash analysis includes the collection, manipulation, and analysis of the crash related data, which can be used effectively and resourcefully. The results of the overlay functions and spatial analysis performed by a GIS can be used as the input and training settings of a neural network, while the results of the neural network may be deployed by a GIS to produce a geospatial output. Each spatial input data and outcome of the neural network can be easily accumulated, normalized, rescaled, re-projected, and overlaid. It may accept different kinds of parameters (e.g., class, ordinal, continuous and categorical) as input or output values, and can handle deficient data [74]. The system is extremely flexible and self-adaptive, and capable of incorporating any improvement new data set. So, a joint approach of DEA-ANN within a GIS environment can provide an easy and an efficient output for decision makers for road safety data analysis and decision making for safety improvement.

Acknowledgments: This research is jointly supported by TITE and IMOB. and sponsored by IMOB for publication. Authors would like to thank HE-Boong Kwon (USA), one of pioneer of DEA-ANN method for his valuable guidance.

Author Contributions: Ali Pirdavani, Tom Brijs and Syyed Adnan Raheel Shah conceived and designed the concept; Syyed Adnan Raheel Shah and Naveed Ahmad performed literature review; Syyed Adnan Raheel Shah, Yongjun Shen and Tom Brijs applied DEA and ANN model; Muhammad Aamir Bashir and Ali Pirdavani contributed in data extraction-GIS application and contributed in analysis tools; Syyed Adnan Raheel Shah wrote the paper.

References

1. Songchitruksa, P.; Tarko, A.P. The extreme value theory approach to safety estimation. *Accid. Anal. Prev.* **2006**, *38*, 811–822. [CrossRef] [PubMed]
2. Golob, T.F.; Recker, W.W.; Alvarez, V.M. Tool to evaluate safety effects of changes in freeway traffic flow. *J. Transp. Eng.* **2004**, *130*, 222–230. [CrossRef]
3. Kwon, H.B. Exploring the predictive potential of artificial neural networks in conjunction with DEA in railroad performance modeling. *Int. J. Prod. Econ.* **2017**, *183*, 159–170. [CrossRef]
4. Hsiang, H.L.; Chen, T.Y.; Chiu, Y.H.; Kuo, F.H. A comparison of three-stage DEA and artificial neural network on the operational efficiency of semi-conductor firms in Taiwan. *Mod. Econ.* **2013**, *4*, 20.
5. Sreekumar, S.; Mahapatra, S. Performance modeling of Indian business schools: A DEA-neural network approach. *Benchmarking* **2011**, *18*, 221–239. [CrossRef]
6. Kwon, H.B. Performance modeling of mobile phone providers: A DEA-ANN combined approach. *Benchmarking* **2014**, *21*, 1120–1144. [CrossRef]

7.   Azadeh, A.; Azadeh, A.; Saberi, M.; Moghaddam, R.T.; Javanmardi, L. An integrated data envelopment analysis–artificial neural network–rough set algorithm for assessment of personnel efficiency. *Expert Syst. Appl.* **2011**, *38*, 1364–1373. [CrossRef]
8.   Mostafa, M.M. Modeling the efficiency of top Arab banks: A DEA–neural network approach. *Expert Syst. Appl.* **2009**, *36*, 309–320. [CrossRef]
9.   Emrouznejad, A.; Anouze, A.L. Data envelopment analysis with classification and regression tree—A case of banking efficiency. *Expert Syst.* **2010**, *27*, 231–246. [CrossRef]
10. Samoilenko, S.; Osei-Bryson, K.M. Using Data Envelopment Analysis (DEA) for monitoring efficiency-based performance of productivity-driven organizations: Design and implementation of a decision support system. *Omega* **2013**, *41*, 131–142. [CrossRef]
11. Çelebi, D.; Bayraktar, D. An integrated neural network and data envelopment analysis for supplier evaluation under incomplete information. *Expert Syst. Appl.* **2008**, *35*, 1698–1710. [CrossRef]
12. Kuo, R.J.; Wang, Y.C.; Tien, F.C. Integration of artificial neural network and MADA methods for green supplier selection. *J. Clean. Prod.* **2010**, *18*, 1161–1170. [CrossRef]
13. Pendharkar, P.C. A hybrid radial basis function and data envelopment analysis neural network for classification. *Comput. Oper. Res.* **2011**, *38*, 256–266. [CrossRef]
14. Al Haji, G. *Towards a Road Safety Development Index (RSDI): Development of an International Index to Measure Road Safety Performance*; Linköping University Electronic Press: Linköping, Sweden, 2005; p. 113.
15. Yannis, G.; Papadimitriou, E.; Lejeune, P.; Treny, V.; Hemdorff, S.; Bergel, R.; Haddak, M.; Holló, P.; Cardoso, J.; Bijleveld, F.; et al. *State of the Art Report on Risk and Exposure Data. SafetyNet, Building the European Road Safety Observatory, Workp 2 Deliv D2*; European Road Safety Observatory: Brussels, Belgium, 2007; p. 120.
16. Elke, H.; Tom, B.; Geert, W.; Koen, V. Benchmarking road safety: Lessons to learn from a data envelopment analysis. *Accid. Anal. Prev.* **2009**, *41*, 174–182.
17. Wegman, F.; Oppe, S. Benchmarking road safety performances of countries. *Saf. Sci.* **2010**, *48*, 1203–1211. [CrossRef]
18. Shen, Y.; Hermans, E.; Bao, Q.; Brijs, T.; Wets, G. Serious injuries: An additional indicator to fatalities for road safety benchmarking. *Traffic Inj. Prev.* **2015**, *16*, 246–253. [CrossRef] [PubMed]
19. Charnes, A.; Cooper, W.W.; Rhodes, E. Measuring the efficiency of decision making units. *Eur. J. Oper. Res.* **1978**, *2*, 429–444. [CrossRef]
20. Shen, Y.; Hermans, E.; Brijs, T.; Wets, G.; Vanhoof, K. Road safety risk evaluation and target setting using data envelopment analysis and its extensions. *Accid. Anal. Prev.* **2012**, *48*, 430–441. [CrossRef] [PubMed]
21. Shen, Y.; Hermans, E.; Ruan, D.; Wets, G.; Brijs, T.; Vanhoof, K. Evaluating trauma management performance in Europe: A multiple-layer data envelopment analysis model. *Transp. Res. Rec.* **2010**, *2148*, 69–75. [CrossRef]
22. Shen, Y.; Hermans, E.; Bao, Q.; Brijs, T.; Wets, G. Road safety development in Europe: A decade of changes (2001–2010). *Accid. Anal. Prev.* **2013**, *60*, 85–94. [CrossRef] [PubMed]
23. Shen, Y.; Shen, Y.; Hermans, E.; Bao, Q.; Brijs, T.; Wets, G.; Wang, W. Inter-national benchmarking of road safety: State of the art. *Transp. Res. Part C* **2015**, *50*, 37–50. [CrossRef]
24. Bastos, J.T.; Shen, Y.; Hermans, E.; Brijs, T.; Wets, G.; Ferraz, A.C.P. Traffic fatality indicators in Brazil: State diagnosis based on data envelopment analysis research. *Accid. Anal. Prev.* **2015**, *81*, 61–73. [CrossRef] [PubMed]
25. Williams, J.; Li, Y. A case study using neural networks algorithms: Horse racing predictions in Jamaica. In Proceedings of the International Conference on Artificial Intelligence (ICAI 2008), Las Vegas, NV, USA, 14–17 July 2008.
26. Abdelwahab, H.; Abdel Aty, M. Development of artificial neural network models to predict driver injury severity in traffic accidents at signalized intersections. *Transp. Res. Rec.* **2001**, *1746*, 6–13. [CrossRef]
27. Chong, M.M.; Abraham, A.; Paprzycki, M. Traffic accident analysis using decision trees and neural networks. *arXiv* **2004**, arXiv:cs/0405050.
28. Yasin Çodur, M.; Tortum, A. An Artificial Neural Network Model for Highway Accident Prediction: A Case Study of Erzurum, *Turkey. Promet-Traffic Transp.* **2015**, *27*, 217–225.
29. Zeng, Q.; Huang, H.; Pei, X.; Wong, S.C. Modeling nonlinear relationship between crash frequency by severity and contributing factors by neural networks. *Anal. Sci Accid. Res.* **2016**, *10*, 12–25. [CrossRef]
30. Athanassopoulos, A.D.; Curram, S.P. A comparison of data envelopment analysis and artificial neural networks as tools for assessing the efficiency of decision making units. *J. Oper. Res. Soc.* **1996**, 1000–1016. [CrossRef]

31. Vaninsky, A. Combining data envelopment analysis with neural networks: Application to analysis of stock prices. *J. Inf. Optim. Sci.* **2004**, *25*, 589–611. [CrossRef]

32. Azadeh, A.; Javanmardi, L.; Saberi, M. The impact of decision-making units features on efficiency by integration of data envelopment analysis, artificial neural network, fuzzy C-means and analysis of variance. *Int. J. Oper. Res.* **2010**, *7*, 387–411. [CrossRef]

33. Ülengin, F.; Kabak, Ö.; Önsel, S.; Aktas, E.; Parker, B.R. The competitiveness of nations and implications for human development. *Socio-Econ. Plan. Sci.* **2011**, *45*, 16–27. [CrossRef]

34. Wu, D.D.; Yang, Z.; Liang, L. Using DEA-neural network approach to evaluate branch efficiency of a large Canadian bank. *Expert Syst. Appl.* **2006**, *31*, 108–115. [CrossRef]

35. Wu, D. Supplier selection: A hybrid model using DEA, decision tree and neural network. *Expert Syst. Appl.* **2009**, *36*, 9105–9112. [CrossRef]

36. Ciobanu, S.M.; Benedek, J. Spatial characteristics and public health consequences of road traffic injuries in Romania. *Environ. Eng. Manag.* **2015**, *14*, 2689–2702.

37. Wang, C.; Quddus, M.A.; Ison, S.G. Impact of traffic congestion on road accidents: A spatial analysis of the M25 motorway in England. *Accid. Anal. Prev.* **2009**, *41*, 798–808. [CrossRef] [PubMed]

38. Chen, C.; Li, T.; Sun, J.; Chen, F. Hotspot Identification for Shanghai Expressways Using the Quantitative Risk Assessment Method. *Int. J. Environ. Res. Public Health* **2016**, *14*, 20. [CrossRef] [PubMed]

39. Zhang, C.; Yan, X.; Ma, L.; An, M. Crash prediction and risk evaluation based on traffic analysis zones. *Math. Probl. Eng.* **2014**, *2014*, 9. [CrossRef]

40. Moradi, A.; Soori, H.; Kavousi, A.; Eshghabadi, F.; Jamshidi, E.; Zeini, S. Spatial analysis to identify high risk areas for traffic crashes resulting in death of pedestrians in Tehran. *Med. J. Islam. Repub. Iran* **2016**, *30*, 450. [PubMed]

41. Steenberghen, T.; Dufays, T.; Thomas, I.; Flahaut, B. Intra-urban location and clustering of road accidents using GIS: a Belgian example. *Int. J. Geogr. Inf. Sci.* **2004**, *18*, 169–181. [CrossRef]

42. Pirdavani, A.; Bellemans, T.; Brijs, T.; Wets, G. Application of geographically weighted regression technique in spatial analysis of fatal and injury crashes. *J. Transp. Eng.* **2014**, *140*, 04014032. [CrossRef]

43. Pirdavani, A.; Bellemans, T.; Brijs, T.; Kochan, B.; Wets, G. Assessing the road safety impacts of a teleworking policy by means of geographically weighted regression method. *J. Saf. Res.* **2014**, *39*, 96–110. [CrossRef]

44. Eksler, V.; Lassarre, S. Evolution of road risk disparities at small-scale level: Example of Belgium. *J. Pet. Sci. Eng.* **2008**, *39*, 417–427. [CrossRef] [PubMed]

45. Yang, Y.; Rosenbaum, M. Artificial neural networks linked to GIS for determining sedimentology in harbours. *J. Pet. Sci. Eng.* **2001**, *29*, 213–220. [CrossRef]

46. Sameen, M.I.; Pradhan, B. Severity Prediction of Traffic Accidents with Recurrent Neural Networks. *Appl. Sci.* **2017**, *7*, 476. [CrossRef]

47. Pradhan, B.; Lee, S.; Buchroithner, M.F. A GIS-based back-propagation neural network model and its cross-application and validation for landslide susceptibility analyses. *Comput. Environ. Urban Syst.* **2010**, *34*, 216–235. [CrossRef]

48. Elsafi, S.H. Artificial neural networks (ANNs) for flood forecasting at Dongola Station in the River Nile, Sudan. *Alex. Eng. J.* **2014**, *53*, 655–662. [CrossRef]

49. Lee, S.; Park, I.; Koo, B.J.; Ryu, J.H.; Choi, J.K.; Woo, H.J. Macrobenthos habitat potential mapping using GIS-based artificial neural network models. *Mar. Pollut. Bull.* **2013**, *67*, 177–186. [CrossRef] [PubMed]

50. Pijanowski, B.C.; Brown, D.G.; Shellito, B.A.; Manik, G.A. Using neural networks and GIS to forecast land use changes: A land transformation model. *Comput. Environ. Urban Syst.* **2002**, *26*, 553–575. [CrossRef]

51. Yoo, C.; Kim, J.M. Tunneling performance prediction using an integrated GIS and neural network. *Comput. Geotech.* **2007**, *34*, 19–30. [CrossRef]

52. Mas, J.F.; Puig, H.; Palacio, J.L.; Sosa López, A. Modelling deforestation using GIS and artificial neural networks. *Environ. Model. Soft* **2004**, *19*, 461–471. [CrossRef]

53. Janssens, D.; Wets, G.; Timmermans, H.J.; Arentze, T.A. Modelling short-term dynamics in activity-travel patterns: Conceptual framework of the Feathers model. In Proceedings of the 11th World Conference on Transport Research, Berkeley, CA, USA, 24–28 June 2007.

54. Avkiran, N.K. An application reference for data envelopment analysis in branch banking: Helping the novice researcher. *Int. J. Bank Mark* **1999**, *17*, 206–220. [CrossRef]

55. Galagedera, D.; Silvapulle, P. Experimental evidence on robustness of data envelopment analysis. *J. Oper. Res. Soc.* **2003**, *54*, 654–660. [CrossRef]

56. Raab, R.L.; Lichty, R.W. Identifying subareas that comprise a greater metropolitan area: The criterion of county relative efficiency. *J. Reg. Sci.* **2002**, *42*, 579–594. [CrossRef]

57. Shmueli, G.; Patel, N.R.; Bruce, P.C. *Data Mining for Business Analytics: Concepts, Techniques and Applications*; John Wiley & Sons: Hoboken, NJ, USA, 2016.

58. Jmp, A.; Proust, M. *Specialized Models*; AS Institute Inc.: Cary, NC, USA, 2013.

59. Tso, G.K.; Yau, K.K. Predicting electricity energy consumption: A comparison of regression analysis, decision tree and neural networks. *Energy* **2007**, *32*, 1761–1768. [CrossRef]

60. Elvik, R. Speed and road safety: Synthesis of evidence from evaluation studies. *Transp. Res. Rec.* **2005**, *1908*, 59–69. [CrossRef]

61. Kweon, Y.J.; Kockelman, K. Safety effects of speed limit changes: Use of panel models, including speed, use, and design variables. *Transp. Res. Rec.* **2005**, *1908*, 148–158. [CrossRef]

62. WHO. *World Report on Road Traffic Injury Prevention*; World Health Organization: Geneva, Switzerland, 2004.

63. Garber, N.; Ehrhart, A. Effect of speed, flow, and geometric characteristics on crash frequency for two-lane highways. *Transp. Res. Rec.* **2000**, *1717*, 76–83. [CrossRef]

64. Golob, T.F.; Recker, W.; Pavlis, Y. Probabilistic models of freeway safety performance using traffic flow data as predictors. *Saf. Sci.* **2008**, *46*, 1306–1333. [CrossRef]

65. Xie, F.; Feng, Q. Research of effects of accident on traffic flow characteristics. In Proceedings of the International Conference on Mechatronic Sciences, Electric Engineering and Computer (MEC), Shengyang, China, 20–22 December 2013.

66. Zhang, Y. Analysis of the Relation between Highway Horizontal Curve and Traffic Safety. In Proceedings of the International Conference on Measuring Technology and Mechatronics Automation (ICMTMA), Zhangjiajie, China, 11–12 April 2009.

67. Vayalamkuzhi, P.; Amirthalingam, V. Influence of geometric design characteristics on safety under heterogeneous traffic flow. *Transp. Res. Rec.* **2016**, *3*, 559–570. [CrossRef]

68. Ma, M.; Yan, X.; Abdel Aty, M.; Huang, H.; Wang, X. Safety analysis of urban arterials under mixed-traffic patterns in Beijing. Transportation Research Record. *Transp. Res. Rec.* **2010**, *2193*, 105–115. [CrossRef]

69. Zero, T. *Towards Zero: Achieving Ambitious Road Safety Targets through a Safe System Approach*; OECD: Paris, France, 2008.

70. Yannis, G.; Evgenikos, P.; Papadimitriou, E. *Best Practice for Cost-Effective Road Safety Infrastructure Investments*; Conference of European Directors of Road (CEDR): Paris, France, 2008.

71. Blumenberg, S. Benchmarking Financial Processes with Data Envelopment Analysis. 2005. Available online: www.is-frankfurt.de/publikationenNeu/BenchmarkingFinancialProcesses1208.pdf (accessed on 20 June 2017).

72. Charnes, A.; Cooper, W.W.; Lewin, A.Y.; Seiford, L.M. *Data Envelopment Analysis: Theory, Methodology, and Applications*; Springer Science & Business Media: New York, NY, USA, 2013.

73. Tu, J.V. Advantages and disadvantages of using artificial neural networks versus logistic regression for predicting medical outcomes. *J. Clin. Epidemiol.* **1996**, *49*, 1225–1231. [CrossRef]

74. Tsangaratos, P.; Benardos, A. Applying artificial neural networks in slope stability related phenomena. In Proceedings of the 13th International Congress-Bulletin of the Geological Society of Greece (BGSG), Chania, Greece, 5–8 September 2013; pp. 1901–1911.

5

A New Damage Assessment Method by Means of Neural Network and Multi-Sensor Satellite Data

Alessandro Piscini *, Vito Romaniello, Christian Bignami and Salvatore Stramondo

Istituto Nazionale di Geofisica e Vulcanologia, 00143 Roma, Italy; vito.romaniello@ingv.it (V.R.); christian.bignami@ingv.it (C.B.); salvatore.stramondo@ingv.it (S.S.)
* Correspondence: alessandro.piscini@ingv.it

Abstract: Artificial Neural Network (ANN) is a valuable and well-established inversion technique for the estimation of geophysical parameters from satellite images. After training, ANNs are able to generate very fast products for several types of applications. Satellite remote sensing is an efficient way to detect and map strong earthquake damage for contributing to post-disaster activities during emergency phases. This work aims at presenting an application of the ANN inversion technique addressed to the evaluation of building collapse ratio (CR), defined as the number of collapsed buildings with respect to the total number of buildings in a city block, by employing optical and SAR satellite data. This is done in order to directly relate changes in images with damage that has occurred during strong earthquakes. Furthermore, once they have been trained, neural networks can be used rapidly at application stage. The goal was to obtain a general tool suitable for re-use in different scenarios. An ANN has been implemented in order to emulate a regression model and to estimate the CR as a continuous function. The adopted ANN has been trained using some features obtained from optical and Synthetic Aperture Radar (SAR) images, as inputs, and the corresponding values of collapse ratio obtained from the survey of the 2010 M7 Haiti Earthquake, i.e., as target output. As regards the optical data, we selected three change parameters: the Normalized Difference Index (NDI), the Kullback–Leibler divergence (KLD), and Mutual Information (MI). Concerning the SAR images, the Intensity Correlation Difference (ICD) and the KLD parameters have been considered. Exploiting an object-oriented approach, a segmentation of the study area into several regions has been performed. In particular, damage maps have been generated by considering a set of polygons (in which satellite parameters have been calculated) extracted from the open source Open Street Map (OSM) geo-database. The trained ANN has been proposed for the M6.0 Amatrice earthquake that occurred on 24 August 2016, in central Italy, by using the features extracted from Sentinel-2 and COSMO-SkyMed images as input. The results show that the ANN is able to retrieve a building collapse ratio with good accuracy. In particular, the fusion approach modelled the collapse ratio characterized by high values of CR (more than 0.5) over the historical center that agrees with observed damages. Since the technique is independent from different typologies of input data (i.e., for radiometric or spatial resolution characteristics), the study demonstrated the strength of the proposed approach for estimating damaged areas and its importance in near real time monitoring activities, owing to its fast application.

Keywords: earthquake; damage assessment; neural networks; satellite data; SAR; Sentinel-2

---

1. Introduction

Artificial neural networks (ANN), computational modelling tools, have found wide acceptance in many disciplines due to their adaptability to complex real world problems.

ANNs have demonstrated their ability to model non-linear physics systems [1], involving complex physical behaviors, and were applied to the analysis of remotely sensed images with promising results. Some examples of applications are: retrieval of soil moisture and agricultural variables from microwave radiometry [2], estimation of snow water equivalent and snow water depth from microwave images [3], retrieval of leaf area index (LAI) and other biophysical variables from the MEdium Resolution Imaging Spectrometer (MERIS) and MODerate-resolution Imaging Spectroradiometer (MODIS) instruments [4], estimation of chlorophyll from MERIS [5], retrieval of volcanic ash and sulphur dioxide from hyperspectral data [6]. ANNs are also used, with good results, for rainfall prediction involving other geophysical data [7,8].

Since an earthquake usually acts in a nonlinear way, the neural network algorithm can be an appropriate method for damage estimation purposes since it demonstrated to be a good non-linear approximator [9]. Recently, ANNs have been applied to detect damaged buildings, following an earthquake, by using high spatial resolution optical images acquired after the seismic event [10]. A neural network based-approach is being implemented to assess the status of buildings after earthquake excitation, predicting the displacement at different floors considering the wave energy propagating only into the ground floor [11].

By using change features from satellite images, accurate and reliable damage mapping can be obtained, exploiting both optical and radar sensors [12–14]. In Romaniello et al., 2016, an unsupervised algorithm for damage classification purposes has been developed [14].

Currently, quantitative estimation of earthquake damage level as a continuous function, using an ANN, has not yet been exploited and the present study represents a first attempt at applying an ANN to both optical high resolution Sentinel-2 and Synthetic Aperture Radar (SAR) remote sensing data for collapse ratio modelling. To the best of our knowledge, the methodology based on ANN has not been utilized in modelling earthquake damage assessment.

In the present study, two different neural networks using different Earth Observation (EO) datasets have been realized in order to model, as a continuous function, building damage. The first neural network (NN) experiment used, as input, only features obtained from optical data, whilst the second one (in a data fusion approach) also the features obtained from SAR images. NDI, KLD and MI features from optical data, and ICD and KLD from SAR data were used. Regarding the SAR data, KLD and MI parameters are very suitable features that can contribute to damage estimation [15]; the ICD demonstrated itself a very good damage proxy [13,16]. Concerning the optical data, the most significant performances are related to the NDI, KLD [17], and MI indexes [18]. These features show very good sensitivity to the collapse ratio.

The case study is the Central Italy strong earthquake, which took place on August 2016. On 24 August 2016 at 1:36 UTC, a M6.0 earthquake occurred in the Apennines of Central (hereafter Amatrice earthquake) Italy at depth 8 km, over a NNW-SSE striking, WSW dipping normal fault [19], destroying the closest towns to the epicenter—Amatrice, Accumoli and Arquata del Tronto—and causing near 300 fatalities. This earthquake revealed the importance of a rapid earthquake damage assessment, right after a seismic event, which can address the civil protection interventions towards the most affected areas. This work allowed to quantitatively evaluate the performance of NNs in terms of CR retrieval accuracy and generalization capability.

2. Neural Network Approach and Employed Features

ANNs are based on the concept of the single artificial neuron, the 'Perceptron', introduced by [20] to solve problems in the area of character recognition [21]. Using supervised learning, with the Error-Correction Learning (ECL) rule for network weights adjustment, those networks can learn to map from one data space to another using examples [22]. One of the most common and reliable learning techniques is the back-propagation (BP) algorithm [23]. BP consists of two phases: in the feedforward pass, an input vector is presented to the network and propagated forward to the output; in the back-propagation phase, the network output is compared to a desired output; network weights are then adjusted in accordance with an ECL rule [23–25]. Cross validation can be used to detect when over-fitting starts during supervised training of a neural network; training is then stopped before convergence to avoid over-fitting (early stopping). Early stopping using cross validation was done by splitting the training data into a training set, a cross-validation set, and a test set, and then training the networks only using the training set and evaluating the per-example error on the test set on a sample basis after a defined number of epochs. Finally, training was stopped when the error—the difference between neural network output and target—on the cross validation set was higher than the previous error value [26].

The performance of a trained ANN is generally assessed by computing the root mean square error (RMSE) between expected values and activation values at the output nodes or, in the case of classification, a percentage of correctly classified examples of the validation set. At the time our study was carried out, the ground truth was unavailable for Amatrice earthquake, so a visual inspection using high resolution images has been adopted.

From studies of past literature and development activities performed in a European research project (APhoRISM, www.aphorism-project.eu), we identified and employed a set of features that demonstrated a good sensitivity to damage at object scale. Regarding optical data, we used the NDI, KLD and MI change indexes. As far as SAR data is concerned, we considered the KLD and ICD parameters. Note that all change indexes have been calculated at object scale, i.e., by considering polygons that refer to city blocks. These latter have been extracted through the free geo-database of the Open Street Map project.

The NDI parameter is defined as

$$NDI_i = \frac{POST_i - PRE_i}{POST_i + PRE_i} \tag{1}$$

where PRE_i and $POST_i$ indicate the mean values of intensity, respectively for pre- and post-seismic images, associated to i-th polygon (see also Figure 2 as example). The intensity values are obtained averaging the Top of Atmosphere (TOA) Red-Green-Blue (RGB) reflectance and corrected by applying the Flat Field procedure for atmospheric correction [27].

The MI index measures the correlation loss between pre- and post-seismic images (see Equation (2)).

$$MI_i = - \ln\left(1 - r_i^2\right) \times 0.5 \tag{2}$$

where r_i^2 is the correlation between the pre- and post-seismic pixels within each polygon. Correlation is obtained from intensity and backscattering values for optical and SAR data, rescpectively. MI is inversely proportional to the damage grade.

The KLD parameter is defined as

$$KLD_i = \frac{(PRE_i - POST_i)^2 + Var(PRE_i) + Var(POST_i)}{2} \times \left(\frac{1}{Var(PRE_i)} + \frac{1}{Var(POST_i)}\right) - 2 \tag{3}$$

where PRE_i and $POST_i$ are the same parameters in the Equation (1), and $Var(PRE_i)$ and $Var(POST_i)$ are their variances within the i-th polygon. The KLD parameter has the same behavior of NDI: KLD increasing values correspond to increasing damage level. The ICD parameter is calculated on the base of the Pearson Correlation coefficient (ρ_i) estimated on the pre-seismic SAR image pair ($ICpre$) and on the co-seismic SAR image pair ($ICcos$). From these two intermediate outputs, we can compute the ICD

$$ICD_i = ICpre - ICcos \tag{4}$$

3. Dataset and NN Training

The training case study was the Mw 7.0 earthquake that hit Haiti on 12 January 2010. The epicenter was located about 25 km west–southwest of Port-au-Prince city. The disastrous shock caused the collapse of a huge number of buildings and widespread damage.

Table 1 describes satellite dataset used which consists of GeoEye-1 optical images (one pre- and one post-seismic), and three TerraSAR-X SAR images (two pre- and one post-seismic).

Table 1. EO data list for NN training.

Datatype	Satellite	Acquisition
Optical	GeoEye-1	2009/10/01
Optical	GeoEye-1	2010/01/13
SAR	TerraSAR-X	2009/05/01
SAR	TerraSAR-X	2009/10/13
SAR	TerraSAR-X	2010/01/20

Optical images inputs, both for pre- and post-, consisted of TOA reflectances at 2 m spatial resolution for RGB spectral bands; starting from these reflectances, the intensity values are derived (as described in Section 2).

As regards SAR data, intensity images have been obtained by multi-looking 3×3 m TerraSAR-X applying a re-sampling at 10×10 m.

The computation of change indexes based on EO imagery has been performed at object scale by considering a set of polygons, extracted from the open source Open Street Map (OSM) geo-database over Port-au-Prince. A total of 1513 polygons corresponding to city blocks of the affected areas have been considered. All the features (NDI, KLD, MI and ICD) are grouped in a unique dataset necessary for the Neural Network approach.

In addition to satellite data, a Ground Truth (GT) survey for Port-au-Prince town, expressed in terms of collapse ratio (CR), has been used (see Figure 1). The CR has been calculated by using GT information collected during a post-earthquake survey and available from the JRC (Joint Research Center) database, and considering the same polygons (city blocks) used for the satellite features calculation.

Figure 1. Collapse ratio (CR) for each polygon obtained from JRC survey data over Port-au-Prince.

In this work, Back-Propagation Neural Network (BPNN) has been used [23]. The BPNN to model CR was implemented using, as input, the features extracted from satellite images (see previous section, e.g., NDI in Figure 2), and CR values as target output. The 1513 samples used for NN training were split in Training, Cross Validation, and Test datasets of 65, 20 and 15%, respectively.

Figure 2. NDI map obtained from pre- and post-seismic GeoEye-1 images over Port-au-Prince.

A first network topology (Exp 1) consisted of only optical features as input (NDI, KLD and MI), five hidden layers with variable number of neurons [5–10–20–10–5], and one output, the CR. Furthermore, in order to perform a data fusion exercise, a different neural network using five inputs was adopted (Exp 2), adding to the optical features also the SAR ones, i.e., IC and KLD.

Figure 3 shows the statistical distributions for CR, train (a), cross-validation (b) and test (c) sets used during NN training phase. Despite of histograms put in evidence that values higher than 0.5 are statistically not well represented, we can consider the dataset a good training ensemble, because it covers the entire range of values.

Figure 3. Distribution histograms for (**a**) training; (**b**) cross validation and (**c**) test of Haiti datasets used in the training phase.

Results of NNs training phase for Exp 1 and Exp 2 are depicted in Figures 4 and 5, respectively. As regard the network using only optical features, the regression coefficient obtained applying the network to the test dataset is 0.67 (Figure 4), whilst NN using both and optical features obtains 0.73 (Figure 5). In this phase, it seems that using more features improves the NN retrieval accuracy.

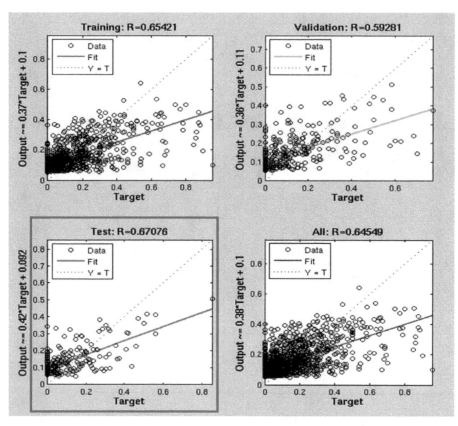

Figure 4. Regression curves for training, cross validation, test and total sets, considering Exp 1. Red squared regression curves describe the result obtained when an independent dataset is applied.

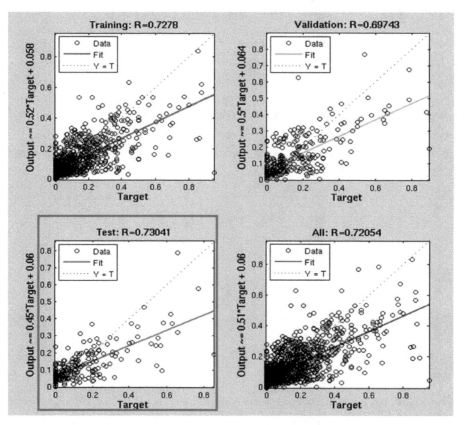

Figure 5. Regression curves for training, validation, test, and total sets, considering Exp 2. Red squared regression curves describe the result obtained when an independent dataset is applied.

4. Application of the Trained NN and Results for the Central Italy Case Study

In order to evaluate the performance of NNs in terms of retrieval accuracy and generalization capability, they were applied to the Amatrice earthquake. The goal is also to evaluate NN capability for modelling the building collapse ratio. The satellite dataset is made up of two Sentinel-2 optical images (1 pre- and 2 post-seismic), three COSMO-Sky SAR images (2 pre- and 1 post-seismic), and a building footprint layer extracted by the Open Street Map service (see Table 2 for EO data characteristics). Both optical and SAR data have a 10 m spatial resolution.

Table 2. EO data list for NN application.

Datatype	Satellite	Acquisition
Optical	Sentinel-2	2016/08/14
Optical	Sentinel-2	2016/09/04
SAR	COSMO-Sky	2016/07/19
SAR	COSMO-Sky	2016/08/20
SAR	COSMO-Sky	2016/08/28

The Amatrice footprint layer extracted from OSM, which is at a single building scale, was modified to obtain polygons surrounding more than one building. In this way, there are more pixels associated to each polygon leading a better estimation of change features over the polygon itself. The resulting layer consists of 112 polygons.

In the Figure 6, histograms for the CR obtained by the NN approach, Exp 1 (a) and Exp 2 (b), are depicted.

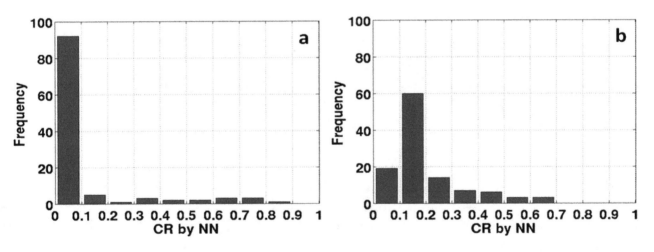

Figure 6. NN Collapse Ratio estimated after the M6.0 earthquake on 24 August 2016. (**a**) NN approach using optical inputs and (**b**) optical and SAR.

The histogram comparison shows that Exp 1 estimates most of polygons with a CR less than 0.1 (92 polygons, 82% of total), which means no damage. Five polygons present a CR between 0.1 and 0.2 and just one between 0.2 and 0.3. Only about 13% of polygons have a CR higher than 0.3, which spans from medium damaging to total collapsing. Specifically, three polygons with CR between 0.3 and 0.4, four between 0.4 and 0.6, six between 0.6 and 0.8, and only one higher than 0.8, indicating total collapse.

The location of polygons is well shown in Figure 7a, which describes the damage CR maps obtained analyzing post 24 August 2016. The map indicates that the severe damage is mainly retrieved in the northeast part of the Amatrice (red blocks), although information from Civil Protection confirms that collapses involved the whole historical center. This is confirmed by looking at an optical very high resolution (VHR) image (Figure 8) acquired after the Amatrice seismic event.

Figure 7. Amatrice CR maps obtained analyzing post 24 August 2016 remote sensing data: (**a**) NN with only optical features as input; (**b**) NN using both optical and SAR features.

Figure 8. Amatrice historical center, post-earthquake damage from RGB VHR DigitalGlobe image.

The fusion of optical and SAR features significantly changes this distribution (Figure 7b), where occurrences of CR less than 0.1 (blue blocks) drastically decrease with respect to the Exp 1 (19 polygons, only 17% of total). About 54% of polygons show a CR between 0.1 and 0.2 (grey blocks), indicating light damage. Looking at map in Figure 7b, these light damages lie partially in the historical center, but most of them are located in the southeast part of the city.

Considering only the historical center and the comparison with the VHR image (Figure 8), we suppose that Exp 2, involving SAR features, is more sensitive to light/medium damage level. Indeed, 14 polygons present a CR between 0.2 and 0.3, which indicate a medium damage level. We also notice that these polygons are located in the southeast part of the historical city center, an area involving severe damage as confirmed by the VHR image depicted in Figure 8. Considering CR higher than 0.3, polygons retrieved by Exp 2 increase by about 27% (19 respects of previous 14 of Exp 1).

The NN employed for the data fusion approach (Figure 7b), better identifies the severely damaged areas in the historical center, also where Exp 1 underestimated the collapses. At the same time, it seems to overestimate the damage in other areas, especially outside of the historical center. Indeed, Exp 2 indicates zones characterized by building collapses also in the southeast and partially in the northwest (Figure 7b). In this sense, the Exp 2 seems to provide a more realistic damage distribution: at first glance the NN regression model using the fusion approach gives the best results better identifying areas affected by collapses, whereas the Exp 1 seems to underestimate them.

5. Conclusions

The present work shows that neural networks, once they have been trained, can be used to rapidly retrieve building collapse ratios from optical and SAR remote sensed data. The implemented ANNs modelled the collapse ratio with quite high accuracy when applied to the post Amatrice earthquake independent dataset. High values of CR (more than 0.5) were retrieved over the historical center that are in agreement with a damage assessment observed by optical VHR imagery. The fusion of optical and SAR derived parameters seem to give a more reliable result using only optical data, even though it probably underestimates the occurrences of collapse ratios higher than 0.5, due mainly to a minor statistical characterization of values during the ANN training phase (only 3%).

Considering that the technique is independent of different types of input data, both for radiometric characteristics and spatial resolution, our work has demonstrated that ANNs are powerful tools able to estimate damaged areas, and they are important in near real time monitoring activities, owing to their fast application.

When looking at using an ANN approach in near real time monitoring, special care has to be taken during the training phase. This is because the neural network needs to be fed and trained continuously also during its operating phase in order to keep phenomena knowledge updated and retrieval performance accurate at the operating stage.

Another aspect is related to the input data characteristics, such as the spatial resolution, which can limit the neural network retrieval accuracy. In this case, a possible improvement could be the use of VHR satellite data, which could provide information at a building scale.

Future work will consider earthquake surveys in order get ground truth datasets to make a quantitative assessment of ANN performance. It will also consider extending the analysis to other areas affected by 2016 Central Italy seismic sequence. Furthermore, because the described approach needs both pre- and post-earthquake images, and in many cases a pre-earthquake image is not available or not up to date, a future goal will be the application of Neural Network approaches only using post-earthquake remote sensed images, available in near real time.

Acknowledgments: The research leading to these results has received funding from the European Union's Seventh Framework Program (FP7/2007-2013) under grant agreement number 606738, APhoRISM project.

Author Contributions: All authors contributed in a substantial way to the writing of the manuscript: Alessandro Piscini conceived the experiments. Alessandro Piscini, Vito Romaniello and Christian Bignami performed the analysis and wrote the manuscript. Salvatore Stramondo supervised the research at all stages.

References

1. Rumelhart, D.E.; Durbin, R.; Golden, R.; Chauvin, Y. Backpropagation: The basic theory. In *Backpropagation: Theory, Architecture, and Applications*; Rumelhart, D.E., Yves, C., Eds.; Lawrence Erlbaum: Hillsdale, NJ, USA, 1995; pp. 1–34.
2. Del Frate, F.; Ferrazzoli, P.; Schiavon, G. Retrieving soil moisture and agricultural variables by microwave radiometry using neural networks. *Remote Sens. Environ.* **2003**, *84*, 174–183. [CrossRef]
3. Tedesco, M.; Pulliainen, J.; Takala, M.; Hallikainen, M.; Pampaloni, P. Artificial neural network-based techniques for the retrieval of SWE and snow depth from SSM/I data. *Remote Sens. Environ.* **2004**, *90*, 76–85. [CrossRef]
4. Verger, A.; Baret, F.; Weiss, M. Performances of neural networks for deriving LAI estimates from existing CYCLOPES and MODIS products. *Remote Sens. Environ.* **2008**, *112*, 2789–2803. [CrossRef]
5. Vilas-González, L.; Spyrakos, E.; Torres-Palenzuela, J.M. Neural network estimation of chlorophyll a from MERIS full resolution data for the coastal waters of Galician rias (NW Spain). *Remote Sens. Environ.* **2011**, *115*, 524–535. [CrossRef]
6. Piscini, A.; Carboni, E.; Del Frate, F.; Grainger, R.G. Simultaneous retrieval of volcanic sulphur dioxide and plume height from hyperspectral data using artificial neural networks. *Geophys. J. Int.* **2014**, *198*, 697–709. [CrossRef]

7.   Wu, C.L.; Chau, K.W.; Fan, C. Prediction of rainfall time series using modular artificial neural networks coupled with data-preprocessing techniques. *J. Hydrol.* **2010**, *389*, 146–167. [CrossRef]
8.   Chau, K.W.; Wu, C.L. A Hybrid Model Coupled with Singular Spectrum Analysis for Daily Rainfall Prediction. *J. Hydroinform.* **2010**, *12*, 458–473. [CrossRef]
9.   Krasnopolsky, V.M.; Breaker, L.C.; Gemmill, W.H. A neural network as a nonlinear transfer function model for retrieving surface wind speeds from the Special Sensor Microwave Imager. *J. Geophys. Res.* **1995**, *100*, 11033–11045. [CrossRef]
10.   Ahadzadeh, S.; Valadanzouj, M.; Sadeghian, S.; Ahmadi, S. Detection of damaged buildings after an earthquake using artificial neural network algorithm. In Proceedings of the International Archives of the Photogrammetry, Remote Sensing and Spatial Information Sciences, Beijing, China, 3–11 July 2008.
11.   Chakraborty, S.; Kumar, P.; Chakraborty, S. Neural Network Approach to Response of Buildings Due to Earthquake Excitation. *Int. J. Geosci.* **2012**, *3*, 630–639. [CrossRef]
12.   Bignami, C.; Chini, M.; Pierdicca, N.; Stramondo, S. Comparing and combining the capability of detecting earthquake damage in urban areas using SAR and optical data. In Proceedings of the IEEE IGARSS, Anchorage, AK, USA, 20–24 September 2004; Volume 1, pp. 55–58.
13.   Stramondo, S.; Bignami, C.; Chini, M.; Pierdicca, N.; Tertulliani, A. Satellite radar and optical remote sensing for earthquake damage detection: Results from different case studies. *Int. J. Remote Sens.* **2006**, *27*, 4433–4447. [CrossRef]
14.   Romaniello, V.; Piscini, A.; Bignami, C.; Anniballe, R.; Stramondo, S. A multisensor approach for the 2016 Amatrice earthquake damage assessment. *Ann. Geophys.* **2016**, *59*. [CrossRef]
15.   Brunner, D.; Lemoine, G.; Bruzzone, L. Earthquake damage assessment of buildings using VHR optical and SAR imagery. *IEEE Trans. Geosci. Remote Sens.* **2010**, *48*, 2403–2420. [CrossRef]
16.   Romaniello, V.; Piscini, A.; Bignami, C.; Anniballe, R.; Stramondo, S. Earthquake damage mapping by using remotely sensed data: The Haiti case study. *J. Appl. Remote Sens.* **2017**, *11*. [CrossRef]
17.   Kullback, S.; Leibler, R.A. On Information and Sufficiency. *Ann. Math. Stat.* **1951**, *22*, 79–86. [CrossRef]
18.   Xie, H.; Pierce, L.E.; Ulaby, F.T. Mutual information based registration of SAR images. In Proceedings of the IEEE International Geoscience and Remote Sensing Symposium, Toulouse, France, 21–25 July 2003; pp. 4028–4031.
19.   Tinti, E.; Scognamiglio, L.; Michelini, A.; Cocco, M. Slip heterogeneity and directivity of the ML 6.0, 2016, Amatrice earthquake estimated with rapid finite-fault inversion. *Geophys. Res. Lett.* **2016**, *43*. [CrossRef]
20.   Rosenblatt, R. *Principles of Neurodynamics*; Spartan Books: New York, NY, USA, 1962.
21.   Hecht-Nielsen, R. *Neurocomputing*; Addison-Wesley: Boston, MA, USA, 1990.
22.   Bishop, C. *Neural Networks for Pattern Recognition*; Oxford University Press: Oxford, UK, 1995.
23.   Rumelhart, D.E.; Hinton, G.E.; Williams, R.J. Learning internal representation by error propagation. In *Parallel Distributed Processing: Exploration in the Microstructure of Cognition*; Rumelhart, D.E., McClleland, J.L., Eds.; MIT Press: Cambridge, MA, USA, 1986; Volume 1, Chapter 8.
24.   Hassoun, M.H. *Fundamentals of Artificial Neural Networks*; MIT Press: Cambridge, MA, USA, 1995.
25.   Haykin, S. *Neural Networks: A Comprehensive Foundation*, 2nd ed.; Prentice-Hall: Upper Saddle River, NJ, USA, 1999.
26.   Prechelt, L. Automatic Early Stopping Using Cross Validation: Quantifying the Criteria. *Neural Netw.* **1998**, *11*, 761–767. [CrossRef]
27.   Rast, M.; Hook, S.J.; Elvidge, C.D.; Alley, R.E. An evaluation of techniques for the extraction of mineral absorption features from high spectral resolution remote sensing data. *Photogramm. Eng. Remote Sens.* **1991**, *57*, 1303–1309.

6

Impacts of Sample Design for Validation Data on the Accuracy of Feedforward Neural Network Classification

Giles M. Foody

School of Geography, University of Nottingham, University Park, Nottingham NG7 2RD, UK;
giles.foody@nottingham.ac.uk

Abstract: Validation data are often used to evaluate the performance of a trained neural network and used in the selection of a network deemed optimal for the task at-hand. Optimality is commonly assessed with a measure, such as overall classification accuracy. The latter is often calculated directly from a confusion matrix showing the counts of cases in the validation set with particular labelling properties. The sample design used to form the validation set can, however, influence the estimated magnitude of the accuracy. Commonly, the validation set is formed with a stratified sample to give balanced classes, but also via random sampling, which reflects class abundance. It is suggested that if the ultimate aim is to accurately classify a dataset in which the classes do vary in abundance, a validation set formed via random, rather than stratified, sampling is preferred. This is illustrated with the classification of simulated and remotely-sensed datasets. With both datasets, statistically significant differences in the accuracy with which the data could be classified arose from the use of validation sets formed via random and stratified sampling ($z = 2.7$ and 1.9 for the simulated and real datasets respectively, for both $p < 0.05\%$). The accuracy of the classifications that used a stratified sample in validation were smaller, a result of cases of an abundant class being commissioned into a rarer class. Simple means to address the issue are suggested.

Keywords: cross-validation; multi-layer perceptron; remote sensing; classification error; sample design; machine learning

1. Introduction

Artificial neural networks are widely used for supervised classification applications. In these applications, cases of known class membership are used to train the neural network in order to allow it to predict the class membership of previously unseen and unlabeled cases. This type of analysis is common in, for example, the production of thematic maps, such as those depicting land cover, from remotely-sensed imagery [1–3]. The imagery contain data on the remotely-sensed response of the land surface that is converted into information on land cover class via the classification analysis and a wide variety of approaches and applications have been investigated, e.g., [2,4,5]. Neural networks have become a popular method for image classification as numerous studies have shown that they can yield more accurate maps than a variety of other alternative approaches to classification [6–8]. The relative performance of neural network classifiers in relation to a range of alternative methods, including standard statistical classifiers, machine learning methods, and decision trees, is discussed in the literature (e.g., [1]). As with all classifiers, the quality of the final classification is, however, in part a function of the classifier and the nature, notably the size and quality, of the ground reference dataset on the class membership used [9–12]. Ground reference data are used to provide data to train the neural network and to evaluate the quality of its predictions. The latter is typically expressed in terms of the overall accuracy of the classification output from the network.

In remote sensing applications, the class membership of the cases in the training set is typically determined by ground-based observation or interpretation of very fine spatial resolution imagery [1,13]. Clearly, the nature and quality of the ground reference data used will impact upon the accuracy of the predictions obtained from the neural network. Classification accuracy is, for example, influenced by the size, composition, and quality of the ground reference dataset used in training [2,9,10,14–16]. The way the ground reference data are used is also important, especially if part of it is used for the purpose of validation [10]. The use of some reference data for validation purposes, to indicate the quality of the trained model generated, is commonplace in defining a supposedly optimal neural network. In common remote sensing applications, the optimal approach would be the one that yielded the most accurate thematic map when applied to the image data.

In a typical remote sensing application, the aim of the classification is to accurately map the land cover classes in the region of study. Throughout there are a set of basic assumptions made in classification analyses. These include the need for the set of classes to be mutually exclusive and exhaustively defined. Failure to satisfy these assumptions will result in errors. Cases of an untrained class will, for example, typically have to be allocated erroneously to one of the defined classes [17]. Although there are instances when a neural network may be used to obtain a non-standard classification, such as soft classification when classes intergrade or are mixed together [17], the standard hard classification in which a case belongs fully to a single class is the focus on this article. In all cases, the ground reference data used in training, validating, and testing the neural network have a key role to play in the production of an accurate map. Hence, the design of these datasets is important.

Good practices for classification accuracy assessment for the evaluation of the quality of thematic maps derived via remote sensing have been defined and include guidance on the construction of the testing sample upon which the assessment is to be based. Typically, for example, the use of a reference dataset acquired following a probabilistic sampling design is recommended [18,19]. This allows a rigorous design-based assessment of classification accuracy, typically based upon the analysis of an error or confusion matrix in which the predicted class label obtained from the neural network is cross-tabulated against the label in the reference dataset for the sample of cases under consideration [1,18].

The goal of training a classifier is different to that of testing its output predictions and, hence, the nature of the ideal training set may be very different to that of the testing set. In training, the aim is, essentially, to guide the neural network to learn the identity of the classes from their remotely-sensed response. Thus, the observed remotely-sensed response for the training sample of known class membership is used in network learning to ultimately form decision rules to accurately label cases of unknown class membership in order to map the region of interest [1]. Much conventional guidance on the design of the training set in remote sensing applications is based upon historical work undertaken with conventional statistical classifiers, such as the maximum likelihood classification. This advice typically calls for the sample size to be estimated following basic sampling theory in order to derive a representative and unbiased description of each class to allow cases of unknown class membership to then be allocated the label of the class they had greatest similarity to. Alternatively, a simple heuristic, such as the use of a sample of cases for each class, the size of which is at least 10 times the number of discriminating variables, such as the spectral wavebands, are used [10,20,21]. In essence, this type of approach is calling for the use of a stratified sample design in the formation of the training set. Despite the development of new classifiers, such approaches are still widely used even though the nature of the ideal training set varies between classifiers.

With neural networks and machine learning methods such as the support vector machine (SVM) and relevance vector machine (RVM) attention in training is focused more on individual cases than broad statistical summary statistics that are central to statistical classifiers, such as the maximum likelihood classification [1]. The individual training cases in the training set can vary greatly in value to an artificial neural network classification [22,23]. In addition, different classifiers may ideally focus upon different cases in the training set. For example, a SVM may require only a very small training sample and, ideally, cases that lie in the boundary region between classes, while a RVM might also require only a small number of training cases, but these are anti-boundary in nature [12]. Considerable research has, therefore, focused on how the properties of the training dataset impact on a classification, typically with a desire to maximize the final mapping accuracy (e.g., [9,10,24–26]). Sometimes it may be possible to predict the location of useful candidate training cases for a classification [27] or, contrary to conventional approaches, even deliberately use mixed cases in training and the analyst can seek to form a training set intelligently for a given application scenario [22,28,29]. As such, there is no single ideal way to define a training set that is universally applicable. Given the popularity of consensus or ensemble methods that use a variety of different classifiers [7,30] it is common to see relatively large training sets acquired following conventional guidance. While this approach may sometimes be inefficient, notably in that the sample may be larger than needed, it has the capacity to provide useful training data for a wide range of classifiers. It is, therefore, common to see either a training sample designed explicitly for a specific task, which could be a small and highly unrepresentative sample [29,31], or the use of a stratified design that seeks to ensure each class can be described well [1]. The design details are, however, important. The nature of the training set has a significant impact on the accuracy of predictions by a neural network [2,15,32]. For example, the size of the training set, notably in relation to the complexity of the network, can have a marked effect on classification by a neural network [9,16,33]. The composition of the training set in terms of relative class abundance is also important [15,34]. Variations in class abundance can yield imbalanced datasets that, as in other classifiers, can have substantial impacts on the final classification.

In a supervised classification with a conventional feedforward neural network it is common for part of the training sample to be used for validation purposes [1,2,10,33–35]. In this data splitting approach part of the training set is used in the normal way to provide examples of the classes upon which the classifier may learn to form rules to classify cases of unknown membership. The remaining part of the training set forms the validation set and is used to evaluate the performance of the network in terms of the accuracy with which the validation set is classified, as well as help determine when to stop network learning [10,36–39]. A variety of approaches exist for the splitting of the training cases to form the training set and the validation set. If reference data are plentiful the training set and validation set could be completely separate and independent samples but if this is not the case other approaches to cross-validation, such as the leave one out approach, may be used [34]; for simplicity, the focus here is on the use of a completely independent validation set. The use of a validation set is important in neural network-based approaches to classification as there is a desire to avoid overfitting to the training data and there are a variety of network parameters that require definition. The latter includes the basic structure of the network (e.g., number of hidden layers and units) and the learning algorithm and parameters (e.g., momentum, learning rate and number of iterations). There is an extensive literature on this topic [2,10]. The basic idea is that a range of different networks can be defined and one that is optimal for the task at hand, defined with the aid of the validation dataset. Thus, for example, a range of different networks may be generated and the one that classifies the validation set to the highest accuracy is selected for the final classification analysis to produce a thematic map. This analysis might also suggest ways to further enhance the classification by, for example, indicating redundant discriminating variables that could be deleted in order to allow more rapid computation [28,40]. It is also useful as some approaches used in training may artificially inflate class separability and be unhelpful [23]. Given that the ultimate aim of the analysis in a typical remote sensing application is the production of an accurate thematic map via a classification analysis the nature of the validation set can be important. Commonly, with classifiers that use validation data, the validation set is formed by simply taking cases randomly from the entire set of data acquired for training activity [38,41] or a separate, often stratified, sample of cases is obtained (e.g., [42]). The size of the validation set is important. As in other aspects of the analysis, the literature contains guidance on the way the reference data should be divided up. For example, Mas et al. [43,44] suggests that half of the labelled cases be used for training, a quarter for validation and the final quarter for testing.

The composition of the sample of cases forming the validation set, notably in terms of the relative abundance of the cases of the classes may vary as a function of the design used to acquire the training dataset. If the validation sample is formed with a simple random sample design it is likely to be imbalanced in composition, with the number of cases of a class reflecting its relative abundance in the region being mapped. The use of a stratified sample design in the generation of the validation sample will act to give a balanced dataset but this need not be an ideal approach. Indeed, the use of a stratified design could result in the selection of a network that was sub-optimal for the task if the classes vary in abundance and separability. For example, standard approaches to the assessment of overall classification accuracy weight errors equally and could inflate the importance of rare classes while deflating that of abundant classes. Sometimes it may be possible to account for the sample design if there is information on class abundance [18]. Alternatively, if the study aim is focused on a single class, which is often the case, the accuracy assessment used in the validation could be focused on that class at the expense of the others. However, for a general purpose map, it may be more appropriate to follow the guidance on sampling that is typically used in the formation of the testing set as this typically allows for variations in class abundance to be accounted for.

The effect of different sample designs for the formation of the validation dataset is explored in this paper. Specifically, the focus is on the use of samples acquired by simple random and stratified random sampling designs (both without replication). Due to the way overall accuracy is typically estimated with a validation sample, it is hypothesized that the overall accuracy of the final classification, evaluated using the testing set, will be larger for a neural network trained using a validation sample generated via simple random as opposed to stratified random sampling. Indeed a series of outcomes may be predicted as having the potential to arise as a function of the validation dataset used. For example, in relation to two classes that overlap, it would be expected that the hyperplane to best separate the classes fitted when a balanced validation set was used would migrate away from the more abundant class if an imbalanced training set was used. As a consequence of this, there is an opportunity for the accuracy with which the more abundant class is classified to rise as fewer cases of it will be omitted from it and fewer of its cases commissioned by the rarer class (es). These trends arise as classification errors are weighted equally in standard assessments of overall accuracy. The overall trend expected would be for the accuracy with which the abundant class is classified to increase while the accuracy of the classification of the rarer class would decline. As such it is hypothesized that the use of a stratified sample design may not be ideal as its use relative to a randomly-defined validation dataset would be associated with a decrease in overall accuracy, arising noticeably through a decrease in the accuracy for the abundant class (es) as a result of an increase in the commission of cases of abundant class (es) by the set of rarer classes.

2. Data and Methods

Two datasets were used. First, a simulated dataset was used to illustrate the issues and also to facilitate, if desired, replication. Second, a real dataset consisting of remotely-sensed data and associated ground reference data on land cover class labels was used.

A very simple multi-class classification scenario was simulated. This comprised data on four classes acquired in two dimensions, or bands. The data for each class were formed using a random number generator using analyst provided values for the class mean and standard deviation on the assumption that the data for each class were normally distributed. For each class the standard deviation was set equal to 5 and the mean values used to generate the data in each band are shown in Table 1. In the scenario generated, class 1 was the most abundant class. Specifically, class 1 was five times more abundant than each of the other classes. Most attention was focused on class 1 and class 2, which exhibited a degree of overlap in their distributions with class 3 and, especially, class 4 was highly separable (Figure 1).

For the analyses of the simulated dataset, training validation and testing sets were generated (Table 2). These were used in two series of analyses, one using a validation dataset formed via simple random sampling and the other formed via a stratified random sample design. In brief, a single training set was generated using a stratified random sample of 400 cases per-class. Similarly, a single testing set was used to evaluate the accuracy of the classifications from the neural networks selected as optimal in each series of analyses undertaken. This testing set was simulated to represent a sample acquired by simple random sampling and comprised 800 cases. Due to the way the scenario was designed, class 1 was five times more abundant in it than the other classes. The two validation datasets used each contained 800 cases, but one was formed via a simple random sample in which class abundance varied as in the testing set, while the other was formed with a stratified random sample in which each class was equally represented. Each series of neural network analyses used a software package that sought to generate an optimal network, with optimality defined as the maximization of the overall accuracy with which the validation set was classified.

The remotely-sensed data were acquired by an airborne thematic mapper (ATM) sensor for a test site near the village of Feltwell, Norfolk, UK. The latter is located approximately 58 km to the northeast of the city of Cambridge. The land around the village was topographically flat and its land cover mosaic was characterized mainly by large agricultural fields. At the time of the ATM data acquisition these fields had also typically been planted with a single crop type (Figure 2).

The ATM used was a basic multispectral scanning system that acquired data in 11 spectral wavebands from blue to thermal infrared wavelengths (Table 3). Given the relatively low altitude of airborne data acquisition (~2000 m), the spatial resolution of the imagery was very much smaller than the typical field size, approximately 5 m. As a result, image pixels tended to represent an area composed of a single class (i.e., pure pixels) and, hence, were appropriate for hard image classification analysis; boundary pixels were ignored. Attention focused here on the six crop classes that dominated the region at the time of the ATM data acquisition. These classes and their approximate coverage (%) of the study area at the time were: sugar beet (S, 30.3%), wheat (W, 30.0%), barley (B, 16.0%), carrot (C, 10.3%), potato (P, 7.8%), and grass (G, 5.4%).

Table 1. The classes in the simulated dataset; note: units are arbitrary.

Band	Class 1	Class 2	Class 3	Class 4
Band 1	50	35	83	20
Band 2	50	60	50	20

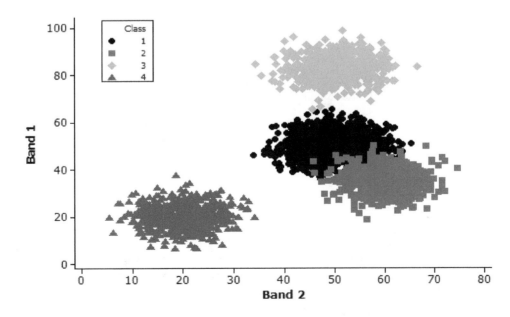

Figure 1. Simulated data (training, validation (random), and testing). Note: class 2 is shown overlaid on top of part of class 1 and the area of overlap may be inferred.

Figure 2. Data for Feltwell. (**a**) The location of Feltwell, © OpenStreetMap contributors; (**b**) airborne thematic mapper (ATM) image extract in waveband 4 and (**c**) ATM image extract in waveband 7.

Table 2. Class composition of the datasets used for analyses of the simulated data.

Dataset	Size and Class Composition
Training	400 cases of each class; total = 1600 cases
Validation (random)	500 class 1 100 class 2 100 class 3 100 class 4; total = 800 cases
Validation (stratified)	200 cases of each class; total = 800 cases
Testing	500 class 1 100 class 2 100 class 3 100 class 4; total = 800 cases

Table 3. The 11 spectral wavebands of the ATM sensor used.

Waveband	Wavelength (μm)
1	0.42–0.45
2	0.45–0.52
3	0.52–0.60
4	0.60–0.63
5	0.63–0.69
6	0.69–0.75
7	0.76–0.90
8	0.91–1.05
9	1.55–1.75
10	2.08–2.35
11	8.50–13.00

Ground reference data for the purposes of network learning (training) and evaluation were acquired. Following [43,44], the ground reference dataset was partitioned such that 50% was used for training, 25% for validation, and 25% for testing. The composition of these datasets, however, sometimes varied (Table 4).

Conventional guidance on the design of the training set was followed with 110 cases of each class obtained for training purposes; meeting the often stated requirement of a sample of at least 10 times the number of discriminating variables (wavebands) used as input. The training set, therefore, contained 660 cases acquired by a stratified random sampling design. This training set was used throughout.

Testing sets should, ideally, be acquired using a probability sampling design [18,19]. Here, a simple random sample (without replication) was used to acquire 330 cases to use for testing. Note that this sample size exceeds that required for accuracy estimation of a map with an accuracy of 85%, a standard if contentious target accuracy in remote sensing, with an allowable error of 4%. This testing set was used in all analyses of the ATM data. This latter issue impacts on comparisons of accuracy estimates and requires the use of a technique suited for related samples [45].

Table 4. Class composition of the datasets used for analyses of the remotely sensed data.

Dataset	Size and Class Composition
Training	110 cases of each class; total = 660 cases
Validation (random)	100 sugar beet 99 wheat 53 barley 34 carrot 26 potato 18 grass; total = 330 cases
Validation (stratified)	55 cases of each class; total = 330 cases
Testing	100 sugar beet 99 wheat 53 barley 34 carrot 26 potato 18 grass; total = 330 cases

As with the analyses of the simulated dataset, the search for an optimal neural network was undertaken twice. In each case the training and testing sets were the same, the only difference was the composition of the validation set used to identify the optimal network from a set of candidate networks generated for the task. In one set of analyses, the validation set was generated by simple random sampling and thus the number of cases of each class tended to reflect the actual abundance of the classes in the region to be mapped. Indeed, here, the sample was selected to ensure that the class composition of the validation set equalled that of the testing set. This sample comprised 330 cases. In the other set of analyses, a validation sample of the same size, but acquired following a stratified random sample such that each class was equally represented was used. The nature of the datasets used is summarized in Table 4.

In the search for an optimal neural network to classify the ATM data, optimality was defined in relation to the maximum overall accuracy of the classification of the validation data. The accuracy of each classification was calculated from the confusion or error matrix generated for it which shows a cross-tabulation predicted and actual class label for each case in the dataset analysed [1,18]. Using the layout and notation defined for the confusion matrix shown in Table 5, which is used throughout the article, overall accuracy, O, was calculated using Equation (1):

$$O = \frac{\sum n_{ii}}{n} \tag{1}$$

In addition to the global estimate of classification quality conveyed by overall accuracy, the producer's, P, and user's accuracy, U, were calculated with reference to each class [1,18]. These were obtained for class i from Equations (2) and (3) respectively:

$$P_i = \frac{n_{ii}}{n_{\cdot i}} \tag{2}$$

$$U_i = \frac{n_{ii}}{n_{i.}} \qquad (3)$$

Note that, as is common, the calculations of these measures of accuracy are based on the raw counts of cases shown in the elements of the confusion matrix. These approaches to accuracy assessment were used to support all analyses, whether based on the training, validation, or testing datasets. Most focus is, however, on the accuracy values arising from analyses of the validation and testing datasets.

To determine if the use of different validation samples impacted on the accuracy of the final thematic classification, the statistical significance of differences in the accuracy of the classifications of the testing set were assessed. Standard approaches for the comparison of accuracy values that are popular in remote sensing projects are unsuitable here as the same sample of testing cases was used throughout. To accommodate for this situation, the statistical significance of differences in accuracy was assessed using the McNemar test [45,46]. The latter is a non-parametric test that is based on a binary confusion matrix which shows the cross-tabulation of the cases that have been labelled correctly and incorrectly by the two classifications being compared. The test focuses on the discordant cases, those which were classified correctly by one classifier, but incorrectly by the other. Without continuity correction, the test is based on the normal curve deviate, z, as expressed as:

$$z = \frac{n_{CI} - n_{IC}}{\sqrt{n_{CI} + n_{IC}}} \qquad (4)$$

where n_{CI} indicates the number of cases in the relevant element of the matrix with the subscript C indicating if the classification was correct in its labelling or I if it was incorrect and order of the subscripts indicates the specific classification from the pair under study. For a standard two-tailed test at the 95% level of confidence, the null hypothesis of no significant difference is rejected if the calculated z exceeds the critical value of $|1.96|$. Similarly, for a one-tailed test, if the hypothesis under test has a directional component, the direction (sign) needs consideration and the magnitude of the critical value of z to indicate that a significant difference exists at the 95% level of confidence is 1.645.

Table 5. The confusion matrix based on raw counts of cases for a classification of q classes. Matrix columns show the label in the reference data and rows the label in the classification.

Class	1	2	...	q	Total
1	n_{11}	n_{12}	...	n_{1q}	$n_{1.}$
2	n_{21}	n_{22}	...	n_{2q}	$n_{2.}$
:	:	:		:	:
q	n_{q1}	n_{q2}	...	n_{qq}	$n_{q.}$
Total	$n_{.1}$	$n_{.2}$	...	$n_{.q}$	n

3. Results and Discussion

With the simulated dataset, two sets of analyses were undertaken to identify optimal neural networks, one using the validation set formed via simple random sampling and the other formed using a stratified random sample. The key properties of the selected networks and their ability to classify the datasets are defined in Table 6 with a full set of confusion matrices for each set of analyses shown in Tables 7 and 8.

Attention focused especially on the accuracy with which the testing set was classified, as this reflects the accuracy of the final product obtained. It was evident that the accuracy of classification obtained with the use of the simple random sample (98.37%) was slightly higher than that arising from the use of the validation set formed via stratified sampling (96.62%). Although small, this difference was significant ($p < 0.05$), with the calculated value of z from Equation (4) given the 26 discordant cases observed being 2.745 (Table 9). It was also evident that the predictions of what might happen as moving from a balanced stratified to random validation sample outlined in the introduction occurred.

Note, for example, that relative to the classification obtained with a stratified validation sample, the use of the validation set formed by random sampling resulted in a higher overall accuracy of the testing set. This arose because of a higher accuracy with which the abundant class was classified due to a reduction in omission error (from 27 to 7 cases) and the accuracy of the rarer class it was confused with declined (Tables 7 and 8). Critically, the results suggest that the use of a balanced training set acquired via a stratified random sample may produce a sub-optimal final output. As hypothesized, the use of a stratified sample resulted in a lower accuracy than was achievable because the accuracy with which the abundant class was classified declined, associated with a large omission error due to cases being commissioned into a rarer class.

A series of analyses was undertaken to define an optimal neural network using the two validation datasets in the analyses of the remote sensing data. The core focus here is on the classification results from the network determined to be optimal from each series of analyses. Although the precise details of the neural networks are relatively unimportant as the classifications that arise from them are the focus of attention, key details on the networks selected are summarized in Table 10 and the confusion matrices for the testing set are shown in Tables 11 and 12.

Table 6. Key characteristics of the networks selected for the analyses of the simulated data; note: network architecture is expressed as input:hidden:output.

Validation Dataset	Architecture	Algorithm and Iterations	Classification Accuracy (%) Training Validation Testing		
Random	2:15:2	Backpropagation 48	97.25	97.62	98.37
Stratified	2:8:2	Conjugate gradient 55	98.25	98.75	96.62

Table 7. Confusion matrices for the network selected using a validation set formed by random sampling.

			Training			
Class	1	2	3	4	Total	User's (%)
1	398	40	2	0	440	90.45
2	2	360	0	0	362	99.44
3	0	0	398	0	398	100
4	0	0	0	400	400	100
Total	400	400	400	400	1600	
Producer's (%)	99.50	90.00	99.50	100		
			Validation			
Class	1	2	3	4	Total	User's (%)
1	490	9	0	0	499	98.19
2	10	91	0	0	101	90.09
3	0	0	100	0	100	100
4	0	0	0	100	100	100
Total	500	100	100	100	800	
Producer's (%)	99.00	91.00	100	100		
			Testing			
Class	1	2	3	4	Total	User's (%)
1	493	6	0	0	499	98.79
2	7	94	0	0	101	93.06
3	0	0	100	0	100	100
4	0	0	0	100	100	100
Total	500	100	100	100	800	
Producer's (%)	98.60	94.00	100	100		

Table 8. Confusion matrices for the network selected using a validation set formed by stratified sampling.

(a)	Training					
Class	1	2	3	4	Total	User's (%)
1	386	14	0	0	400	96.50
2	14	386	0	0	400	96.50
3	0	0	400	0	400	100
4	0	0	0	400	400	100
Total	400	400	400	400	1600	
Producer's (%)	96.50	96.50	100	100		

(b)	Validation					
Class	1	2	3	4	Total	User's (%)
1	195	5	0	0	200	96.50
2	5	195	0	0	200	96.50
3	0	0	200	0	200	100
4	0	0	0	200	200	100
Total	200	200	200	200	800	
Producer's (%)	97.50	97.50	100	100		

(c)	Testing					
Class	1	2	3	4	Total	User's (%)
1	473	0	0	0	473	100
2	27	100	0	0	127	78.74
3	0	0	100	0	100	100
4	0	0	0	100	100	100
Total	500	100	100	100	800	
Producer's (%)	94.60	100	100	100		

Table 9. Cross-tabulation of labelling from classifiers using random (columns) and stratified (rows) validation sets for the simulated data.

	Correct	Incorrect	Total
Correct	767	6	773
Incorrect	20	7	27
Total	787	13	800

Table 10. Key characteristics of the networks selected for the analyses of the remotely-sensed data.

Validation Dataset	Architecture	Algorithm and Iterations	Classification Accuracy (%) Training Validation Testing		
Random	9:11:6	Conjugate gradient, 435	97.27	98.18	97.87
Stratified	11:16:6	Conjugate gradient, 205	96.21	97.27	96.36

The selected neural networks were able to classify the data accurately. In each case, however, it was evident that classification accuracy was slightly, no more than 1.51%, less accurate when the stratified, rather than random, validation dataset had been used in network selection. Although small, these differences can still be significant. Indeed, a key result was that the accuracy of the classification of the testing dataset, which indicates the accuracy of the land cover map obtainable, was higher when the validation set formed using random (overall accuracy = 97.87%), rather than stratified, sampling (overall accuracy = 96.36%). Although small, a test of the significance of the difference in overall accuracy, using a one-sided McNemar test to recognize the directional nature of the test and the use of the same testing set, revealed it to be statistically significant at the 95% level of confidence (Table 13). Specifically, given the seven discordant cases observed (Table 13), Equation (4) yields $z = 1.889$.

The difference in accuracy between the classifications of the testing sets by the two selected neural networks (Tables 11 and 12) was attributable mostly to seven cases of sugar beet being commissioned into the potato class when the validation set formed using a stratified sample was used. As a result of these errors the producer's accuracy for the sugar beet class declined from 99.00% when the validation sample acquired with a random sample was used to 93.00% when the validation sample had been generated with a stratified sample. The user's accuracy for the potato class also differed for the classifications of the testing set obtained when using the validation set defined with random and stratified sampling, the accuracies being 96.00% and 77.42%, respectively.

Given that the testing set had been generated using a simple random sample design the variation in the number of cases per-class reflected the relative abundance of the classes in the region to be mapped. Critically, the size of the sample of cases for the potato class was approximately one quarter of that for the sugar beet class. As hypothesized, the overall accuracy of the testing set decreased when the stratified rather than random sample was used in validation because cases of an abundant class (sugar beet) were commissioned by a relatively rare class (potato). Thus, even in a situation such as that encountered here, in which the classes are very highly separable, the sample design used in the formation of the validation dataset can have a statistically significant effect on the overall accuracy of the final land cover map, as reflected in the accuracy of the classifications of the testing set (Tables 11 and 12).

Table 11. Confusion matrix for the testing set from the network selected using a validation set formed by random sampling.

Class	S	W	B	C	P	G	Total	User's (%)
S	99	0	1	0	0	0	100	99.00
W	0	96	0	0	1	0	97	98.97
B	0	2	52	0	0	0	54	96.30
C	0	0	0	34	1	0	35	97.14
P	1	0	0	0	24	0	25	96.00
G	0	1	0	0	0	18	19	94.74
Total	100	99	53	34	26	18	330	
Producer's (%)	99.00	96.97	98.11	100	92.31	100		

Table 12. Confusion matrix for the testing set from the network selected using a validation set formed by stratified sampling.

Class	S	W	B	C	P	G	Total	User's (%)
S	93	0	1	0	0	0	94	98.94
W	0	97	0	0	1	0	98	98.98
B	0	2	52	0	0	0	54	96.30
C	0	0	0	34	1	0	35	97.14
P	7	0	0	0	24	0	31	77.42
G	0	0	0	0	0	18	18	100
Total	100	99	53	34	26	18	330	
Producer's (%)	93.00	97.98	98.11	100	92.31	100		

Table 13. Cross-tabulation of labelling from classifiers using random (columns) and stratified (rows) validation sets for the remotely sensed data.

	Correct	Incorrect	Total
Correct	317	1	318
Incorrect	6	6	12
Total	020	7	330

Finally, it was also evident that the networks selected for the analyses of the remotely-sensed data differed, most notably in terms of architecture (Table 10). When the validation set constructed with simple random sampling had been used, the data acquired in wavebands 1 and 8 were deemed unnecessary and, hence, only nine input units used. In addition this latter network also had fewer hidden units than the network selected when the validation set had been formed with a stratified sample. Overall, the network formed with the use of the validation sample acquired by random sampling was smaller and less complex than that selected when the validation set formed with a stratified sample was used. As such the network might be expected to be less likely to over-train and have a higher ability to generalize than that selected from the use of the validation sample formed via stratified sampling. Slightly different trends were observed for the simulated dataset. Here, however, it should be noted that a large number of networks of very different size, but very similar performance in terms of ability to classify the data, were generated, limiting the ability to comment on the issue; note, for example, that some candidate networks that yielded the same accuracy for the classification of the testing set after use of the random validation set had the smallest number of hidden units.

The results show that the design of the validation sample has a significant effect on classification by a neural network. Thus, the sample design used to form the validation set should be considered carefully when using neural networks. If, for example, there are constraints that limit design possibilities, it may be possible to make simple adaptations to standard practice. For example, if a stratified sample must be used for the validation sample, then this feature of the dataset should be accounted for in the assessment of the accuracy of the classification of the validation set. Thus, rather than use a standard confusion matrix, as indicated in Table 5, the elements of the matrix could be converted from raw counts to proportions via $p_{ij} = W_i \frac{n_{ij}}{n_{i.}}$ where W_i is the proportion of the area mapped as class i [18]. The use of this approach, for example, shows that the estimated accuracy with which the validation set formed by stratified sampling of the simulated data was slightly less (98.12%) than the naïve assessment of the matrix (98.75%) and may indicate that the network it is associated with is, therefore, less attractive as a candidate for the task at hand than it first appears from the naïve assessment. Alternatively, if the focus of the application is on a subset of the classes it may be sensible to weight errors differentially or focus on only on the classes of interest rather than use overall accuracy. Critically, consideration needs to be given to the design of the validation sample in classification by a neural network. It should also be noted that this is only one small part of a set of broader validation issues that should be considered in the use of neural networks [47].

4. Conclusions

Feedforward neural networks are often constructed with the aid of a validation dataset. The latter data are typically used to indicate the accuracy of the neural network on a dataset independent of that used in its training phase. Commonly, the optimal network is selected on the basis of the overall accuracy with which the validation dataset is classified based on the analysis of a raw confusion matrix. A problem with this approach is that all classification errors are typically treated equally and the magnitude of the overall accuracy can be distorted by the sample design used to form the validation sample. Here, it was shown that the use of a stratified, rather than random, sample of cases as a validation set resulted in a statistically significant reduction of the accuracy with which an independent test set was classified. This difference in accuracy arose because of the commission of cases of an abundant class by a relatively rare class. Moreover, these results were obtained for analyses of a datasets in which the classes were very highly separable within the feature space of both the simulated and the remotely-sensed datasets used. Assuming that the desired aim is the production of a map with high overall accuracy, simple ways to address issues connected with the design of the validation dataset are to the use of a validation sample acquired by simple random sampling or to ensure the accuracy of its classification is based on proportions calculated on the basis of class abundance rather than basic raw counts.

Acknowledgments: The ATM data were acquired through the European AgriSAR campaign and the research benefits from earlier work with colleagues, notably Manoj Arora, which is gratefully acknowledged. The neural network analyses were undertaken with the Trajan package.

References

1. Tso, B.; Mather, P.M. *Classification Methods for Remotely Sensed Data*, 2nd ed.; Taylor & Francis: London, UK; New York, NY, USA, 2001.

2. Mas, J.F.; Flores, J.J. The application of artificial neural networks to the analysis of remotely sensed data. *Int. J. Remote Sens.* **2008**, *29*, 617–663. [CrossRef]

3. Jensen, R.R.; Hardin, P.J.; Yu, G. Artificial neural networks and remote sensing. *Geogr. Compass* **2009**, *3*, 630–646. [CrossRef]

4. Yue, J.; Zhao, W.; Mao, S.; Liu, H. Spectral-spatial classification of hyperspectral images using deep convolutional neural networks. *Remote Sens. Lett.* **2015**, *6*, 468–477. [CrossRef]

5. Li, L.; Chen, Y.; Xu, T.; Huang, C.; Liu, R.; Shi, K. Integration of Bayesian regulation back-propagation neural network and particle swarm optimization for enhancing sub-pixel mapping of flood inundation in river basins. *Remote Sens. Lett.* **2016**, *7*, 631–640. [CrossRef]

6. Peddle, D.R.; Foody, G.M.; Zhang, A.; Franklin, S.E.; LeDrew, E.F. Multi-source image classification II: An empirical comparison of evidential reasoning and neural network approaches. *Can. J. Remote Sens.* **1994**, *20*, 396–407. [CrossRef]

7. Lu, D.; Weng, Q. A survey of image classification methods and techniques for improving classification performance. *Int. J. Remote Sens.* **2007**, *28*, 823–870. [CrossRef]

8. Serpico, S.B.; Bruzzone, L.; Roli, F. An experimental comparison of neural and statistical non-parametric algorithms for supervised classification of remote-sensing images. *Pattern Recognit. Lett.* **1996**, *17*, 1331–1341. [CrossRef]

9. Paola, J.D.; Schowengerdt, R.A. A review and analysis of backpropagation neural networks for classification of remotely-sensed multi-spectral imagery. *Int. J. Remote Sens.* **1995**, *16*, 3033–3058. [CrossRef]

10. Kavzoglu, T.; Mather, P.M. The use of backpropagating artificial neural networks in land cover classification. *Int. J. Remote Sens.* **2003**, *24*, 4907–4938. [CrossRef]

11. Pal, M.; Foody, G.M. Evaluation of SVM, RVM and SMLR for accurate image classification with limited ground data. *IEEE J. Sel. Top. Appl. Earth Obs. Remote Sens.* **2012**, *5*, 1344–1355. [CrossRef]

12. Foody, G.M.; Pal, M.; Rocchini, D.; Garzon-Lopez, C.X.; Bastin, L. The sensitivity of mapping methods to reference data quality: Training supervised image classifications with imperfect reference data. *ISPRS Int. J. Geo Inf.* **2016**, *5*, 199. [CrossRef]

13. Antoniou, V.; Fonte, C.C.; See, L.; Estima, J.; Arsanjani, J.J.; Lupia, F.; Minghini, M.; Foody, G.; Fritz, S. Investigating the feasibility of geo-tagged photographs as sources of land cover input data. *ISPRS Int. J. Geo Inf.* **2016**, *5*, 64. [CrossRef]

14. Kavzoglu, T. Increasing the accuracy of neural network classification using refined training data. *Environ. Model. Softw.* **2009**, *24*, 850–858. [CrossRef]

15. Foody, G.M.; McCulloch, M.B.; Yates, W.B. The effect of training set size and composition on artificial neural network classification. *Int. J. Remote Sens.* **1995**, *16*, 1707–1723. [CrossRef]

16. Zhuang, X.; Engel, B.A.; Lozano-Garcia, D.F.; Fernandez, R.N.; Johannsen, C.J. Optimisation of training data required for neuro-classification. *Int. J. Remote Sens.* **1994**, *15*, 3271–3277. [CrossRef]

17. Foody, G.M. Hard and soft classifications by a neural network with a non-exhaustively defined set of classes. *Int. J. Remote Sens.* **2002**, *23*, 3853–3864. [CrossRef]

18. Olofsson, P.; Foody, G.M.; Herold, M.; Stehman, S.V.; Woodcock, C.E.; Wulder, M.A. Good practices for estimating area and assessing accuracy of land change. *Remote Sens. Environ.* **2014**, *148*, 42–57. [CrossRef]

19. Stehman, S.V. Basic probability sampling designs for thematic map accuracy assessment. *Int. J. Remote Sens.* **1999**, *20*, 2423–2441. [CrossRef]

20. Piper, J. Variability and bias in experimentally measured classifier error rates. *Pattern Recognit. Lett.* **1992**, *13*, 685–692. [CrossRef]

21. Garson, G.D. *Neural Networks: An Introductory Guide for Social Scientists*; Sage: London, UK, 1998.

22. Ahmad, S.; Tesauro, G. Scaling and generalisation in neural networks: A case study. In *Proceedings 1988 Connectionist Models Summer School*; Touretzky, D., Hinton, G., Sejnowsjki, T., Eds.; Morgan Kaufmann: San Mateo, CA, USA, 1989; pp. 3–10.

23. Foody, G.M. The significance of border training patterns in classification by a feedforward neural network using back propagation learning. *Int. J. Remote Sens.* **1999**, *20*, 3549–3562. [CrossRef]

24. Li, C.; Wang, J.; Wang, L.; Hu, L.; Gong, P. Comparison of classification algorithms and training sample sizes in urban land classification with Landsat thematic mapper imagery. *Remote Sens.* **2014**, *6*, 964–983. [CrossRef]

25. Silva, J.; Bacao, F.; Dieng, M.; Foody, G.M.; Caetano, M. Improving specific class mapping from remotely sensed data by cost-sensitive learning. *Int. J. Remote Sens.* **2017**, *38*, 3294–3316. [CrossRef]

26. Ma, X.; Tong, X.; Liu, S.; Luo, X.; Xie, H.; Li, C. Optimized sample selection in SVM classification by combining with DMSP-OLS, Landsat NDVI and GlobeLand30 products for extracting urban built-up areas. *Remote Sens.* **2017**, *9*, 236. [CrossRef]

27. Foody, G.M.; Mathur, A. Toward intelligent training of supervised image classifications: Directing training data acquisition for SVM classification. *Remote Sens. Environ.* **2014**, *93*, 107–117. [CrossRef]

28. Chang, E.I.; Lippmann, R.P. Using genetic algorithms to improve pattern classification performance. In *Advances in Neural Information Processing Systems*; Lippmann, R.P., Moody, J.E., Touretzky, D.S., Eds.; Morgan Kaufmann: San Mateo, CA, USA, 1991; Volume 3, pp. 797–803.

29. Mathur, A.; Foody, G.M. Crop classification by support vector machine with intelligently selected training data for an operational application. *Int. J. Remote Sens.* **2008**, *29*, 2227–2240. [CrossRef]

30. Du, P.; Xia, J.; Zhang, W.; Tan, K.; Liu, Y.; Liu, S. Multiple classifier system for remote sensing image classification: A review. *Sensors* **2012**, *12*, 4764–4792. [CrossRef] [PubMed]

31. Foody, G.M.; Mathur, A. The use of small training sets containing mixed pixels for accurate hard image classification: Training on mixed spectral responses for classification by a SVM. *Remote Sens. Environ.* **2006**, *103*, 179–189. [CrossRef]

32. Mueller, A.V.; Hemond, H.F. Statistical generation of training sets for measuring NO^{3-}, NH^{4+} and major ions in natural waters using an ion selective electrode array. *Environ. Sci. Process. Impacts* **2016**, *18*, 590–599. [CrossRef] [PubMed]

33. Bishop, C.M. *Neural Networks for Pattern Recognition*; Oxford University Press: Oxford, UK, 1995.

34. Lek, S.; Giraudel, J.L.; Guegan, J.-F. Neuronal networks: Algorithms and architectures for ecologists and evolutionary ecologists. In *Artificial Neuronal Networks. Application to Ecology and Evolution*; Lek, S., Guegan, J.-F., Eds.; Springer: Berlin, Germany, 2000; pp. 3–27.

35. Fardanesh, M.T.; Ersoy, O.K. Classification accuracy improvement of neural network classifiers by using unlabeled data. *IEEE Trans. Geosci. Remote Sens.* **1998**, *36*, 1020–1025. [CrossRef]

36. Twomey, J.M.; Smith, A.E. Bias and variance of validation methods for function approximation neural networks under conditions of sparse data. *IEEE Trans. Syst. Man Cybern. Part C* **1998**, *28*, 417–430. [CrossRef]

37. Prechelt, L. Automatic early stopping using cross validation: Quantifying the criteria. *Neural Netw.* **1998**, *11*, 761–767. [CrossRef]

38. Setiono, R. Feedforward neural network construction using cross validation. *Neural Comput.* **2001**, *13*, 2865–2877. [CrossRef] [PubMed]

39. Huynh, T.Q.; Setiono, R. Effective neural network pruning using cross-validation. In Proceedings of the IEEE International Joint Conference on Neural Networks, Montreal, QC, Canada, 31 July–4 August 2005; Volume 2, pp. 972–977.

40. Lee, C.; Landgrebe, D.A. Decision boundary feature extraction for neural networks. *IEEE Trans. Neural Netw.* **1997**, *8*, 75–83. [CrossRef] [PubMed]

41. Zhang, G.; Hu, M.Y.; Patuwo, B.E.; Indro, D.C. Artificial neural networks in bankruptcy prediction: General framework and cross-validation analysis. *Eur. J. Oper. Res.* **1999**, *116*, 16–32. [CrossRef]

42. Pal, M.; Mather, P.M. Support vector machines for classification in remote sensing. *Int. J. Remote Sens.* **2005**, *26*, 1007–1011. [CrossRef]

43. Mas, J.F. Mapping land use/cover in a tropical coastal area using satellite sensor data, GIS and artificial neural networks. *Estuar. Coast. Shelf Sci.* **2004**, *59*, 219–230. [CrossRef]

44. Mas, J.F.; Puig, H.; Palacio, J.L.; Sosa-López, A. Modelling deforestation using GIS and artificial neural networks. *Environ. Model. Softw.* **2004**, *19*, 461–471. [CrossRef]
45. Foody, G.M. Thematic map comparison: Evaluating the statistical significance of differences in classification accuracy. *Photogramm. Eng. Remote Sens.* **2004**, *70*, 627–633. [CrossRef]
46. Agresti, A. *Categorical Data Analysis*, 2nd ed.; Wiley: New York, NY, USA, 2002.
47. Humphrey, G.B.; Maier, H.R.; Wu, W.; Mount, N.J.; Dandy, G.C.; Abrahart, R.J.; Dawson, C.W. Improved validation framework and R-package for artificial neural network models. *Environ. Model. Softw.* **2017**, *92*, 82–106. [CrossRef]

7

Characterization of Surface Ozone Behavior at Different Regimes

Nádia F. Afonso and José C. M. Pires *

Laboratório de Engenharia de Processos, Ambiente Biotecnologia e Energia (LEPABE), Departamento de Engenharia Química, Faculdade de Engenharia, Universidade do Porto, Rua Dr. Roberto Frias, 4200-465 Porto, Portugal; nadia.afonso@iol.pt
* Correspondence: jcpires@fe.up.pt

Abstract: Previous studies showed that the influence of meteorological variables and concentrations of other air pollutants on O_3 concentrations changes at different O_3 concentration levels. In this study, threshold models with artificial neural networks (ANNs) were applied to characterize the O_3 behavior at an urban site (Porto, Portugal), describing the effect of environmental and meteorological variables on O_3 concentrations. ANN characteristics, and the threshold variable and value, were defined by genetic algorithms (GAs). The considered predictors were hourly average concentrations of NO, NO_2, and O_3, and meteorological variables (temperature, relative humidity, and wind speed) measured from January 2012 to December 2013. Seven simulations were performed and the achieved models considered wind speed (at 4.9 m·s^{-1}), temperature (at 17.5 °C) and NO_2 (at 26.6 μg·m^{-3}) as the variables that determine the change of O_3 behavior. All the achieved models presented a similar fitting performance: $R^2 = 0.71$–0.72, $RMSE = 14.5$–14.7 μg·m^{-3}, and the index of agreement of the second order of 0.91. The combined effect of these variables on O_3 concentration was also analyzed. This statistical model was shown to be a powerful tool for interpreting O_3 behavior, which is useful for defining policy strategies for human health protection concerning this air pollutant.

Keywords: air pollution; artificial neural network; genetic algorithms; surface ozone; threshold models

1. Introduction

Surface ozone (O_3) is considered one of the most concerning air pollutants in Europe. It is a secondary pollutant (it is not directly emitted), generated by chemical reactions that occur in the atmosphere between primary air pollutants (nitrogen oxides—NO_x—and volatile organic compounds—VOCs) catalyzed by sunlight [1]. The impact of this air pollutant has been studied in different areas [2–5]. Concerning human health, O_3 can cause injuries to airway epithelial cells (and lung diseases such as asthma), hyperplasia, headaches, and nausea, particularly in sensitive people, such as children and elderly [6–10]. Regarding vegetation, O_3 can damage plant leaves (decreases in both leaf photosynthesis and leaf area), reducing crop yields associated with a high negative economic impact [11–13]. Moreover, as a strong oxidant, it is responsible for the degradation of material via corrosion [14].

As already mentioned, O_3 is produced by chemical reactions between primary pollutants present in the atmosphere. It can also be transported from other locations by the wind (horizontal transport) and from the stratosphere (vertical transport) [15,16]. Thus, the atmosphere works as an open chemical reactor, in which reaction kinetics depend on the concentrations of reactants (primary air pollutants), mixture (wind speed and direction), temperature (influencing exponentially kinetic reaction constants—Arrhenius equation), and solar radiation. Thus, the O_3 behavior is highly dependent on the environment (urban, rural, or background) and meteorological variables.

In urban areas, O_3 concentrations are usually lower than the values observed in rural areas [17–20]. This phenomenon occurs mainly due to high NO_x concentrations. Photochemical equilibrium is defined by the following equations [21–23]:

$$NO_2 + h\nu \; (\lambda < 420 \text{ nm}) \rightarrow NO + O \qquad \text{(R1)}$$

$$O + O_2 + M \rightarrow O_3 + M \qquad \text{(R2)}$$

$$NO + O_3 \rightarrow NO_2 + O_2. \qquad \text{(R3)}$$

The photochemical reaction of NO_2 decomposition (chemical reaction R1) leads to the production of NO molecules and oxygen atoms that combine with molecular oxygen to produce ozone (chemical reaction R2). O_3 can also react with NO, forming NO_2 (chemical reaction R3). The complexity of phenomena associated with O_3 formation makes it hard to understand and predict its concentration in ambient air. Moreover, there are insufficient data (e.g., inventories of air pollutant emissions) to develop consistent phenomenological models to describe O_3 behavior.

Alternatively, statistical models have the capability of characterizing the relationship between variables using collected data and they involve mere pattern-recognition using mathematical operations. One of the most applied statistical models is the artificial neural network (ANN). ANNs are nonlinear models, which are inspired in the biological neural processing system [24,25]. These models are composed by artificial neurons (grouped in layers; three layers—input, hidden, and output—are often applied) that receive an input value and converts to an output through a selected function (activation function). Additionally, ANNs are characterized by a high fitting performance. The rapid development of computer hardware has increased the processing capabilities, which have led to achievement of ANN models with less computation time [26]. Therefore, these models have been used in a wide range of applications, including classification, regression, and mapping [27–29]. However, there are too many variables that need to be defined before the model parameters can be determined, including (i) the number of processing neurons in the hidden layer, and (ii) the activation function for each neuron. In recent years, genetic algorithms (GAs) have been applied to help in the definition of these variables. GAs are commonly applied to generate high-quality solutions for optimization and search problems, based on bio-inspired operators, such as mutation, crossover, and selection [30,31]. In GAs, a set of candidate solutions (called population—a group of individuals) are iteratively modified though the mentioned genetic operators in order to find a group of better solutions for the next generation (new iteration). GAs present the following advantages: (i) continuous or discrete variables can be optimized; (ii) a derivative function is not required; (iii) multivariable problems can be optimized; (iv) extremely complex cost surfaces can be dealt with; and (v) a list of optimal solutions (and not just a single one) is provided.

In Porto (selected area in this study—Portugal), an increasing trend of O_3 concentrations (147% higher) has been observed since the 19th century due to the photochemical production of this pollutant, associated with the increase in anthropogenic emissions mainly due to traffic [32]. Additionally, in the north of Portugal, Lamas d'Olo is a rural site where the highest O_3 concentrations are usually measured. Consequently, this site is often selected to evaluate O_3 behavior. Russo et al. [33] mentioned that high-ozone episodes can be explained by several factors: (i) atmospheric stagnation; (ii) horizontal transport by the wind of ozone-rich air masses; (iii) high solar radiation and temperature; and (iv) the influence of local winds (sea breezes and valley winds). Carvalho et al. [19] observed a positive correlation between O_3 concentrations with temperature and a negative correlation with relative humidity. Regarding the effect of wind field, the northeast flow from Spain (Galicia and Asturias) was observed, and this can be associated with the long-range transport of atmospheric pollutants to Portugal. Fernández-Guisuraga et al. [34] compared O_3 trends at urban and rural sites. At the rural site, O_3 concentrations were mainly influenced by the wind (transport), showing low variability with the concentrations of other pollutants. On the other hand, at the urban site, most of the variance was explained by the NO_2/NO_x ratio. Several research studies can be found where ANN models were applied to determine O_3 trends and to predict its concentration (to provide early warning to the population when high O_3 concentration episodes occur). Comrie [35] compared the performance of an ANN model with a multiple linear regression (MLR) model to predict daily average O_3 concentrations in different cities with distinct climate and O_3 regimes. ANN models presented slightly better performance than MLR. Abdul-Wahab and Al-Alawi [36] developed ANN models to predict O_3 concentrations through meteorological and environmental data. The contribution of the meteorological data was defined between 33% and 41%, while the remaining variation was attributed to chemical pollutants. NO, SO_2, relative humidity (the highest contribution), non-methane hydrocarbon, and NO_2 were the variables that most influenced the O_3 concentrations. Additionally, temperature also presents an important role, while solar radiation had a lower effect than expected. Pires et al. [37] compared threshold autoregressive (TAR) models, autoregressive (AR) models, and ANN in the prediction of the next day hourly average O_3 concentrations. In the training period, ANN presented a higher performance. However, in the test period, TAR models presented more accurate results and the distinction became greater when the evaluation was performed for the prediction of extreme values.

In recent studies, O_3 concentrations have shown different behaviors regarding certain explanatory variables [25,37], which can be classified as O_3 regimes. This observation can be justified by the chemical reactions associated with O_3 formation/destruction that are influenced by certain variables, such as temperature, solar radiation, and wind speed [32]. To take these regimes account, threshold regression models were considered in this study [38]. Thus, GAs were used to define the threshold variable and value (the value of the explanatory variable corresponding to the change of the regime; two regimes were selected), the number of hidden neurons, and the activation function in the hidden and output layers. In this study, hourly average O_3 concentrations were modeled using threshold models with an ANN, whose structure was iteratively optimized by GAs. The achieved models enable the characterization of O_3 variability with selected meteorological and environmental variables in different regimes.

2. Materials and Methods

2.1. Data

Air quality data were obtained from an urban background site (*Sobreiras—Lordelo do Ouro*, see Figure 1) of the Air Quality Monitoring Network (AQMN) of Porto, Portugal. The AQMN is managed by the Regional Commission of Coordination and Development of Northern Portugal (*Comissão de Coordenação e Desenvolvimento Regional do Norte*), under the responsibility of the Ministry of Environment. Hourly average concentrations of NO, NO_2, and O_3 from the period from January 2012 to December 2013 (8760 hourly average values in 2012 and 7481 in 2013) were used to develop the proposed models. NO and NO_2 were obtained through the chemiluminescence method according to European Union (EU) Directive 1999/30/EC (European Community). According to EU Directive 2002/3/EC, O_3 measurements were performed through UV-absorption photometry using the equipment 41 M UV Photometric Ozone Analyzer (Environment S.A., Poissy, France). This monitoring equipment was subject to a rigid maintenance program, calibrated every 4 weeks. Measurements were continuously registered, and hourly average concentrations (in $\mu g \cdot m^{-3}$) were recorded.

The meteorological data were collected in a meteorological station located at *Pedras Rubras*, which is managed by *Instituto Português do Mar e da Atmosfera* (IPMA, I.P.); these values are considered representative for the entire Metropolitan Area of Porto. In this study, hourly averages of temperature (T, °C), relative humidity (RH, %), and wind speed (WS, $m \cdot s^{-1}$) were used to analyze the influence of meteorological conditions on O_3 concentrations.

Figure 1. Monitoring site of *Sobreiras—Lordelo do Ouro*.

2.2. Statistical Model

In this study, threshold models with ANNs were defined with GAs, aiming to evaluate the effect of environmental and meteorological variables in O_3 concentrations. The applied model is defined as the following:

$$y = \begin{cases} \text{net}_1(x_i), \text{ if } x_d \leq r \\ \text{net}_2(x_i), \text{ if } x_d > r \end{cases} \tag{1}$$

where y is the output variable, net_1 and net_2 are ANN models, x_i are the exploratory variables, x_d is the threshold variable, and r is the threshold value. Applied feedforward ANN models had three layers (input, hidden, and output) and considered eight input variables (hourly average data): NO concentration, NO_2 concentration (due to the chemical reactions R1 and R3), the ratio NO_2/NO (due to the equilibrium constant of the chemical reaction R3), T, RH, $1/RH$ (as RH usually shows a negative effect on O_3 levels), WS, and $1/WS$ (the same as the RH effect). The output variable was the hourly average O_3 concentrations measured at the same time of the input data to infer the direct influence of these variables on O_3 chemistry. Regarding the activation functions, the linear function was considered for the output neuron and four functions were selected by GAs: sigmoid, hyperbolic tangent, inverse, and radial basis. The data were divided in training (75%) and validation (25%) sets and the early stopping method (ANN training procedure is stopped when an increase in validation error is observed) was applied to avoid overfitting. The division of the data was performed by time: 75% for training (January 2012 to 25 May 2013); 25% for validation (25 May 2013 to 19 December 2013). In the training set, O_3 concentrations ranged from 0 to 161 $\mu g \cdot m^{-3}$, while O_3 concentrations ranged from 0 to 170 $\mu g \cdot m^{-3}$ in the validation set.

GAs are a search and optimization technique based on Darwin principles of evolution and natural genetics [30,31]. This procedure begins with a set of individuals (population) that is randomly generated. Each individual (also called chromosome) is a binary code string and contains information about a set of parameters, which is a potential solution to a given problem. To evaluate the quality of the proposed solution (to rank the individuals in the population), a fitness function should be defined. To create new chromosomes for the next generation, the fittest chromosomes are submitted to the genetic operations [30]: (i) selection; (ii) crossover; (iii) mutation. These new chromosomes are then evaluated according to the fitness function, and the ones with the highest performance were selected. The repetition of this procedure generates a sequence of populations containing better solutions. The termination criteria can be (i) to stop after a previously defined maximum number of generations is achieved, or (ii) to stop when a desired fitness value is achieved. In this study, GAs were used to define the threshold variable and value, the number of hidden neurons, and the activation function in the hidden layer, and to select the explanatory variables to be used in each ANN model. The determination of the models was coded by the authors with MATLAB® software (R2014a, MathWorks, Natick, MA, USA, 2014) using the following specifications:

- a population size of 100;
- a selection probability of 0.20 (proportion of the individuals of the new generation obtained by selection operator);
- a selection criterion based on elitism (a small proportion of the fittest candidates is copied unchanged into the next generation);
- a crossover probability of 0.70 (proportion of the individuals of the new generation obtained by crossover operator);
- a mutation probability of 0.1 (proportion of the individuals of the new generation obtained by mutation operator);
- an evaluation of root mean squared error (*RMSE*) in training and validation sets;
- a stopping criterion based on the maximum number of generations.

Figure 2 shows an example of chromosome (37 bits). It is divided in 8 sets of bits (SB_i). SB_1 (3 bits) defines the threshold variable (from the explanatory variables; the maximum number of 8) through the conversion from binary to decimal numbers (MATLAB function *bin2dec*). SB_2 (8 bits) defines the threshold value. With the threshold variable already defined, the maximum (x_{max}) and minimum (x_{min}) values of this variable are determined. Threshold value is calculated based on Equation (2).

$$r = \frac{bin2dec(SB_2)}{255} \times (x_{max} - x_{min}) + x_{min}. \tag{2}$$

SB_3 and SB_6 (2 bits) define the activation function for the hidden layer of each ANN: 00—log-sigmoid (*logsig*); 01—hyperbolic tangent sigmoid (*tansig*); 10—inverse (*netinv*); 11—radial basis (*radbas*). SB_4 and SB_7 (3 bits) define the number of neurons in the hidden layer through the conversion from binary to decimal number (1 to 8). SB_5 and SB_8 (8 bits) define the explanatory variables that are used in each ANN (1 bit for each explanatory variable): 0—not selected; 1—selected.

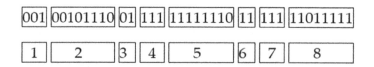

Figure 2. Example of a chromosome.

3. Results and Discussion

3.1. *Air Quality and Meteorological Data Characterization*

During the analyzed period, the hourly average O_3 concentrations were between 0 and 170 $\mu g \cdot m^{-3}$ (not exceeding the information neither the alert threshold—180 and 240 $\mu g \cdot m^{-3}$, respectively). Regarding O_3 exceedances to EU limits for the protection of human health, the 8 h average O_3 concentrations were higher than 120 $\mu g \cdot m^{-3}$ twice in September 2012, twice in July 2013, and once in August 2013. Figure 3 shows the average daily profile of O_3 concentrations. As a photochemical pollutant, its concentration increases during the daylight period, presenting a maximum between 14 and 15 h and a minimum at night time. The observed profile is characteristic of an urban site, as it does not present a high amplitude of concentrations (due to the presence of high NO_x concentrations).

Figure 4 shows the monthly average values of NO, NO_2, and O_3 concentrations, as well as the analyzed meteorological variables (temperature, relative humidity, and wind speed). High O_3 concentrations were observed in April 2012 (63.7 $\mu g \cdot m^{-3}$) and from March to July 2013 (59.9–71.3 $\mu g \cdot m^{-3}$). In this period, low concentrations of NO and NO_2 were also measured.

High temperatures and low relative humidity were also observed. On the other hand, lower O_3 concentrations were measured for periods with high NO and NO_2 concentrations and lower temperatures. These observations are in agreement with other research studies in which the behavior of O_3 was analyzed [18,39–41]. Pires et al. [40] compared several linear models to predict O_3 concentrations at an urban site in Porto. The correlation analysis performed between O_3 concentrations and meteorological variables showed also negative correlations with NO, NO_2, and RH and a positive correlation with T. In another study focusing on the same region [41], O_3 concentrations were negatively correlated with NO, NO_2, and RH and positively correlated with T and WS. Zhang, Wang, Park, and Deng [18] analyzed high O_3 concentration episodes and related them with meteorological variables. O_3 concentrations were highly correlated with maximum temperature and minimum relative humidity. The effect of minimum WS was also analyzed at urban, suburban, and rural sites. O_3 concentrations were positively (negatively) correlated with minimum WS at urban (suburban and rural) sites. Shan, Yin, Zhang, Ji, and Deng [39] analyzed the effect of meteorological variables on O_3 concentrations at an urban site in China. Daily average O_3 concentrations were negatively correlated with pressure and RH, and positively correlated with temperature, solar radiation, sunshine duration, and wind speed.

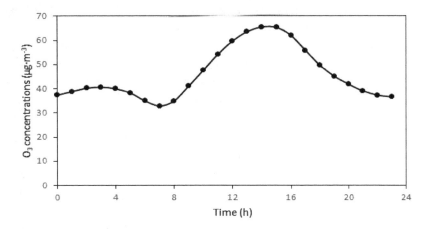

Figure 3. Daily average profile of O_3 concentrations at the monitoring site.

Figure 4. *Cont.*

(e)   (f)

Figure 4. Monthly average values of (**a**) NO concentrations, (**b**) temperature, (**c**) NO_2 concentrations, (**d**) relative humidity, (**e**) O_3 concentrations, and (**f**) wind speed.

3.2. Linear Correlation Analysis

Figure 5 shows the variation in linear correlation between O_3 and meteorological parameters on a monthly basis. Negative correlations were observed for NO (-0.547 to -0.296) and NO_2 (-0.807 to -0.276) concentrations. The effect of these air pollutants was more significant in winter periods than in summer periods. Regarding the effect of meteorological variables, temperature was usually positively correlated with O_3, which was in agreement with what was expected. The highest value ($R = 0.661$) was determined in September 2013 and an unusual negative correlation ($R = -0.376$) was determined in July 2013. *RH* was negatively correlated in almost all periods. The highest impact was also observed in September 2013 ($R = -0.685$) and an unusual positive correlation was determined in July 2013 ($R = 0.419$). Chen et al. [42] demonstrated that *RH* favors O_3 decomposition, justifying the associated negative effect. Regarding *WS*, this variable can have two different effects on O_3 concentrations. Low *WS* can promote the accumulation of O_3 produced in the region (increasing its concentration), while high values reduce the levels of other air pollutants (such as NO_x) that influence the O_3 chemistry (in the case of NO_x, its concentration decrease leads to the increase in O_3 levels). Thus, the effect of *WS* on O_3 concentrations depends on the studied environment: urban or rural. In this study (urban environment), *WS* was positively correlated with O_3 concentrations, with the highest value ($R = 0.593$) in February 2013.

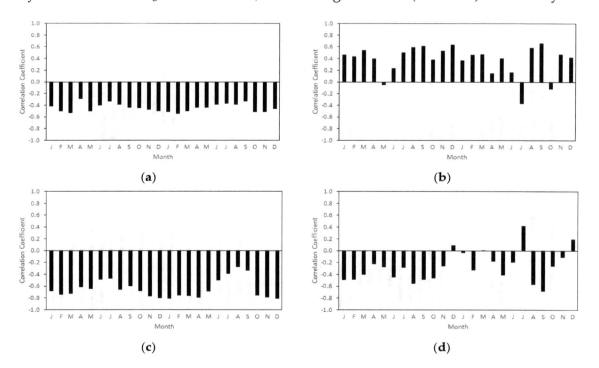

(a)   (b)

(c)   (d)

Figure 5. *Cont.*

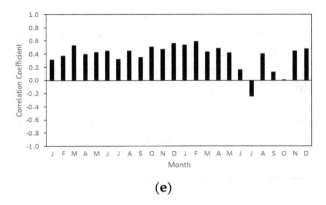

(e)

Figure 5. Temporal variation of linear correlation between O_3 concentrations and the following: (**a**) NO concentrations, (**b**) temperature, (**c**) NO_2 concentrations, (**d**) relative humidity, and (**e**) wind speed.

3.3. ANN Models and Interpretation

Seven simulations were performed to determine the models able to describe the relationship between O_3 concentrations with NO, NO_2, *T*, *RH*, and *WS* (measured at same time). These models are threshold models, considering two O_3 regimes where the relationship between output and input variables are different. The change from one regime to another depends on the value (threshold value) of a specific input variable (threshold variable). GAs were used to optimize the ANN characteristics (the number of hidden neurons, the activation function in the hidden layer, and the input variables), and the threshold variable and value. The models were evaluated according their fitting performance in training and validation sets. Table 1 shows the best models for the seven simulations. All of them presented similar fitting performance: (i) R^2 = 0.71–0.72; (ii) *RMSE* between 14.5 and 14.7 $\mu g \cdot m^{-3}$; and (iii) the index of agreement of the second order of 0.91. Three explanatory variables were selected, each one with a specific threshold value: (i) *WS* with 4.9 $m \cdot s^{-1}$; (ii) *T* with 17.5 °C; and (iii) NO_2 with 26.6 $\mu g \cdot m^{-3}$. Generally, hyperbolic tangent and radial basis were the functions selected for the hidden layer, composed by 7 or 8 neurons. In almost all models, all input variables were selected as ANN inputs in both O_3 regimes.

The analysis of the combined effect of input variables (two variables) was performed for the three threshold variables, considering the two regimes determined by the best models in Simulations I, II, and III. Figure 6 shows the combined effect of NO_2, *T*, and *WS* on O_3 concentrations for *WS* $\leq$ 4.9 $m \cdot s^{-1}$ and for *WS* > 4.9 $m \cdot s^{-1}$. For *WS* $\leq$ 4.9 $m \cdot s^{-1}$, O_3 concentrations (i) decreased with NO_2 except when *T* > 17 °C (without significant variation), (ii) increased with *T* except when *WS* > 2.8 $m \cdot s^{-1}$ (O_3 presented a maximum between 17 and 20 °C), and (iii) did not change significantly with *WS* except when *T* > 26 °C (presenting a decreasing tendency). For *WS* > 4.9 $m \cdot s^{-1}$, O_3 concentrations (i) decreased with NO_2 except when *T* > 17 °C (presenting a slight increase), (ii) presented high values for high *T* with all tested ranges of NO_2 concentrations (presenting a local maximum—$\approx$ 87.4 $\mu g \cdot m^{-3}$—for *T* $\approx$ 11 °C and NO_2 $\approx$ 6 $\mu g \cdot m^{-3}$), (iii) increased with *T* for the tested range of *WS*, (iv) did not change significantly with *WS*, and (v) presented higher values than those where *WS* $\leq$ 4.9 $m \cdot s^{-1}$. The combined effect of *T*-NO_2 is in agreement with what was concluded in linear correlation analysis. The effect of NO_2 is more significant in the winter period, in which temperatures are usually low and NO_2 concentrations are high (see Figure 4). With high NO_2 concentrations, the chemical equilibrium given by Equation (R3) limits the increase in O_3 concentrations. In addition, based on a comparison of the two regimes defined by *WS*, the combined effect of these two variables presented similar behavior; however, O_3 concentrations were higher when *WS* > 4.9 $m \cdot s^{-1}$ (49–104 $\mu g \cdot m^{-3}$) than they were when *WS* $\leq$ 4.9 $m \cdot s^{-1}$ (11–41 $\mu g \cdot m^{-3}$). High values of *WS* are associated with the dispersion of air pollutants, reducing their concentration. As NO_2 concentrations decrease, O_3 concentrations can achieve higher values [17,19,20,22]. The combined effect of *T*-*WS* showed the highest variability of O_3 with *T*, showing a positive correlation between these variables. The O_3 variability with *WS* is almost insignificant.

Regarding the combined effect NO_2-WS, similar conclusions were drawn: O_3 concentrations presented a decreasing tendency with NO_2 concentrations, and their variability with WS is almost insignificant. Figures S1 and S2 present the effect of the same combinations of variables on O_3 concentrations considering T (simulation II) and NO_2 (simulation III) as threshold variables, respectively. Similar analysis can be performed through these figures.

Table 1. ANN models: their input variables, activation functions (AF), number of hidden neurons (HN), and performance indexes (R^2, RMSE (root mean squared error) and d_2) for each performed simulation (Sim).

Sim	Model	AF	HN	R^2/RMSE/d_2
I	$O_{3\|t} = \begin{cases} net_1\left(NO, NO_2, \frac{NO_2}{NO}, T, RH, \frac{1}{RH}, WS, \frac{1}{WS}\right), \text{ if } WS \leq 4.9 \\ net_2\left(NO, NO_2, \frac{NO_2}{NO}, T, RH, \frac{1}{RH}, WS, \frac{1}{WS}\right), \text{ if } WS > 4.9 \end{cases}$	tansig radbas	8 8	0.71/14.7/0.91
II	$O_{3\|t} = \begin{cases} net_1\left(NO, NO_2, \frac{NO_2}{NO}, T, RH, \frac{1}{RH}, WS, \frac{1}{WS}\right), \text{ if } T \leq 17.5 \\ net_2\left(NO, NO_2, \frac{NO_2}{NO}, T, RH, \frac{1}{RH}, WS, \frac{1}{WS}\right), \text{ if } T > 17.5 \end{cases}$	tansig tansig	7 7	0.72/14.5/0.91
III	$O_{3\|t} = \begin{cases} net_1\left(NO, NO_2, \frac{NO_2}{NO}, T, RH, \frac{1}{RH}, WS\right), \text{ if } NO_2 \leq 26.6 \\ net_2\left(NO, NO_2, \frac{NO_2}{NO}, T, RH, \frac{1}{RH}, WS, \frac{1}{WS}\right), \text{ if } NO_2 > 26.6 \end{cases}$	tansig radbas	8 7	0.71/14.7/0.91
IV	$O_{3\|t} = \begin{cases} net_1\left(NO, NO_2, \frac{NO_2}{NO}, T, RH, \frac{1}{RH}, WS, \frac{1}{WS}\right), \text{ if } WS \leq 4.9 \\ net_2\left(NO, NO_2, \frac{NO_2}{NO}, T, RH, \frac{1}{RH}, WS, \frac{1}{WS}\right), \text{ if } WS > 4.9 \end{cases}$	tansig radbas	8 8	0.71/14.7/0.91
V	$O_{3\|t} = \begin{cases} net_1\left(NO, NO_2, \frac{NO_2}{NO}, T, RH, \frac{1}{RH}, WS, \frac{1}{WS}\right), \text{ if } T \leq 17.5 \\ net_2\left(NO, NO_2, \frac{NO_2}{NO}, T, RH, \frac{1}{RH}, WS, \frac{1}{WS}\right), \text{ if } T > 17.5 \end{cases}$	tansig tansig	7 7	0.72/14.5/0.91
VI	$O_{3\|t} = \begin{cases} net_1\left(NO, NO_2, \frac{NO_2}{NO}, T, RH, \frac{1}{RH}, WS\right), \text{ if } NO_2 \leq 26.6 \\ net_2\left(NO, NO_2, T, RH, \frac{1}{RH}, WS, \frac{1}{WS}\right), \text{ if } NO_2 > 26.6 \end{cases}$	tansig radbas	8 8	0.71/14.7/0.91
VII	$O_{3\|t} = \begin{cases} net_1\left(NO, NO_2, \frac{NO_2}{NO}, T, RH, WS\right), \text{ if } T \leq 17.5 \\ net_2\left(NO_2, \frac{NO_2}{NO}, T, RH, \frac{1}{RH}, WS, \frac{1}{WS}\right), \text{ if } T > 17.5 \end{cases}$	tansig tansig	7 7	0.72/14.5/0.91

The application of this statistical methodology allows for the determination of the influence of environmental and meteorological variables on O_3 concentration. Consequently, it is possible to develop more accurate predictive models for this secondary pollutant, which is important for the definition of policy measures for human health protection.

Figure 6. *Cont.*

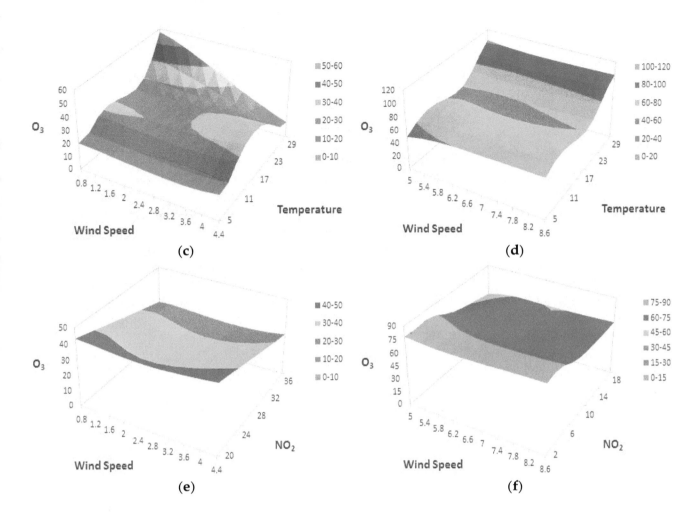

Figure 6. The combined effect of NO_2 concentrations, temperature, and wind speed on O_3 concentrations based on model determined in Simulation I where WS (wind speed) ≤ 4.9 m·s^{-1} (**a,c,e**) and where $WS > 4.9$ m·s^{-1} (**b,d,f**).

4. Conclusions

Linear correlation analysis showed a positive relationship between O_3 concentrations with T and WS, while NO, NO_2, and RH showed a negative effect. In the studied period, the highest O_3 concentrations were observed for low NO_x concentrations and high wind speed. Threshold models with ANNs and those defined by genetic algorithms define three important variables that could define different O_3 regimes: (i) a wind speed of 4.9 m·s^{-1}; (ii) a temperature of 17.5 °C; and (iii) an NO_2 concentration of 26.6 μg·m^{-3}. The achieved models enabled the evaluation of the combined effect of two input variables in different O_3 regimes. This information may be useful for defining policy strategies for human health protection concerning surface ozone.

Acknowledgments: This work was financially supported by Project POCI-01-0145-FEDER-006939 (LEPABE) funded by FEDER funds through COMPETE2020—Programa Operacional Competitividade e Internacionalização (POCI)—and by national funds through FCT—Fundação para a Ciência e a Tecnologia. J.C.M.P. acknowledges the FCT Investigator 2015 Programme (IF/01341/2015).

Author Contributions: N.F.A. collected the data, ran the simulations, analyzed the results, and wrote the paper; J.C.M.P. conceived the study, developed the model code, and revised the paper.

References

1. Lin, M.Y.; Horowitz, L.W.; Payton, R.; Fiore, A.M.; Tonnesen, G. Us surface ozone trends and extremes from 1980 to 2014: Quantifying the roles of rising asian emissions, domestic controls, wildfires, and climate. *Atmos. Chem. Phys.* **2017**, *17*, 2943–2970. [CrossRef]

2. Lefohn, A.S.; Malley, C.S.; Simon, H.; Wells, B.; Xu, X.; Zhang, L.; Wang, T. Responses of human health and vegetation exposure metrics to changes in ozone concentration distributions in the European Union, United States, and China. *Atmos. Environ.* **2017**, *152*, 123–145. [CrossRef]

3. Hollaway, M.J.; Arnold, S.R.; Collins, W.J.; Folberth, G.; Rap, A. Sensitivity of midnineteenth century tropospheric ozone to atmospheric chemistry-vegetation interactions. *J. Geophys. Res.-Atmos.* **2017**, *122*, 2452–2473. [CrossRef]

4. Proietti, C.; Anav, A.; De Marco, A.; Sicard, P.; Vitale, M. A multi-sites analysis on the ozone effects on gross primary production of European forests. *Sci. Total Environ.* **2016**, *556*, 1–11. [CrossRef] [PubMed]

5. Jiang, Z.; Miyazaki, K.; Worden, J.R.; Liu, J.J.; Jones, D.B.A.; Henze, D.K. Impacts of anthropogenic and natural sources on free tropospheric ozone over the Middle East. *Atmos. Chem. Phys.* **2016**, *16*, 6537–6546. [CrossRef]

6. Jorres, R.A.; Holz, O.; Zachgo, W.; Timm, P.; Koschyk, S.; Muller, B.; Grimminger, F.; Seeger, W.; Kelly, F.J.; Dunster, C.; et al. The effect of repeated ozone exposures on inflammatory markers in bronchoalveolar lavage fluid and mucosal biopsies. *Am. J. Respir. Crit. Care Med.* **2000**, *161*, 1855–1861. [CrossRef] [PubMed]

7. Frank, R.; Liu, M.C.; Spannhake, E.W.; Mlynarek, S.; Macri, K.; Weinmann, G.G. Repetitive ozone exposure of young adults—Evidence of persistent small airway dysfunction. *Am. J. Respir. Crit. Care Med.* **2001**, *164*, 1253–1260. [CrossRef] [PubMed]

8. Holz, O.; Mucke, M.; Paasch, K.; Bohme, S.; Timm, P.; Richter, K.; Magnussen, H.; Jorres, R.A. Repeated ozone exposures enhance bronchial allergen responses in subjects with rhinitis or asthma. *Clin. Exp. Allergy* **2002**, *32*, 681–689. [CrossRef] [PubMed]

9. McConnell, R.; Berhane, K.; Gilliland, F.; London, S.J.; Islam, T.; Gauderman, W.J.; Avol, E.; Margolis, H.G.; Peters, J.M. Asthma in exercising children exposed to ozone: A cohort study. *Lancet* **2002**, *359*, 386–391. [CrossRef]

10. Goldberg, M.S.; Burnett, R.T.; Brook, J.; Bailar, J.C.; Valois, M.F.; Vincent, R. Associations between daily cause-specific mortality and concentrations of ground-level ozone in Montreal, Quebec. *Am. J. Epidemiol.* **2001**, *154*, 817–826. [CrossRef] [PubMed]

11. Carter, C.A.; Cui, X.; Ding, A.; Ghanem, D.; Jiang, F.; Yi, F.; Zhong, F. Stage-specific, nonlinear surface ozone damage to rice production in China. *Sci. Rep.* **2017**, *7*, 44224. [CrossRef] [PubMed]

12. Morgan, P.B.; Ainsworth, E.A.; Long, S.P. How does elevated ozone impact soybean? A meta-analysis of photosynthesis, growth and yield. *Plant Cell Environ.* **2003**, *26*, 1317–1328. [CrossRef]

13. Sadiq, M.; Tai, A.P.K.; Lombardozzi, D.; Val Martin, M. Effects of ozone-vegetation coupling on surface ozone air quality via biogeochemical and meteorological feedbacks. *Atmos. Chem. Phys.* **2017**, *17*, 3055–3066. [CrossRef]

14. Christodoulakis, J.; Tzanis, C.G.; Varotsos, C.A.; Ferm, M.; Tidblad, J. Impacts of air pollution and climate on materials in Athens, Greece. *Atmos. Chem. Phys.* **2017**, *17*, 439–448. [CrossRef]

15. Banta, R.M.; Senff, C.J.; White, A.B.; Trainer, M.; McNider, R.T.; Valente, R.J.; Mayor, S.D.; Alvarez, R.J.; Hardesty, R.M.; Parrish, D.; et al. Daytime buildup and nighttime transport of urban ozone in the boundary layer during a stagnation episode. *J. Geophys. Res.-Atmos.* **1998**, *103*, 22519–22544. [CrossRef]

16. Monier, E.; Weare, B.C. Climatology and trends in the forcing of the stratospheric ozone transport. *Atmos. Chem. Phys.* **2011**, *11*, 6311–6323. [CrossRef]

17. Tong, L.; Zhang, H.L.; Yu, J.; He, M.M.; Xu, N.B.; Zhang, J.J.; Qian, F.Z.; Feng, J.Y.; Xiao, H. Characteristics of surface ozone and nitrogen oxides at urban, suburban and rural sites in Ningbo, China. *Atmos. Res.* **2017**, *187*, 57–68. [CrossRef]

18. Zhang, H.; Wang, Y.H.; Park, T.W.; Deng, Y. Quantifying the relationship between extreme air pollution events and extreme weather events. *Atmos. Res.* **2017**, *188*, 64–79. [CrossRef]

19. Carvalho, A.; Monteiro, A.; Ribeiro, I.; Tchepel, O.; Miranda, A.I.; Borrego, C.; Saavedra, S.; Souto, J.A.; Casares, J.J. High ozone levels in the northeast of Portugal: Analysis and characterization. *Atmo. Environ.* **2010**, *44*, 1020–1031. [CrossRef]

20. Pires, J.C.M.; Alvim-Ferraz, M.C.M.; Martins, F.G. Surface ozone behaviour at rural sites in Portugal. *Atmos. Res.* **2012**, *104*, 164–171. [CrossRef]

21. Sanchez, B.; Santiago, J.L.; Martilli, A.; Palacios, M.; Kirchner, F. Cfd modeling of reactive pollutant dispersion in simplified urban configurations with different chemical mechanisms. *Atmos. Chem. Phys.* **2016**, *16*, 12143–12157. [CrossRef]

22. Pires, J.C.M. Ozone weekend effect analysis in three European urban areas. *Clean-Soil Air Water* **2012**, *40*, 790–797. [CrossRef]

23. Gressent, A.; Sauvage, B.; Cariolle, D.; Evans, M.; Leriche, M.; Mari, C.; Thouret, V. Modeling lightning-nox chemistry on a sub-grid scale in a global chemical transport model. *Atmos. Chem. Phys.* **2016**, *16*, 5867–5889. [CrossRef]

24. Agatonovic-Kustrin, S.; Beresford, R. Basic concepts of artificial neural network (ann) modeling and its application in pharmaceutical research. *J. Pharm. Biomed.* **2000**, *22*, 717–727. [CrossRef]

25. Pires, J.C.M.; Goncalves, B.; Azevedo, F.G.; Carneiro, A.P.; Rego, N.; Assembleia, A.J.B.; Lima, J.F.B.; Silva, P.A.; Alves, C.; Martins, F.G. Optimization of artificial neural network models through genetic algorithms for surface ozone concentration forecasting. *Environ. Sci. Pollut. Res.* **2012**, *19*, 3228–3234. [CrossRef] [PubMed]

26. Tawadrous, A.S.; Katsabanis, P.D. Prediction of surface crown pillar stability using artificial neural networks. *Int. J. Numer. Anal. Methods* **2007**, *31*, 917–931. [CrossRef]

27. Pasini, A. Artificial neural networks for small dataset analysis. *J. Thorac. Dis.* **2015**, *7*, 953–960. [PubMed]

28. Keshavarzi, A.; Sarmadian, F.; Omran, E.-S.E.; Iqbal, M. A neural network model for estimating soil phosphorus using terrain analysis. *Egypt. J. Remote Sens. Space Sci.* **2015**, *18*, 127–135. [CrossRef]

29. Zare, M.; Pourghasemi, H.R.; Vafakhah, M.; Pradhan, B. Landslide susceptibility mapping at vaz watershed (iran) using an artificial neural network model: A comparison between multilayer perceptron (mlp) and radial basic function (rbf) algorithms. *Arab. J. Geosci.* **2013**, *6*, 2873–2888. [CrossRef]

30. Goldberg, D.E. *Genetic Algorithms in Search, Optimization and Machine Learning*, 1st ed.; Addison-Wesley Professional: Boston, MA, USA, 1989.

31. Holland, J.H. *Adaptation in Natural and Artificial Systems*; University of Michigan Press: Ann Arbor, MI, USA, 1975.

32. Alvim-Ferraz, M.C.M.; Sousa, S.I.V.; Pereira, M.C.; Martins, F.G. Contribution of anthropogenic pollutants to the increase of tropospheric ozone levels in the oporto metropolitan area, Portugal since the 19th century. *Environ. Pollut.* **2006**, *140*, 516–524. [CrossRef] [PubMed]

33. Russo, A.; Trigo, R.M.; Martins, H.; Mendes, M.T. NO_2, PM10 and O_3 urban concentrations and its association with circulation weather types in Portugal. *Atmos. Environ.* **2014**, *89*, 768–785. [CrossRef]

34. Fernandez-Guisuraga, J.M.; Castro, A.; Alves, C.; Calvo, A.; Alonso-Blanco, E.; Blanco-Alegre, C.; Rocha, A.; Fraile, R. Nitrogen oxides and ozone in Portugal: Trends and ozone estimation in an urban and a rural site. *Environ. Sci. Pollut. Res.* **2016**, *23*, 17171–17182. [CrossRef] [PubMed]

35. Comrie, A.C. Comparing neural networks and regression models for ozone forecasting. *J. Air Waste Manag.* **1997**, *47*, 653–663. [CrossRef]

36. Abdul-Wahab, S.A.; Al-Alawi, S.M. Assessment and prediction of tropospheric ozone concentration levels using artificial neural networks. *Environ. Model. Softw.* **2002**, *17*, 219–228. [CrossRef]

37. Pires, J.C.M.; Alvim-Ferraz, M.C.M.; Pereira, M.C.; Martins, F.G. Evolutionary procedure based model to predict ground-level ozone concentrations. *Atmos. Pollut. Res.* **2010**, *1*, 215–219. [CrossRef]

38. Terui, N.; van Dijk, H.K. Combined forecasts from linear and nonlinear time series models. *Int. J. Forecast.* **2002**, *18*, 421–438. [CrossRef]

39. Shan, W.P.; Yin, Y.Q.; Zhang, J.D.; Ji, X.; Deng, X.Y. Surface ozone and meteorological condition in a single year at an urban site in central-eastern China. *Environ. Monit. Assess.* **2009**, *151*, 127–141. [CrossRef] [PubMed]

40. Pires, J.C.M.; Alvim-Ferraz, M.C.M.; Pereira, M.C.; Martins, F.G. Comparison of several linear statistical models to predict tropospheric ozone concentrations. *J. Stat. Comput. Simul.* **2012**, *82*, 183–192. [CrossRef]

41. Pires, J.C.M.; Martins, F.G.; Sousa, S.I.V.; Alvim-Ferraz, M.C.M.; Pereira, M.C. Selection and validation of parameters in multiple linear and principal component regressions. *Environ. Model. Softw.* **2008**, *23*, 50–55. [CrossRef]

8

Data Mining Approaches for Landslide Susceptibility Mapping

Sunmin Lee [1,2], Moung-Jin Lee [2,*] and Hyung-Sup Jung [1,*]

[1]   Department of Geoinformatics, University of Seoul, 163 Seoulsiripdaero, Dongdaemun-gu, Seoul 02504, Korea; smlee@kei.re.kr

[2]   Center for Environmental Assessment Monitoring, Environmental Assessment Group, Korea Environment Institute (KEI), Sejong-si 30147, Korea

*   Correspondence: leemj@kei.re.kr (M.-J.L.); hsjung@uos.ac.kr (H.-S.J.)

Academic Editor: Saro Lee

Abstract: The application of data mining models has become increasingly popular in recent years in assessments of a variety of natural hazards such as landslides and floods. Data mining techniques are useful for understanding the relationships between events and their influencing variables. Because landslides are influenced by a combination of factors including geomorphological and meteorological factors, data mining techniques are helpful in elucidating the mechanisms by which these complex factors affect landslide events. In this study, spatial data mining approaches based on data on landslide locations in the geographic information system environment were investigated. The topographical factors of slope, aspect, curvature, topographic wetness index, stream power index, slope length factor, standardized height, valley depth, and downslope distance gradient were determined using topographical maps. Additional soil and forest variables using information obtained from national soil and forest maps were also investigated. A total of 17 variables affecting the frequency of landslide occurrence were selected to construct a spatial database, and support vector machine (SVM) and artificial neural network (ANN) models were applied to predict landslide susceptibility from the selected factors. In the SVM model, linear, polynomial, radial base function, and sigmoid kernels were applied in sequence; the model yielded 72.41%, 72.83%, 77.17% and 72.79% accuracy, respectively. The ANN model yielded a validity accuracy of 78.41%. The results of this study are useful in guiding effective strategies for the prevention and management of landslides in urban areas.

Keywords: spatial data mining; SVM; ANN; validation; ROC

1. Introduction

The rapid growth in data due to the development of information and communication technology (ICT) has spurred the demand for data mining, which is generally considered to be the most useful analytical tool in the analysis of large data sets. Data mining has been defined in various ways previously, and is commonly referred to as an exploratory analysis tool for large amounts of data [1]. Therefore, data mining is a process of finding useful information that is not easily exposed in a large amount of data, most recently known as "big data".

With the rapid growth of ICT since the 1980s, countries and corporations have made significant efforts to build databases to store and manage vast amounts of data. Rational and rapid decision-making is necessary to increase the utilization of large-scale databases. In this environment, it has become important to find meaningful new information that can support optimal decision-making. In this process, an efficient methodology for data mining that extracts not only known information from previously established databases but also information that was not previously known has gained popularity. Therefore, the data mining methodology makes it possible to effectively analyze large amounts of data in a database and the relationships between parameters.

In recent years, storms bringing localized heavy rains have been reported worldwide, and precipitation in the mid-latitude regions has increased since the 20th century [2]. In addition, rapid climate change and abnormal climate conditions have led to extreme weather phenomena, which provide the appropriate conditions for the creation of landslides, and the frequency of landslides has increased accordingly [3]. A landslide is the movement of soil down a slope when the stress exceeds the strength of the soil material due to rapid changes in the natural environment such as heavy rains or earthquakes [4]. Landslides are caused by various factors, such as topography, geology, soil characteristics, forest conditions, and climate variables [5–8]. The main purpose of this study was to generate and validate a landslide susceptibility map using data mining models for Umyeonsan, South Korea. Therefore, data mining techniques should be used appropriately for the analysis of the relationships between landslides and impact factors on landslides in landslide susceptibility analyses. Additionally, given that the data mining models derive their results from the database, the input data represent a key part in the model. Thus, the database should be constructed using reliable data, such as nationally distributed government data. Remote sensing data is one of the main sources to be used to generate the reliable data [9–11].

Due to the complex factors influencing landslides, it is important to perform an appropriate assessment of the potential for landslide susceptibility using data mining techniques. A susceptibility analysis based on data on previous damage should be performed to define landslide susceptibility in areas where landslides are anticipated so that the damage can be predicted in advance and appropriate mitigation measurements adopted. Over the past several decades, a number of data mining approaches have been developed for mapping landslide susceptibility [12,13]. The most commonly used methods are soft computing or statistical techniques such as artificial neural network (ANN) models [14,15], fuzzy logic methods [16,17] or logistic regression models [18,19]. Landslides are caused by multiple interactions between various factors, including geometry, geology, soil characteristics, and vegetation conditions [20,21], and are highly dependent on terrain features [22,23].

As geological hazards, landslides can limit the sustainable development of urban areas, causing environmental damage, serious casualties, and loss of property [24,25]. An appropriate assessment of the susceptibility for landslide damage is required to minimize the complex array of losses. In South Korea, according to an analysis of the scale of landslides by the Korea Forest Service (KFS), the average annual area affected by landslides increased markedly from 231 ha in the 1980s, to 349 ha in the 1990s, and 713 ha in 2000 [26]. Typhoons accompanied by intense storms caused by local climatic conditions often cause significant damage; and landslides are concentrated in the rainy season every summer. Because much of South Korea is mountainous, most studies on landslides in the country have focused on mountainous areas [27–29]; e.g., in Jangheung [23] or Yongjin [6,30]. Furthermore, due to the frequency of landslides in mountainous areas, most of the studies in Southeast Asia have been performed in non-urban areas [31–33]. Assessment of landslide susceptibility in urban areas could provide an important contribution to minimizing additional damage from these natural disasters; it could be also used to planning and multi hazard assessment [34–37].

Seoul, the capital of South Korea, also has urban areas occupying mountainous land. Unlike landslides that occur in sparsely populated mountainous areas, landslides that occur in mountainous areas in cities result in significant casualties/fatalities and structural damage. For example, landslides in Umyeonsan located in central Seoul in July 2011 caused unprecedented damage to the city. Following the 2011 Umyeonsan landslide, the downtown area, which had previously been considered safe, was no longer regarded as a landslide-safe zone. Various aspects of Korea's landslide policy have been re-evaluated and are expected to be revised in the near future. Landslides are more dangerous in urban areas than non-urban mountainous areas. Policies have shifted to preventative and resident-oriented initiatives, with a focus on minimizing injuries to citizens and restoration of their properties after landslides. Therefore, it is important to analyze potential landslide hazards in urban areas.

To apply the model for this type of data mining analysis, a spatial database should first be constructed containing information on the area of landslide damage. Because landslides are difficult to access, the collection of field survey data to build a landslide spatial database is costly and time-consuming. An effective alternative is to build a database using aerial photography and geographic information systems (GISs). Landslides leave traces that last for months or years; hence, aerial imagery can be used to extract data on areas damaged by landslides after the event. Using these data, the susceptibility for landslides can be analyzed by data mining modeling techniques.

2. Study Area

Seoul is the largest city in South Korea and is located in the center of the Korean Peninsula. The total urban area covered by Seoul (approximately 605 km^2) is less than 1% of the total land area of South Korea. However, Seoul is one of the most heavily populated cities in the world; over 10 million citizens were recorded in 2011. Thus, the city has a dense and complex transportation system. Natural disasters, including landslides, have the potential to cause serious damage in the city [2]. The study area, Umyeonsan, is an urbanized mountain rising 321.6 m above sea level in the southern sub-central area of Seoul (Figure 1). Due to the fact that it is not very high or steep, the mountain is a popular site for Seoul residents and it has a large transient population

Figure 1. Map showing the location of the study area of Umyeonsan: (**a**) South Korea; (**b**) Seoul; and (**c**) Umyeonsan.

Umyeonsan is comprised of a low mountainous area with a slope of approximately 30° near the top of the mountain. The area is mainly comprised of gneissic rock formed by geomorphological movement and weathering. The composition of biotite gneiss, which is prone to weathering and fault formation, increases its susceptibility to landslides. The foliation structure of the gneiss is sparsely generated due to multiple flexures. The main soil type distributed in the study area is brown forest soil formed from the base materials of biotite gneiss and granite gneiss, which are metamorphic rocks, and the soil texture is sandy loam to loam, which has good drainage. Deep soils and high organic content brown forest soil is distributed in some valley areas in the study area, where the soil texture is silt loam in silt and sand. Due to the soil textures present, the Umyeonsan area is prone to severe weathering, causing the unstable condition of outcrops.

The main forest type in Umyeonsan is temperate deciduous broad-leaved forest. Broad-leaved forest, pine forest, poplar forest, and broad-leaved plantation forest are all present in the study area. The principal species in the forests are similar to those in the central Korean Peninsula: *Quercus mongolica*, *Robinia pseudoacacia*, *Quercus variabilis*, oak trees and *Asian black birch*.

On 27 July 2011, heavy rains occurred in central Korea and a large landslide occurred in Umyeonsan. The heavy rainfall lasted from 26–28 July. The cumulative precipitation in the Seoul metropolitan area was 587.5 mm, which is the largest three-day cumulative precipitation recorded since precipitation monitoring began in Seoul. Based on data from Seocho automatic weather system, the maximum precipitation per hour at the time of the landslide was 85.5 mm [38]. The landslide from the heavy rains affected an area of 73.23 ha. Sixteen people were killed, including 6 residents from the village of Namtaeryeong and 7 residents from the southern area of the southern ring roads; 2 people remained missing, and approximately 400 people were evacuated [39]. The water supply was cut off. Thousands of apartments near Gangnam and Umyeonsan were damaged; and over 20,000 households experienced water problems. Therefore, we chose this landslide to obtain data for our study (Figure 2d,e). The study area was determined by the road and river boundaries in the Umyeonsan area (Figure 2).

3. Data

Given that data mining models are a data-driven methodology, accurate data are essential to build a suitable database. Accurate data on the state of the earth's surface can be obtained rapidly using aerial photographs, which also effectively capture complicated urban structures. Moreover, despite the high cost of aerial photographs, it is cost-effective to acquire the images of and continuously monitor urban areas due to their dense nature. To acquire accurate data on the landslide occurrence, aerial photographs from before and after the Umyeonsan landslides were used (Figure 2a,b). Location data for the Umyeonsan landslide area were acquired through comparative analysis, as traces of landslides are visible for several years after the events.

In addition, a digital topographic map (1:5000) generated by the National Geographic Information Institute (NGII) was obtained, and geometric correction was performed. Overlay and comparative analysis of the aerial photographs and the 1:5000 digital topographic map indicated the landslide occurrence area (Figure 2c). The mapped landslide occurrence data were finally confirmed by comparison with data surveyed in the field by the Seoul metropolitan government [39].

Figure 2. *Cont.*

Figure 2. Data of landslide occurrence locations from aerial photographs: (**a**) before landslides in 2011; (**b**) after landslides in 2011; (**c**) training and validation data; (**d**) damage situation in Southern Circular Road [40]; (**e**) Gyungnam apartment buildings in Seocho-gu [41].

Slopes are established by erosion and sedimentation of terrain surfaces and have a key effect on landslide occurrence as they affect the flow of water and the soil characteristics [42]. Therefore, among the factors influencing landslides, topographic factors were preferentially considered and were used as input data to identify the relationships with landslide occurrence (Figure 3). Prior to calculation of the topographic factors, a triangulated irregular network (TIN) was produced from a pre-generated topographic map, and a digital elevation model (DEM) was created using ArcGIS 10.1 (ESRI, Redlands, LA, USA).

All topographic factors were calculated from the DEM using ArcGIS software (ESRI, Redlands, LA, USA). The slope, indicated as the angle to the horizontal plane or the ratio of the vertical height to the horizontal distance as a percentage, was calculated using the SLOPE tool. The ASPECT tool was used to calculate the aspect, the compass direction of the slope faces. The curvature values of the terrain surface were calculated using the CURVATURE tool. The curvature represents the morphological characteristics of the study area; an upwardly convex surface possesses a positive value, while an upwardly concave surface possesses a negative value.

Figure 3. Topographical factors related to landslides used to construct a spatial database: (**a**) slope gradient; (**b**) slope aspect; (**c**) curvature; (**d**) topographic wetness index (TWI); (**e**) stream power index (SPI); (**f**) slope length factor (SLF); (**g**) standardized height; (**h**) valley depth; (**i**) downslope distance gradient.

The topographic wetness index (TWI) is defined for a steady state. The index is commonly used for hydrological processes to quantify the disposition of water flow and the impact on the topography [43]. The TWI is calculated by the run-off model as follows, which reflects the accumulation of water at a specific point in the catchment area and the force of gravity on the water in the area [43].

$$TWI = In\left(\frac{a}{\tan \beta}\right) \tag{1}$$

where α is the cumulative upslope area drained per unit contour length in a cell, and β is the local slope angle of the cell.

The stream power index (SPI) estimates the erosion power of a stream, which affects the stability of the area. Therefore, it is generally used to determine the locations for performing soil conservation measures to reduce the erosive effects of concentrated surface runoff. SPI can be calculated as follows [44]:

$$SPI = As \times \tan(\beta) \tag{2}$$

where As is the area of the target point of the study area, and β is the slope angle.

After calculating the slope, the slope length factor (SLF) for average erosion is calculated using the Revised Universal Soil Loss Equation (RUSLE) [45]. The SLF for slope of length λ is represented as:

$$SLF = \left(\frac{\lambda}{72.6}\right)^m \tag{3}$$

where the constant value 72.6 of the RUSLE is in feet, and the variable, m, is the slope-length exponent calculated from the local slope angle β [45]. The SLF is influenced by the ratio of rill erosion to interrill erosion, β. Rill erosion is caused by flow, while interrill erosion is caused by the effects from rainfall.

System for Automated Geoscientific Analyses (SAGA) GIS were used to calculate additional topographic factors [46]. The standardized height is the absolute height multiplied by the normalized height of the study area; normalized height is the normalization of the height of the study area between 0 and 1. The standardized height was calculated using linear regression to yield the suitable parameters for prediction of the soil attributes by simplifying the complex terrain states [47]. The valley depth is the vertical distance of the specific point from the base level of the channel network [46]. The elevation of the channel network base level is interpolated, and the valley depth is calculated by subtracting the base level of the channel network from the DEM [46]. The downslope distance gradient is a quantitative estimation of the hydraulic gradient. The downslope distance gradient is obtained by calculating the downhill distance when water loses a determined quantity of energy from precipitation [48]. $\tan a_d$ identifies the downslope distance gradient:

$$\tan a_d = \frac{d}{L_d} \tag{4}$$

where d and L_d are the elevation and the horizontal distance to the reference point, respectively.

The type and composition of the soil influence the occurrence of landslides as they determine the degree of erosion and saturation [49]. The slope stability can be strengthened by the roots of vegetation and the degree of strengthening depends on the specific attributes of the vegetation; hence, the impact of heavy rainfall on the slope can be mitigated by vegetation [50]. The vegetation map was provided by the Korea Forest Research Institute and was in polygon format with a scale of 1:25,000 (Figure 4). The attributes of timber diameter, type, density, and age were extracted from the vegetation map. In addition to the vegetation variables, the soil type and condition also influence the occurrence of landslides. The soil maps used in this study were provided by the National Academy of Agricultural Science (RDA) in polygon format with a scale of 1:25,000. Soil depth, drainage, topography, and texture were derived from the soil maps (Figure 5).

Figure 4. Landslide-related factors obtained from the vegetation map: (**a**) timber type; (**b**) timber diameter; (**c**) timber density; (**d**) timber age.

Figure 5. Landslide-related factors obtained from the soil map: (**a**) soil depth; (**b**) soil drainage; (**c**) soil topography; (**d**) soil texture.

All variables used in this study were resampled into a grid format of 5-m spatial resolution to create a spatial database. The total size of the spatial database was 1015 × 874, and the study area contained 457,133 cells. The 103 landslide occurrence locations in point format were also converted into the 5-m grid format to be included in the data in the spatial database. Half of the 103 landslide location data points (51 points) were used as training data, and the other half were used as validation data. All of the collected data were prepared in ASCII format from the grid. Finally, a spatial database was established to apply the model to (Table 1).

Table 1. Landslide-related factors used to construct spatial database.

Original Data	Factors	Data Type	Scale
Aerial photograph	Landslide location	Point	1:1000
Topographical map [a]	Slope gradient [°] Slope aspect Curvature Topographic Wetness Index (TWI) Stream Power Index (SPI) Slope Length Factor (SLF) Standardized height [m] Valley depth [m] Downslope distance gradient [rad]	GRID	1:1000
Forest map [b]	Timber diameter Timber type Timber density Timber age	Polygon	1:25,000
Soil map [c]	Soil depth Soil drainage Soil Topography Soil texture	Polygon	1:25,000

[a] Topographical factors were extracted from digital topographic map by National Geographic Information Institute; [b] The forest map produced by Korea Forest Service; [c] The detailed soil map produced by Rural Development Administration

4. Methodology

Landslides are affected by various factors and each element is simplified as input data in modeling studies. Given that natural phenomena such as landslides are composite events of each element, they should be analyzed using a modeling technique based on the relationships between data in a vast dataset such as a data mining model. Using data mining models, the relationships among all of the predisposing factors and the effects of all of the landslide-related factors on the model results can be analyzed. Therefore, in this study, the factors related to landslide susceptibility were used as the input data of the data mining models. The input variables were stored in a spatial database; 17 landslide-related variables were chosen from the spatial database, including half of the randomly selected landslide locations used as training data. Landslide susceptibility maps were created using two data mining models, specifically support vector machine (SVM) (Boulder, CO, USA) and ANN models, which are based on linear classifiers. Additionally, the weight of each factor was calculated in the data mining modeling process, and a validation was performed (Figure 6). A detailed description of the data mining models used in this study is given below.

Figure 6. Flowchart of the spatial prediction of landslide susceptibility.

4.1. Support Vector Machine

4.1.1. Basic Principles of SVM

The SVM method is a training algorithm based on nonlinear transformations that uses a classification based on the principle of structural risk minimization, which has performed well in the test phase [51]. When training data are represented in multidimensional space, several hyperplanes can be used to distinguish the data into two classes; however, there is only one optimum linear separating hyperplane [52]. This optimal hyperplane should maximize the distance between the data closest to the hyperplane. Thus, the objective of the SVM method is to determine a multi-dimensional hyperplane that differentiates the two classes with the maximum margin. The SVM model performs this process according to three main concepts: margin, support vector, and kernel.

The margin is the distance between a specific data point and the hyperplane that separates the data patterns. It is the shortest distance from the data vector of each class to the given hyperplane. The larger the margin, the more accurately the SVM model performs the classification; hence, the optimal hyperplane maximizes the margin when $\vec{\omega}$ is the normal vector of the hyperplane, and is the norm of the normal of the hyperplane. $||\vec{\omega}||$ is the norm of the normal of the hyperplane.

$$margin = \frac{2}{||\vec{\omega}||} \tag{5}$$

The hyperplane, which defines the two classes that maximize the margins when the data are linearly separable, can be defined by Equation (6). Under these conditions, the training data on these two hyperplanes constitute the support vector. Therefore, the support vector maintains the shortest distance to the hyperplane. The support vectors are dependable on the boundary of the hyperplane and they are time-efficient when a new dataset is tested with a support vector. The dataset of linear separable training vectors x_i ($i = 1, 2, \ldots, n$) and the training vectors consisting of two classes are represented by y_i as the value of ± 1 [53,54].

$$y_i\left((\vec{\omega} \cdot \vec{x_i}) + b\right) - 1 \geq 0 \tag{6}$$

where b is the scalar base, and the scalar product operation is denoted by $(\cdot)$.

In addition, because the two hyperplanes must maximize the margin between them, the optimization of objective Equation (7) with the constraints of Equation (6) yields the planes of the two classes:

$$\min(\frac{1}{2}||\vec{\omega}||^2) \tag{7}$$

However, most cases do not satisfy the constraint as the data are not linearly separable. A formula including the slack variable ξ_i indicates the distance from the hyperplane to the data located on the wrong side [52]; the penalty term C introduced to account for misclassification [55,56] was developed to address this problem.

$$y_i\left((\vec{\omega}\cdot\vec{x_i})+b\right) \geq 1-\xi_i \tag{8}$$

$$\min(\frac{1}{2}||\vec{\omega}||^2 + C\sum_{i=1}^{n}\xi_i) \tag{9}$$

4.1.2. SVM for Nonlinear Classification

In the preceding case of nonlinear boundary data, the SVM model is unable to generate an appropriate hyperplane. Therefore, the data can be classified into a linear problem as a hyperplane by mapping them to a high-dimensional space value from the low-dimensional data that cannot be linearly separated. To solve Equations (8) and (9) when the dataset is linearly separable, the training data are expressed as a vector inner product in the optimization process. This enables the inner product in the training algorithm to be expressed by the inner product of a function in a feature space when nonlinear data in the input space are mapped to a high-dimensional feature space by a specific function.

The hyperplane of the data with nonlinear data patterns can be derived by transforming the data space, but x cannot be computed directly; an arbitrary kernel function should be used to satisfy x [53]. The kernel functions, $K(x_i, x_j)$, convert the original data into linearly separable data in high-dimensional feature space [52]. The kernel function should satisfy Mercer's condition. The following four functions are the main kernel functions used in the SVM model:

$$\text{Linear} : K(x_i, y_i) = x_i^T \cdot x_j \tag{10}$$

$$\text{Polynomial} : K(x_i, y_i) = \left(\gamma\cdot\left(x_i^T\cdot x_j\right)+r\right)^d, \gamma > 0 \tag{11}$$

where γ and d are parameters of the kernel functions.

$$\text{Radial basis function} : K(x_i, y_i) = e^{-\gamma||x_i-x_j||^2}, \gamma > 0 \tag{12}$$

$$\text{Sigmond} : K(x_i, y_i) = \tanh\left(\gamma\cdot x_i^T\cdot x_j + r\right) \tag{13}$$

In radial basis function (RBF) kernel, $||x_i - x_j||$ is the squared Euclidean distance between the vectors, and γ is defined as follows:

$$\gamma = \frac{1}{2\sigma^2} \tag{14}$$

where σ is a freely adjustable parameter that controls the performance of the kernel, and r is a parameter of the sigmoid kernel.

In this study, the Environment for Visualizing Images (ENVI 5.0, Exelis Visual Information Solutions, Boulder, CO, USA) software program was used for the application of the SVM model (Boulder, CO, USA). The software provides four types of kernels in the SVM classifier as mentioned above: linear, polynomial, RBF, and sigmoid. Each kernel function has its own parameters, which should be adjusted appropriately according to the characteristics of the purpose. The default kernel of the ENVI software is the RBF kernel, which is known to provide more accurate prediction results in most classifications, particularly in nonlinear problems. After applying the kernel and classifying each input factor, the SVM model results were mapped using the process of overlapping and summation.

$$SVM_{all} = \sum_{i=1}^{n} SVM_i \ (n = \text{number of input factors}) \tag{15}$$

SVM_{all} is the sum of all classification results by the SVM model for all variables, and SVM_i is the single result of a given variable; in this study, the total number of input factors, n was 17. The results were mapped and validated using the landslide locations not used for training.

4.2. Artificial Neural Network

An ANN is a statistical learning algorithm that describes a neuronal signaling system. The human brain regulates the connection strength of synapses between neurons and determines whether signals are delivered or not; the ANN learns to derive the results with minimum error by adjusting the weight between input data and output data. The unit of the ANN is the perceptron, which is divided into an input layer and an output layer. The input layers are represented by nodes, which are linked to an edge having the weight of each node [57]. The model outputs 1 when the sum of the values of each layer multiplied by the weight is greater than or equal to the threshold; otherwise, the output is -1. This type of perceptron can only be applied to a linear classification and is not applicable in most cases.

In this study, multi-layered perceptrons allowing classification of nonlinear data were used with a hidden layer between the input and output layers. A feed-forward network was used for the map learning. The learning of neural networks progressed via a backpropagation algorithm that propagates the errors generated in the output layer to each layer and corrects the weights. In other words, the backpropagation algorithm learns the neural network by inverting the step, and the result most similar to the output layer value is derived based on the input layer. The process can be simplified as a combination of nodes and edges [58]. The backpropagation algorithm can be expressed by the following equation:

$$\omega_k^{up} = \omega_k - \eta * \frac{\partial E_{total}}{\omega_k} \tag{16}$$

where ω_k is the updated weight for connection k and η is the learning rate. The initial value of weight ω_0 is given randomly. E_{total} is the error of the final output result when the expected output value is assumed to be T, according to the input data [59].

$$E_{total} = \frac{1}{2} \sum_{j}^{n} \left(T_j - out_{o_j} \right)^2 \tag{17}$$

where j is the number assigned to the node in each layer, and n is the number of output layers. Because the backpropagation algorithm updates the weights while propagating the error generated from the output to the input direction, it updates the weights starting from the portion from the output. E_{total} is calculated repeatedly using the updated weight so as to be minimized; the differentiation of E_{total} is fundamental. Therefore, when the calculated value is passed to the next layer or is expected to be the final result, an activation function is used to convert it to a range of values. In this study, a unipolar sigmoid function was used. The equation of the unipolar sigmoid function is as follows:

$$f(x) = \frac{1}{1 + e^{-x}} \tag{18}$$

Using the activation function, the output value of each layer is adjusted from 0 to 1.

In this study, the ANN model was processed using MATLAB R2015a software (MathWorks, Massachusetts, MA, USA). Areas where the slope data were 0 were set as areas not prone to landslide occurrence and used for training data; landslide occurrence point data were also allocated to the training set. The backpropagation algorithm of the ANN was applied to estimate the weights between the input and the hidden layers, and between the hidden and the output layers by updating the hidden nodes and the learning rate. In this study, the input data were normalized to the range of 0.1–0.9 and a $17 \times 34 \times 1$ structure was selected for the networks. The learning rate was set to 0.01 with a randomly selected initial weight. The number of epochs was set to 2000, and the root mean square error value was set to 0.1, which was used for the stopping criterion.

5. Results

Figure 7a–d shows the landslide susceptibility maps produced by applying the SVM models with four kernels; and Figure 7e shows the map of the results from the ANN models. The values of the landslide susceptibility results were sorted in descending order and categorized into five classes for easy visual interpretation. For intuitive and efficient analysis, the results for the landslide susceptible areas were classified into five groups of approximately the same size: very high, high, medium, low, and very low susceptibility. The comparative analysis between the results and existing landslide occurrence locations was also made easier by representing the landslide susceptibility with the five categories. The spatial distribution of the landslide locations was concentrated in high-susceptibility areas.

In Figure 7, the results map of the linear, polynomial, and sigmoid kernels of the SVM models showed a similar pattern and represented the influence of the input factors of Figures 4 and 5. In addition, the RBF kernel showed a similar pattern to the ANN map in that the center of the map had relatively higher susceptibility. In general, highly susceptible areas were principally distributed in the northeast direction from the center of the study area. Areas with high susceptibility were located in dense forest areas of broadleaved forest, above-average soil depth, and steep slopes. The dominant rocks constituting the Umyeonsan area are gneissic rocks, which have high landslide susceptibility because they are heavily eroded; landslides occur in areas of high soil drainage.

Figure 7. *Cont.*

Figure 7. Landslide susceptibility maps generated using data mining models: Using an (**a**) support vector machine (SVM) model with linear kernel; (**b**) SVM model with polynomial kernel; (**c**) SVM model with radial basis function kernel; (**d**) SVM model with sigmoid kernel; (**e**) and the artificial neural network (ANN) model.

In Table 2, the area of each class is represented by percentage. The areas with very high indices in the SVM model were 2.2599, 2.2599, 2.2889 and 2.28369 km^2 for the linear, polynomial, RBF, and sigmoid kernels, respectively. Among the results from the SVM models, the RBF kernel accounted for a slightly higher percentage (20.03%) of the areas with a very high index compared with the linear, polynomial, and sigmoid kernels. The areas with a very high index in the ANN model covered 3.9262 km^2, which had the highest percentage (34.35%). In contrast, the areas with high index accounted for only a small proportion (8.65%). The area with a very low index accounted for similar percentages of 20.00, 20.00, 20.00 and 19.99% for the linear, polynomial, and sigmoid kernels, and ANN, respectively, while the RBF accounted for a lower percentage of 19.89%.

Table 3 shows the relative weights per variable, which were calculated from randomly selected training data for 20 epochs of 200 cycles. The weights between layers represent the contribution of each variable to the prediction of the landslide susceptibility. Timber age showed the lowest average value of weight, 0.04828. For the comparison analysis of the relativeness of the weights, the values were divided by the weight of the timber age and were normalized with respect to the average weight of the timber age. The highest average weight of 0.09615 for the standardized height was normalized to 2.00164, which was approximately twice the value of the weight of the timber age. The second major variable contributing to landslides was timber diameter (1.46886), followed by curvature (1.28335), soil depth (1.27746), and aspect (1.25804). The standard deviation ranged from 0.00022 to 0.01132.

To investigate the reliability of the proposed landslide susceptibility map using the previously described SVM and ANN data mining models, we performed a validation. The 2011 landslide occurrence data not used for the training were used for validation. After 51 points of training data had been extracted using random sampling from the landslide occurrence data from the 2011 Umyeonsan landslide and applied to the models, 52 points were used for the validation. Because the predicted landslide susceptibility map was an indicator of the potential possibility of landslides in the future, the data obtained after the occurrence of the landslides were a suitable choice for the validation data. However, an erosion control project implemented after the 2011 event has significantly reduced the possibility of landslides. Therefore, it is important to obtain information on landslide susceptibility in areas where landslides have not occurred. In addition, implementation of precautionary measures to prevent landslides from occurring in the future is necessary. Thus, half of the landslide occurrence points were used as training data, and the remaining points were used as validation data.

Table 2. Area and percentage of each category.

Landslide Susceptibility Index	SVM								Backpropagation ANN (km²)	Percentage (%)
	Linear (km²)	Percentage (%)	Polynomial (km²)	Percentage (%)	RBF (km²)	Percentage (%)	Sigmoid (km²)	Percentage (%)		
Very high	2.2599	19.77%	2.2599	19.77%	2.2889	20.03%	2.2836	19.98%	3.9262	34.35%
High	2.3081	20.20%	2.3081	20.20%	2.1835	19.11%	2.2866	20.01%	0.9881	8.65%
Moderate	2.2878	20.02%	2.2878	20.02%	2.2359	19.56%	2.2863	20.01%	1.9451	17.02%
Low	2.2866	20.01%	2.2866	20.01%	2.4474	21.41%	2.2859	20.00%	2.2849	19.99%
Very low	2.2858	20.00%	2.2858	20.00%	2.2727	19.89%	2.2859	20.00%	2.2841	19.99%
Total	11.4283	100.00	11.4283	100.00	11.4283	100.00	11.4283	100.00	11.4283	100.00

Table 3. Neural network weight between landslide and related factors.

200 Cycles for 1 epoch	1	2	3	4	5	6	7	8	9	10	Average	Standard Deviation	Normalized Weight with Respect to Timber Age
Slope gradient	0.0555	0.0548	0.0546	0.0546	0.0549	0.0552	0.0557	0.0561	0.0564	0.0567	0.05545	0.00072	1.14851
Slope aspect	0.0616	0.0611	0.0608	0.0607	0.0607	0.0606	0.0606	0.0605	0.0604	0.0603	0.06073	0.00036	1.25787
Curvature	0.0623	0.0615	0.0614	0.0615	0.0616	0.0618	0.0620	0.0623	0.0625	0.0629	0.06198	0.00048	1.28376
TWI	0.0512	0.0520	0.0512	0.0506	0.0500	0.0495	0.0490	0.0485	0.0480	0.0476	0.04976	0.00141	1.03065
SPI	0.0531	0.0520	0.0518	0.0517	0.0517	0.0517	0.0517	0.0517	0.0516	0.0516	0.05186	0.00043	1.07415
SLF	0.0579	0.0594	0.0590	0.0584	0.0578	0.0572	0.0567	0.0564	0.0561	0.0559	0.05748	0.00116	1.19056
Standardized height	0.0699	0.083	0.0909	0.0955	0.0987	0.1014	0.1036	0.1052	0.1063	0.1070	0.09615	0.01132	1.99151
Valley depth	0.0521	0.0517	0.0514	0.0512	0.0510	0.0509	0.0509	0.0508	0.0508	0.0509	0.05117	0.00042	1.05986
Downslope distance gradient	0.0626	0.0579	0.0559	0.0551	0.0547	0.0546	0.0548	0.0552	0.0557	0.0563	0.05628	0.00231	1.16570
Timber diameter	0.072	0.0717	0.0712	0.071	0.0708	0.0707	0.0705	0.0704	0.0704	0.0703	0.07090	0.00055	1.46852
Timber type	0.0546	0.0515	0.0514	0.0513	0.0512	0.0512	0.0513	0.0513	0.0512	0.0512	0.05162	0.00100	1.06918
Timber density	0.0607	0.0582	0.0577	0.0577	0.0579	0.0581	0.0583	0.0584	0.0584	0.0583	0.05837	0.00082	1.20899
Timber age	0.0489	0.0484	0.0482	0.0482	0.0481	0.0482	0.0482	0.0482	0.0482	0.0482	0.04828	0.00022	1.00000
Soil depth	0.0682	0.0686	0.0666	0.0646	0.0626	0.0606	0.0586	0.0566	0.0553	0.0543	0.06160	0.00506	1.27589
Soil drainage	0.0546	0.0544	0.0547	0.0554	0.0557	0.0560	0.0561	0.0562	0.0563	0.0564	0.05558	0.00072	1.15120
Soil Topography	0.0606	0.0607	0.0603	0.0600	0.0598	0.0597	0.0597	0.0596	0.0595	0.0594	0.05993	0.00043	1.24130
Soil texture	0.0542	0.0531	0.0529	0.0527	0.0526	0.0526	0.0526	0.0527	0.0528	0.0530	0.05292	0.00046	1.09611

The AUC of the receiver operating characteristic (ROC) curve was used for the validation of the models. The ROC consists of specificity, the x-axis, which shows the percentage of the areas where landslides are expected to occur but have not occurred, and sensitivity, the y-axis, which represents the predicted landslide areas according to training data of landslide occurrences. The ROC graph was drawn using the following procedure. To show the rank relative to the predicted results, the values of the results of landslide susceptibility from the models were sorted in descending order. The ranked results were represented in ROC curves using 100 classes; a single class comprised an area that was approximately 1% of the total study area. Figure 8a shows the AUC of the landslide susceptibility map for the SVM model; the linear, polynomial, RBF, and sigmoid kernel had 72.41 (0.7241), 72.83 (0.7283), 77.17 (0.7717) and 72.79% (0.7279) accuracy versus 78.41% (0.7841) accuracy in the ANN model (Figure 8b).

Figure 8. Validation results using data mining models: (**a**) SVM model; (**b**) ANN model.

A higher AUC value indicates better model performance, but it is important to perform additional empirical analyses based on input and ground truth data. Our validation results indicated that the five landslide sensitivity assessments performed well; the AUC values were all greater than 0.7 [60]. In the SVM model, the AUC value using the RBF kernel was 0.7717, which was superior to the other three kernels in the study area. The ANN model yielded an AUC value of 0.7841, which was approximately 6% higher than that of the SVM model with the linear kernel, and 0.7% higher than with the RBF kernel. The graphs of the validation results are depicted in Figure 8; in all of the models, 60% of the study area incorporated all of the landslide locations.

6. Discussion

In this study, two types of data mining models were used to create landslide susceptibility maps. Because the results of data mining models are derived from their data, the reliability of the input data is fundamental. Landslide location data were acquired from aerial photographs to provide accurate training datasets. The data mining models are suitable for observing the rapid changes in an urban area with its compactness and complex characteristics. In addition, other topographical-, forest-, and soil-related variables were obtained, and topographic-related factors were calculated from the topographical map produced from the aerial photographs.

The results from each model in Figure 7 were not statistically different in the spatial pattern of the landslide susceptibility area. This can be interpreted that all of the four prediction results from the SVM model were based on the same input factors, and all landslide predisposing factors were regarded as having equal importance so that the same weight was assigned to each predisposing factor. However, for the results from the RBF kernel, which had the highest accuracy among the SVM models, the susceptibility on the west, southwest, and southeast slopes was lower than that from the other kernels, which was similar to the trend for the ANN model. In general, the northern area of the mountain showed high susceptibility, and had the same pattern as the existing damaged area of the southern ring roads.

The percentages of the classes of five indices were also calculated. The linear, polynomial, and sigmoid kernels were distributed fairly evenly, while the RBF kernel showed a slightly higher percentage in the very high index class, and a lower percentage in the very low index class. This suggests that a high percentage in the high susceptibility index yields good validation results. Figure 8a shows that the RBF kernel had a percentage of 77.71%, which was approximately 5% higher than that of the other kernels. Similarly, the results from the ANN model yielded a high percentage (more than 30% of the total) in the very high class, which represented high risk areas, and this also led to a high validation percentage for the ANN model, 78.41%. In addition, the ANN model results in Table 3 show that the standardized height contributed most to landslide susceptibility with a normalized weight of greater than 2. This can be seen in Figure 7e, which shows the effect of standardized height with the high relative weight of the ANN model. Standardized height was the most important factor among all of the landslide-related factors. The factors related to landslide susceptibility should be managed based on their relative weight.

Because this study investigated the susceptibility of landslides in urban areas, the study area was fairly small, with an area of approximately 11.4 km^2 and a spatial resolution of 5 m along road and river boundaries rather than a rectangular boundary. It should be noted that the ROC was characterized using a percentage of the landslide susceptibility rates. Thus, in this study, the validation rate was relatively low compared with most previous studies, which were based on rectangles. Moreover, we could use the slope unit instead of common raster unit, which is closely correlated to the topography, for the further steps; the slope unit is defined as the unit between the valleys and ridges. Most of the mountains in the city are managed in the framework of city parks in which the various economic and social aspects of Umyeonsan are also considered, rather than the area being covered solely by forests that are more resistant to landslides. Given that landslide damage in urban areas can lead to human casualties, it is important to investigate the soil and topographical characteristics of the areas where landslides are likely to occur. Furthermore, landslide risk should be incorporated in city-level policies.

7. Conclusions

In this study, we applied a spatial data mining approach to identify areas where landslides are likely to occur using aerial photographs and GIS. Landslide-prone locations were determined using the interpretation of aerial photographs and field survey results to construct spatial datasets. The topographical predisposing factors were extracted from a digital topographical map constructed from aerial photographs, and the soil and forest predisposing factors were extracted from publicly available maps. A spatial database was constructed using the extracted and calculated factors including randomly extracted training data for the landslide area, and landslide susceptibility was mapped using SVM and ANN models. Finally, the results map was validated using the half of the landslide location data not used for training.

In urban areas such as Seoul, landslides have previously not been considered a serious concern, compared with floods and storms. Seoul has undergone rapid urbanization that has taken place with little consideration of natural disasters. In large cities such as Seoul, global warming, rapid urbanization, extreme precipitation, and population pressures could potentially lead to complex events, including landslides occurring in the city, which could have serious consequences. Therefore, it is important to create a scientifically valid landslide susceptibility map for urban areas such as Seoul.

The predicted rate of landslides was high on the steep mountain tops and similar to the general landslide pattern on the ridges. The map created in this study showed that susceptibility to landslides decreased with slope gradient. Because landslides are caused by differences in gravity due to terrain shape, it is logical that their occurrence is affected by the geographical structure of the area.

The relationship between landslide occurrence and forested land cover showed that broadleaved forests had a high susceptibility, and this type of forest occupied the largest proportion of the study area. In addition, large trunk-size, high-density forests, and middle-aged forests had a high risk of landslides. Among the soil factors, red and yellow soils had higher landslide susceptibility, as well as areas with large soil depth; the occurrence rate was higher in areas with good soil drainage. The performance of the SVM and ANN models was validated using AUC analysis. The SVM model yielded 72.41%, 72.83%, 77.17% and 72.79% accuracy for the linear, polynomial, RBF, and sigmoid kernels, respectively; while the ANN model had 78.41% accuracy. The results of this study are expected to support spatial decision-making from relevant agencies to formulate policies related to landslide disaster risk reduction in urban areas.

Acknowledgments: This research was conducted by Korea Environment Institute (KEI) with support of a grant (16CTAP-C114629-01) from Technology Advancement Research Program (TARP) funded by Ministry of Land, Infrastructure and Transport of Korean government. This work was supported by the Space Core Technology Development Program through the National Research Foundation of Korea, funded by the Ministry of Science, ICT and Future Planning, under Grant (NRF-2014M1A3A3A03034798).

Author Contributions: Sunmin Lee applied the SVM and ANN models and wrote the paper. Hyung-Sup Jung designed the experiments. Also, he interpreted the input data and the results. Moung-Jin Lee suggested the idea and organized the paperwork. In addition, he collected data and conducted pre-processing of the input data. All of authors contributed to the writing of each part.

References

1. Berry, M.J.; Linoff, G. *Data Mining Techniques: For Marketing, Sales, and Customer Support*; John Wiley & Sons, Inc.: Hoboken, NJ, USA, 1997.
2. Pachauri, R.K.; Allen, M.R.; Barros, V.R.; Broome, J.; Cramer, W.; Christ, R.; Church, J.A.; Clarke, L.; Dahe, Q.; Dasgupta, P. Contribution of working groups I, II and III to the fifth assessment report of the intergovernmental panel on climate change. In *Climate Change 2014: Synthesis Report*; IPCC: Geneva, Switzerland, 2014.
3. Centre for Research on the Epidemiology of Disasters, UN Office for Disaster Risk Reduction. *The Human Cost of Natural Disasters: A Global Perspective*; Centre for Research on the Epidemiology of Disaster (CRED): Brussels, Belgium, 2015.
4. Cruden, D.; Varnes, D. Landslide Types and Processes. In *Landslides: Investigation and Mitigation*; Turner, A.K., Schuster, R.L., Eds.; Special Report 247; Transportation Research Board: Washington, DC, USA, 1996; pp. 36–75.
5. Choi, J.; Oh, H.J.; Lee, H.J.; Lee, C.; Lee, S. Combining landslide susceptibility maps obtained from frequency ratio, logistic regression, and artificial neural network models using aster images and gis. *Eng. Geol.* **2012**, *124*, 12–23. [CrossRef]
6. Lee, S.; Ryu, J.H.; Won, J.S.; Park, H.J. Determination and application of the weights for landslide susceptibility mapping using an artificial neural network. *Eng. Geol.* **2004**, *71*, 289–302. [CrossRef]
7. Pradhan, B.; Lee, S. Landslide susceptibility assessment and factor effect analysis: Backpropagation artificial neural networks and their comparison with frequency ratio and bivariate logistic regression modelling. *Environ. Model. Softw.* **2010**, *25*, 747–759. [CrossRef]
8. Pradhan, B.; Lee, S. Utilization of optical remote sensing data and gis tools for regional landslide hazard analysis using an artificial neural network model. *Earth Sci. Front.* **2007**, *14*, 143–151.
9. Ahmed, B.; Dewan, A. Application of bivariate and multivariate statistical techniques in landslide susceptibility modeling in chittagong city corporation, bangladesh. *Remote Sens.* **2017**, *9*. [CrossRef]
10. Lee, S. Application of logistic regression model and its validation for landslide susceptibility mapping using gis and remote sensing data. *Int. J. Remote Sens.* **2005**, *26*, 1477–1491. [CrossRef]
11. Scaioni, M.; Longoni, L.; Melillo, V.; Papini, M. Remote sensing for landslide investigations: An overview of recent achievements and perspectives. *Remote Sens.* **2014**, *6*, 9600–9652. [CrossRef]

12. Akcay, O. Landslide fissure inference assessment by anfis and logistic regression using uas-based photogrammetry. *ISPRS Int. J. Geo-Inf.* **2015**, *4*, 2131–2158. [CrossRef]

13. Bui, D.T.; Pradhan, B.; Lofman, O.; Revhaug, I.; Dick, O.B. Landslide susceptibility assessment in the hoa binh province of vietnam: A comparison of the levenberg–marquardt and bayesian regularized neural networks. *Geomorphology* **2012**, *171*, 12–29.

14. Yilmaz, I. Landslide susceptibility mapping using frequency ratio, logistic regression, artificial neural networks and their comparison: A case study from kat landslides (Tokat—Turkey). *Comput. Geosci.* **2009**, *35*, 1125–1138. [CrossRef]

15. Ermini, L.; Catani, F.; Casagli, N. Artificial neural networks applied to landslide susceptibility assessment. *Geomorphology* **2005**, *66*, 327–343. [CrossRef]

16. Pourghasemi, H.R.; Pradhan, B.; Gokceoglu, C. Application of fuzzy logic and analytical hierarchy process (ahp) to landslide susceptibility mapping at haraz watershed, iran. *Nat. Hazards* **2012**, *63*, 965–996. [CrossRef]

17. Pradhan, B. Use of gis-based fuzzy logic relations and its cross application to produce landslide susceptibility maps in three test areas in malaysia. *Environ. Earth Sci.* **2011**, *63*, 329–349. [CrossRef]

18. Bai, S.; Lü, G.; Wang, J.; Zhou, P.; Ding, L. Gis-based rare events logistic regression for landslide-susceptibility mapping of lianyungang, china. *Environ. Earth Sci.* **2011**, *62*, 139–149. [CrossRef]

19. Gorsevski, P.V.; Gessler, P.E.; Foltz, R.B.; Elliot, W.J. Spatial prediction of landslide hazard using logistic regression and roc analysis. *Trans. GIS* **2006**, *10*, 395–415. [CrossRef]

20. Pradhan, B.; Mansor, S.; Pirasteh, S.; Buchroithner, M.F. Landslide hazard and risk analyses at a landslide prone catchment area using statistical based geospatial model. *Int. J. Remote Sens.* **2011**, *32*, 4075–4087. [CrossRef]

21. Rawat, M.; Joshi, V.; Rawat, B.; Kumar, K. Landslide movement monitoring using gps technology: A case study of bakthang landslide, gangtok, east sikkim, india. *J. Dev. Agric. Econ.* **2011**, *3*, 194–200.

22. Hess, D.M.; Leshchinsky, B.A.; Bunn, M.; Mason, H.B.; Olsen, M.J. A simplified three-dimensional shallow landslide susceptibility framework considering topography and seismicity. *Landslides* **2017**, *08*, 1–21. [CrossRef]

23. Lee, S.; Chwae, U.; Min, K. Landslide susceptibility mapping by correlation between topography and geological structure: The janghung area, korea. *Geomorphology* **2002**, *46*, 149–162. [CrossRef]

24. Zhou, S.; Chen, G.; Fang, L. Distribution pattern of landslides triggered by the 2014 ludian earthquake of china: Implications for regional threshold topography and the seismogenic fault identification. *ISPRS Int. J. Geo-Inf.* **2016**, *5*. [CrossRef]

25. Devkota, K.C.; Regmi, A.D.; Pourghasemi, H.R.; Yoshida, K.; Pradhan, B.; Ryu, I.C.; Dhital, M.R.; Althuwaynee, O.F. Landslide susceptibility mapping using certainty factor, index of entropy and logistic regression models in gis and their comparison at mugling–narayanghat road section in nepal himalaya. *Nat. Hazards* **2013**, *65*, 135–165. [CrossRef]

26. Korea Forest Service. *2013 Detailed Strategy for Primary Policy*; Korea Forest Service: Seoul, Korea, 2013; pp. 12–16.

27. Lee, M.; Park, I.; Won, J.; Lee, S. Landslide hazard mapping considering rainfall probability in inje, Korea. *Geomat. Nat. Hazards Risk* **2016**, *7*, 424–446. [CrossRef]

28. Lee, S.; Hong, S.M.; Jung, H.S. A support vector machine for landslide susceptibility mapping in Gangwon Province, Korea. *Sustainability* **2017**, *9*. [CrossRef]

29. Park, I.; Lee, S. Spatial prediction of landslide susceptibility using a decision tree approach: A case study of the pyeongchang area, Korea. *Int. J. Remote Sens.* **2014**, *35*, 6089–6112. [CrossRef]

30. Lee, S.; Min, K. Statistical analysis of landslide susceptibility at yongin, Korea. *Environ. Geol.* **2001**, *40*, 1095–1113. [CrossRef]

31. Lee, S.; Evangelista, D. *Landslide Susceptibility Mapping Using Probability and Statistics Models in Baguio City, Philippines*; Department of Environment and Natural Resources: Quezon City, Philippines, 2008.

32. Lee, S.; Pradhan, B. Landslide hazard assessment at cameron highland malaysia using frequency ratio and logistic regression models. *Geophys. Res. Abstr.* **2006**, *8*. SRef-ID: 1607-7962/gra/EGU06-A-003241.

33. Oh, H.-J.; Lee, S.; Soedradjat, G.M. Quantitative landslide susceptibility mapping at pemalang area, Indonesia. *Environ. Earth Sci.* **2010**, *60*, 1317–1328. [CrossRef]

34. Bathrellos, G.D.; Gaki-Papanastassiou, K.; Skilodimou, H.D.; Papanastassiou, D.; Chousianitis, K.G. Potential suitability for urban planning and industry development using natural hazard maps and geological–geomorphological parameters. *Environ. Earth Sci.* **2012**, *66*, 537–548. [CrossRef]

35. Bathrellos, G.D.; Skilodimou, H.D.; Chousianitis, K.; Youssef, A.M.; Pradhan, B. Suitability estimation for urban development using multi-hazard assessment map. *Sci. Total Environ.* **2017**, *575*, 119–134. [CrossRef] [PubMed]

36. Chousianitis, K.; Del Gaudio, V.; Sabatakakis, N.; Kavoura, K.; Drakatos, G.; Bathrellos, G.D.; Skilodimou, H.D. Assessment of earthquake-induced landslide hazard in greece: From arias intensity to spatial distribution of slope resistance demand. *Bull. Seismol. Soc. Am.* **2016**, *106*, 174–188. [CrossRef]

37. Papadopoulou-Vrynioti, K.; Bathrellos, G.D.; Skilodimou, H.D.; Kaviris, G.; Makropoulos, K. Karst collapse susceptibility mapping considering peak ground acceleration in a rapidly growing urban area. *Eng. Geol.* **2013**, *158*, 77–88. [CrossRef]

38. Korean Geotechnical Society. *Cause Investigation of Umyeonsan Landslide and Recovery Measure Establishment Service Final Report*; Korean Geotechnical Society: Seoul, Korea, 2011; p. 3.

39. Seoul Special City. *Supplemental Investigation Report of Causes of Umyeonsan Landslides*; Seoul Special City: Seoul, Korea, 2014; p. 447.

40. Nocutnews. A Part of Umyeonsan. Available online: http://news.Naver.Com/main/read.Nhn?Mode=lsd& mid=sec&sid1=102&oid=079&aid=0002273090 (accessed on 20 June 2017).

41. SBS. An Approval of a mAn-Made Disaster, Not a Natural Disaster, Umyeonsan Landslide. Available online: http://news.Sbs.Co.Kr/news/endpage.Do?News_id=n1002292872&plink=oldurl (accessed on 20 June 2017).

42. Jones, D.; Brunsden, D.; Goudie, A. A preliminary geomorphological assessment of part of the karakoram highway. *Q. J. Eng. Geol. Hydrogeol.* **1983**, *16*, 331–355. [CrossRef]

43. Beven, K.; Kirkby, M.J. A physically based, variable contributing area model of basin hydrology/un modèle à base physique de zone d'appel variable de l'hydrologie du bassin versant. *Hydrol. Sci. J.* **1979**, *24*, 43–69. [CrossRef]

44. Moore, I.D.; Grayson, R.; Ladson, A. Digital terrain modelling: A review of hydrological, geomorphological, and biological applications. *Hydrol. Process.* **1991**, *5*, 3–30. [CrossRef]

45. Wischmeier, W.H.; Smith, D.D. *Predicting Rainfall Erosion Losses—A Guide to Conservation Planning*; U.S. Department of Agriculture: Washington, DC, USA, 1978.

46. Conrad, O.; Bechtel, B.; Bock, M.; Dietrich, H.; Fischer, E.; Gerlitz, L.; Wehberg, J.; Wichmann, V.; Böhner, J. System for automated geoscientific analyses (saga) v. 2.1.4. *Geosci. Model Dev.* **2015**, *8*, 1991–2007. [CrossRef]

47. Böhner, J.; Selige, T. Spatial prediction of soil attributes using terrain analysis and climate regionalisation. *Gottinger Geographische Abhandlungen* **2006**, *115*, 13–28.

48. Hjerdt, K.; McDonnell, J.; Seibert, J.; Rodhe, A. A new topographic index to quantify downslope controls on local drainage. *Water Resour. Res.* **2004**, *40*, W05601–W05606. [CrossRef]

49. Doetterl, S.; Berhe, A.A.; Nadeu, E.; Wang, Z.; Sommer, M.; Fiener, P. Erosion, deposition and soil carbon: A review of process-level controls, experimental tools and models to address c cycling in dynamic landscapes. *Earth-Sci. Rev.* **2016**, *154*, 102–122. [CrossRef]

50. Anthes, R.A. Enhancement of convective precipitation by mesoscale variations in vegetative covering in semiarid regions. *J. Clim. Appl. Meteorol.* **1984**, *23*, 541–554. [CrossRef]

51. Choi, J.W.; Byun, Y.G.; Kim, Y.I.; Yu, K.Y. Support vector machine classification of hyperspectral image using spectral similarity kernel. *J. Korea Soc. Geospat. Inf. Sci.* **2006**, *14*, 71–77.

52. Cortes, C.; Vapnik, V. Support-vector networks. *Mach. Learn.* **1995**, *20*, 273–297. [CrossRef]

53. Yao, X.; Tham, L.; Dai, F. Landslide susceptibility mapping based on support vector machine: A case study on natural slopes of Hong Kong, China. *Geomorphology* **2008**, *101*, 572–582. [CrossRef]

54. Xu, C.; Dai, F.; Xu, X.; Lee, Y.H. Gis-based support vector machine modeling of earthquake-triggered landslide susceptibility in the jianjiang river watershed, China. *Geomorphology* **2012**, *145*, 70–80. [CrossRef]

55. Zhu, J.; Hastie, T. Kernel logistic regression and the import vector machine. *J. Comput. Graph. Stat.* **2005**, *14*, 185–205. [CrossRef]

56. Schölkopf, B.; Smola, A.J.; Williamson, R.C.; Bartlett, P.L. New support vector algorithms. *Neural Comput.* **2000**, *12*, 1207–1245. [CrossRef] [PubMed]

57. Atkinson, P.M.; Tatnall, A. Introduction neural networks in remote sensing. *Int. J. Remote Sens.* **1997**, *18*, 699–709. [CrossRef]

58.   Magoulas, G.D.; Vrahatis, M.N.; Androulakis, G.S. Effective backpropagation training with variable stepsize. *Neural Netw.* **1997**, *10*, 69–82. [CrossRef]

59.   Hines, J.; Tsoukalas, L.H.; Uhrig, R.E. *Matlab Supplement to Fuzzy and Neural Approaches in Engineering*; John Wiley & Sons, Inc.: Hoboken, NJ, USA, 1997.

60.   Fawcett, T. An introduction to roc analysis. *Pattern Recognit. Lett.* **2006**, *27*, 861–874. [CrossRef]

9

Analysis of the Pyroclastic Flow Deposits of Mount Sinabung and Merapi using Landsat Imagery and the Artificial Neural Networks Approach

Prima Riza Kadavi [1], Won-Jin Lee [2] and Chang-Wook Lee [1,*]

[1]   Division of Science Education, Kangwon National University, 1 Kangwondaehak-gil, Chuncheon-si, Gangwon-do 24341, Korea; rizakadavi@gmail.com

[2]   Earthquake Volcano Research Center, Korea Meteorological Administration, 61 16-Gil, Yeouidaebang-ro, Dongjak-Gu, Seoul 07062, Korea; wjleeleo@korea.kr

*   Correspondence: cwlee@kangwon.ac.kr

Abstract: Volcanic eruptions cause pyroclastic flows, which can destroy plantations and settlements. We used image data from Landsat 7 Bands 7, 4 and 2 and Landsat 8 Bands 7, 5 and 3 to observe and analyze the distribution of pyroclastic flow deposits for two volcanos, Mount Sinabung and Merapi, over a period of 10 years (2001–2017). The satellite data are used in conjunction with an artificial neural network method to produce maps of pyroclastic precipitation for Landsat 7 and 8, then we calculated the pyroclastic precipitation area using an artificial neural network method after dividing the images into four classes based on color. Red, green, blue and yellow were used to indicate pyroclastic deposits, vegetation and forest, water and cloud, and farmland, respectively. The area affected by a volcanic eruption was deduced from the neural network processing, including calculating the area of pyroclastic deposits. The main differences between the pyroclastic flow deposits of Mount Sinabung and Mount Merapi are: the sediment deposits of the pyroclastic flows of Mount Sinabung tend to widen, whereas those of Merapi elongated; the direction of pyroclastic flow differed; and the area affected by an eruption was greater for Mount Merapi than Mount Sinabung because the VEI (Volcanic Explosivity Index) during the last 10 years of Mount Merapi was larger than Mount Sinabung.

Keywords: Sinabung eruption; Merapi eruption; pyroclastic flow deposits; Landsat imagery; artificial neural network

1. Introduction

Remote sensing research has used multispectral remote sensing imagery to provide additional data, proving to be a valuable source of spatio-temporal data for some applications. Applications that are widely used are: land cover classification, detection catch archeology, extracting spatial features and classification in a residential area, the extraction of the street, the estimation of urban sprawl automatic mapping feature flow, classification and feature extraction transport, mapping snow cover and the evaluation of geomorphological features [1,2].

Land cover assessment (LC) is very important in planning, monitoring and sustaining the utilization of natural resources. LC has a direct impact on water, atmospheric and soil erosion and is therefore directly related to many globally-important environmental issues [3]. Appropriate knowledge, updated and temporal, about LC is very important to address the issue of unplanned development, environmental degradation, loss of wildlife habitat and depletion of primary agricultural and forest land [4]. Therefore, it is important to evaluate and monitor the LC dynamics resulting from anthropogenic activities and natural phenomena to plan, monitor and sustain the utilization of natural resources [5].

In this study, land cover assessment is used to see the distribution of volcanic eruptions' deposits that may be dangerous to the surrounding area when the eruption occurs. The volcano is a naturally-formed entity on the surface of the Earth that occupies an area and displays volcanism. An eruption is the discharge of magma from within the Earth and can be divided into three types: explosive, effusive and hot spot eruptions. The type of eruption that occurs is influenced by many factors such as magma viscosity, the gas content of the magma, the influence of groundwater and the depth of the magma chamber [6]. The products ejected by volcanic eruptions, which are often catastrophic, can be captured by satellite, optical and radar sensors. Remote sensing images can be utilized to detect the spread of eruptive fumes associated with volcanic eruptions that spread in the atmosphere, pyroclastic deposits, incandescent lava, lahar distribution and dome deformation [7].

We analyze the evaluation of volcano hazard, especially pyroclastic deposit, using land classification with Landsat imagery data and the artificial neural network approach. Several different approaches to the evaluation of volcano hazards can be found in the current literature, including direct and indirect heuristic approaches and deterministic, probabilistic and statistical approaches. Lee et al. (2015) summarized the many analyses of volcano activity based on remote sensing and GIS techniques [8]. Recently, studies on volcano activity assessment made use of remote sensing, and many applied probabilistic models such as satellite imagery. We used land classification in satellite imagery using an artificial neural network method to analyze the area affected by an eruption. Lee at al. (2003) showed that the most frequently-used neural network method is the backpropagation learning algorithm, a learning algorithm of a multilayered neural network that consists of an input layer, hidden layers and an output layer [9]. Many experiments have shown that multilayered neural networks are more accurate for land cover classification than traditional statistical methods [10–13].

2. Study Area

Indonesia is a country located in the ring of fire, an area of frequent geological disasters, including earthquakes, volcanoes, flash floods, landslides and tsunamis. These disasters are harmful, destructive and result in a huge loss of life. The most frequent disaster in Indonesia is a volcanic eruption. Indonesia currently has 129 active volcanoes, and 70 volcanoes have erupted in the last 400 years. The area of land threatened by a volcanic eruption in Indonesia is 16,670 km^2, and around 5,000,000 lives are threatened by volcanic eruptions [14].

There have been several volcanic eruptions in Indonesia (Figure 1a) in the last 10 years. The focus of this research is Mount Sinabung (Figure 1b) and Merapi (Figure 1c), the eruptions of which have caused many casualties. The eruption of Merapi in 2010 was the largest eruption in the last 10 years and killed 347 people. Mount Sinabung has erupted two times in the last 10 years in 2010 and 2013–2017. The period of eruption of 2013 is still ongoing and has killed 16 people [14].

Mount Sinabung is in the Karo Highlands, Karo District, North Sumatra Province, Indonesia. The geographical position of the peak of Mount Sinabung is 3°10' north latitude and 98°23.5' east longitude. The height of Mount Sinabung is 2460 m above sea level, and its volcano type is a stratovolcano [14]. Mount Sinabung was dormant for 400 years, from ~1600–2010 CE. Mount Sinabung finally erupted in 2010, an activity that was predicted from the three earthquakes that struck Sumatra. The earthquakes measured 8.8, 7.9 and 8.4 on the Richter scale and struck in 2005, 2007 and 2007, respectively [15].

Mount Merapi is located on the border of four districts; Sleman, Yogyakarta and Magelang, Boyolali and Klaten, in Central Java Province. Its geographical position lies at 7°32'30" north latitude and 110°26'30" east longitude. Based on its tectonic order, the mountain is located in a subduction zone, where the meeting of the Indo-Australian and Eurasian Plates controls the volcanism in Sumatra, Java, Bali and Nusa Tenggara. The height of Mount Merapi is 2986 m above sea level, and its type is a stratovolcano [14]. Mount Merapi is a very active volcano mountain. Since 1600, Merapi volcano has erupted more than 80 times, and an eruption occurs on average every 4–5 years. Merapi is a volcano cone composed of andesitic-basaltic magma with silica (SiO_2) content ranging from 52–56%. The top morphology is characterized by a horseshoe-shaped crater, in which lava domes grow [16–19].

Figure 1. Map of Indonesia using Google Maps (**a**); location of Mt. Sinabung (**b**) and Merapi (**c**) on Sumatra and Java Islands from USGS earth explorer website [20].

3. Data

A pyroclastic flow inundation map can be generated by field surveys from the crater rim to the furthest extent of the pyroclastic flow after a volcanic eruption. However, a field survey in an active volcano can be dangerous because of the exposure to hazardous gases and sudden activity. In contrast, remote sensing techniques are a useful tool for generating pyroclastic flow deposit maps, which provide a safe, cost-effective alternative to field mapping.

A Landsat image is a picture of the surface of the Earth taken from outside the atmosphere at an altitude of approximately 818 km from the Earth's surface, on a scale of 1:250,000. Each Landsat image covers an area of 185 km^2 so that the aspect of a large object can be identified without exploring all of the surveyed areas [21]. Each color in the satellite image has a meaning; a color on the image indicates whether a reflection value corresponds to vegetation, aquatic bodies or a body of the Earth's surface rock [22]. Therefore, geological interpretation of the Landsat image is based on the difference between the reflection values [23].

This study used satellite data from Landsat 7 ETM+ and Landsat 8. Landsat 7 ETM+ was used to see the changes of pyroclastic deposits from 2007 to 2012, while Landsat 8 was used to see changes in pyroclastic deposits from 2013 to 2017. However, the Landsat 7 ETM+ data suffer from the scan line corrector (SLC)-off phenomenon because of the failure of the SLC in the ETM+ instrument. These images can be restored by gap-filling the scan line using one-dimensional interpolation without any other supplementary data with the gap interpolation and filtering technique [24].

The Landsat 7 and Landsat 8 image data from Mount Sinabung and Mount Merapi were taken from USGS earth explorer website [20] using RGB Bands 7, 4 and 2 to see the pyroclastic deposit in Landsat 7 and RGB Bands 7, 5 and 3 to see the pyroclastic deposits in Landsat 8 (Figures 2 and 3). Landsat image data are taken annually to analyze changes in pyroclastic deposit. Landsat images were selected based on their image quality. Several parameters were used to assess the quality of Landsat images, including sunlight and cloud cover in the image of the object of research. Based on these parameters, data selection is performed in different months and dates in each year.

We select the Landsat image data representing data before eruption and after eruption to see the difference in pyroclastic deposit area of Mount Sinabung and Merapi. For Mount Sinabung, we chose Landsat image data taken 7 February 2007 as data before eruption, 17 November 2011 as data after the first eruption, 1 July 2016 as data after the second eruption and 28 July 2017 as current pyroclastic deposit data. For Mount Merapi, we chose Landsat image data taken 16 June 2007 as data before eruption, 10 May 2011 and 2 November 2014 as data after the big eruption in 2010 and 23 May 2017 as current pyroclastic deposit data.

Figure 2. Landsat 7 (Bands 7, 4 and 2) and Landsat 8 (bands 7, 5 and 3) imagery data of Mount Sinabung taken in the year shown below each image. We can see the changes of pyroclastic deposits of Mount Sinabung from before eruption (**a**) 7 February 2007, after the first eruption on (**b**) 17 November 2011, after the second eruption on (**c**) 7 July 2016, and current pyroclastic deposits on (**d**) 28 July 2017, marked with grey color in the middle of the image.

Figure 3. Landsat 7 (Bands 7, 4, 2) and Landsat 8 (Bands 7, 5, 3) imagery data of Mount Merapi taken in the year shown below each image. We can see the changes of pyroclastic deposits of Mount Merapi from before eruption on (**a**) 16 June 2007, and also during the time after the big eruption in 2010, which can see on (**b**) 10 May 2011 and (**c**) 2 November 2014, and current pyroclastic deposits on (**d**) 23 May 2017, marked with dark grey color in the middle of the image.

4. Methodology

A neural network model belongs to the branch of artificial intelligence generally referred to as artificial neural networks (ANNs). ANNs teach a system to execute a task, instead of using a computational programming system to do defined tasks. To perform such tasks, an artificial intelligence system (AI) is generated; a pragmatic model that can quickly and precisely find the patterns buried in data that represent useful knowledge. Neural networks are one example of these AI models. In the area of medical diagnosis relationships with dissimilar data, artificial intelligence is one of the most available techniques [25].

In this method, the analyst first determines some training area (sample area) on the image as a class of the appearance of a particular object. This determination is based on the analyst's knowledge of the region of land cover areas. The pixel values in the sample area are then used by computer software as the key to recognizing other pixels. Areas that have similar pixel values are entered into predefined classes [10]. Therefore, in this method, the system first identifies the class of information, and this class is then used to determine the spectral class that represents the class of information.

An ANN is a mathematical model that has been applied for the identification, modeling, optimization, forecasting, prediction and control of complex systems [26,27]. It can be trained for performing a particular task based on available empirical data [28]. ANN is a nonparametric approach that has advantages over statistical classification techniques and has been widely used in different LC studies in recent years [29,30]. The NN classifier adopted in this study is a nonlinear layered feed-forward model with standard backpropagation for supervised learning. To perform the ANN using ENVI software, the logistic activation method was used, and one hidden layer was selected. The training threshold contribution and training momentum fields were set to a value of 0.9. Finally, the training rate field, training RMS exit criteria field and number of training iterations were set to values of 0.2, 0.1 and 100, respectively.

The links with the neurons located in the so-called hidden neuron layer then take different weights and are trained depending on the required output, thus modeling complex relationships among variables [31]. The system requires feed-forward and backpropagation processes to allow the network to be trained. The visualization of this stage is accomplished through error analysis. If the error becomes smaller and asymptotic, the network will be ready to receive new input data and predict the output [32].

The ANN models used in this study are of the multilayer perceptron ANN type [33]. The architecture is as shown in Figure 4. The input layer is Region 1 (red color as pyroclastic deposits), Region 2 (blue color as water), Region 3 (green color as forest) and Region 4 (yellow color as farmland). In each case, the training of the proposed network was performed with a backpropagation algorithm, which is a supervised learning procedure. It uses the method of gradient descent for minimizing the global quadratic error of the output calculated by the network [32].

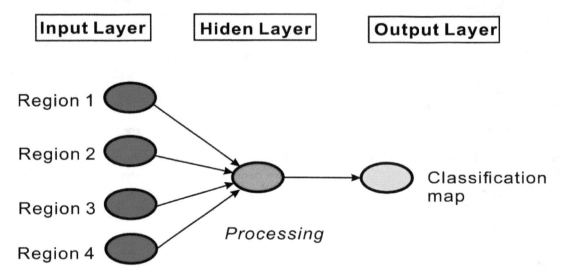

Figure 4. Artificial neural network model used for data analysis. Input layer data Regions 1–4, which are select in ENVI. The output layer is the classification map.

5. Result

5.1. Classification of Landsat Images

The pyroclastic dimension area was analyzed using supervised classification using the artificial neural network approach in the ENVI program. We used the gap-filled Landsat data to delineate the pyroclastic deposit surrounding areas through supervised classification. We established four classes for Mount Sinabung and Mount Merapi. The focus of this study is the pyroclastic deposits, which are marked as red. The classes used on Mt. Sinabung are forest (green), farmland (yellow), water and shadow (blue) and pyroclastic deposit (red). Mt. Merapi is divided into forest (green), farmland (yellow), cloud (blue) and pyroclastic deposit (red).

Land classification on Mount Sinabung using the artificial neural network approach is shown in Figure 5. The land classification is divided into four classes based on its color: green as forest, yellow as farmland, blue as water and shadow and red as pyroclastic precipitate. We selected Landsat image data in this study taken on 7 February 2007 as data before eruption, 17 November 2011 as data after the first eruption, 1 July 2016 as data after the second eruption and 28 July 2017 as current pyroclastic deposit data. Based on the supervised classification maps of Mt. Sinabung (Figure 5), the pyroclastic flow depositional area was primarily in the southern region of Mt. Sinabung before the first eruption in 2010. However, the pyroclastic flow deposit migrated to the eastern part of the volcano after the 2010 eruption (Figure 5a,b). The next eruption was in 2013–2017 (Figure 5b,c), and from the land classification data, it can be seen that after the second eruption on 1 July 2016, there was a very big change in the pyroclastic deposits of Mount Sinabung. The pyroclastic precipitate flows to the southeast and extends downward with a shape resembling a landslide. On 28 July 2017, it was seen that pyroclastic deposits did not change too much from the previous year.

Figure 5. Supervised classification maps from ANN processing in the ENVI program of the Mount Sinabung Landsat Imagery taken on: (**a**) 7 February 2007; (**b**) 17 November 2011; (**c**) 1 July 2016; and (**d**) 28 July 2017.

Land classification on Mount Merapi using the artificial neural network approach is shown in Figure 6. The land classification is divided into four classes based on its color: green as forest, yellow as farmland, blue as cloud and red as pyroclastic precipitate. We selected Landsat image data in this study taken on 16 June Landsat 2007 as data before eruption, 10 May 2011 and 2 November 2014 as data after the big eruption in 2010 and 23 May 2017 as current pyroclastic deposits. Based on the supervised classification maps of Mount Merapi (Figure 6), the pyroclastic flow depositional area primarily existed in the southwestern region before the eruption in 2010. However, the pyroclastic flow deposit migrated to the southern part of the volcano after the 2010 eruption.

Figure 6. Supervised classification maps using ANN processing in the ENVI program of Mount Merapi Landsat imagery data taken on: (**a**) 16 June 2007; (**b**) 10 May 2011; (**c**) 2 November 2014; and (**d**) 23 May 2017.

5.2. Time Series Analysis of Pyroclastic Deposit

Pyroclastic deposits area are constantly changing after an eruption, and therefore, time series analysis is needed to compare the area of pyroclastic deposits in each year. Time series analysis of pyroclastic deposit land classification with Landsat imagery data is used to calculate the area of pyroclastic deposits. We calculate the area of the pyroclastic deposits using ArcMap. The data of the land classification map with a multilayer format are used to calculate pixels in each class [34], because the concern of this research is the pyroclastic deposit, so we only calculate red pixels, which means pyroclastic deposit on the land classification map. The Landsat image has a resolution of 30 × 30 m, so we just multiply the number of pixels by the resolution of the Landsat image to get the area of pyroclastic deposits.

The pyroclastic deposits on Mt. Sinabung increased after the eruption in 2010. Since Mt. Sinabung has been highly active since 2010, the pyroclastic deposit area changes every year. We can see the change of the pyroclastic deposit of Mt. Sinabung each year based on Table 1. Before the eruption, pyroclastic deposits were calculated as 0.3807 km^2 in 2007. The area of pyroclastic deposits increased in 2011 due to the eruption in 2010 to 1.5066 km^2. The increase of pyroclastic precipitate in 2011 was not too big because the eruption that occurred in 2010 was a small eruption. Then, in 2013–2017, the second eruption occurred, and in 2016, the pyroclastic deposit area due to the second eruption of Mount Sinabung in the last ten years increased very significantly to 20.5911 km^2; this can be seen in Figure 7. The current pyroclastic deposit area of Mount Sinabung is estimated at 18.4572 km^2.

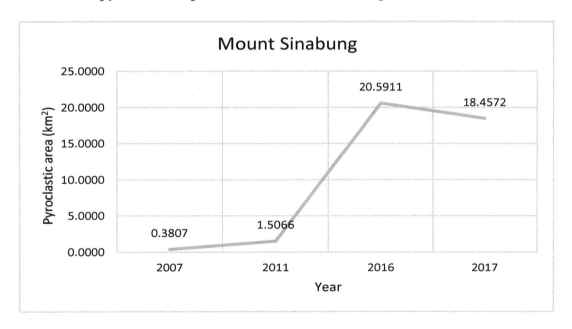

Figure 7. Time series of pyroclastic flow deposits for Mount Sinabung.

Table 1. Time series of pyroclastic flow deposits.

Mount Sinabung No.	Year	Pyroclastic Deposit Area	
		Pixel	km^2
1	2007	423	0.3807
2	2011	1674	1.5066
3	2016	22,879	20.5911
4	2017	20,508	18.4572

Mount Merapi erupted in 2006; thus, the pyroclastic deposit of Mount Merapi was still high in 2007, as it was calculated to be 16.443 km^2 in Table 2. In 2010, Merapi erupted with VEI (Volcanic Explosivity

Index) 4. Therefore, much volcanic material was ejected and caused damage. The pyroclastic deposits caused by the eruption in 2010 were very big, increasing two-fold from previous eruptions (Figure 8). We can calculate the pyroclastic deposit of by the 2010 eruption by 2011 data; it is 38.2536 km^2. After the big eruption of 2010, the pyroclastic deposits area was decreased, calculated for Landsat 2014 as 20.5911 km^2. The current pyroclastic deposit area of Mount Sinabung is estimated at 18.4572 km^2.

Figure 8. Time series of pyroclastic flow deposit for Mount Merapi.

Table 2. Time series of pyroclastic flow deposit.

Mount Merapi. No	Year	Pyroclastic Deposit Area	
		Pixel	km^2
1	2007	18,270	16.443
2	2011	42,504	38.2536
3	2014	34,708	31.2372
4	2017	28,543	25.6887

6. Discussion

The supervised classification neural network analysis was used to view and calculate the area of pyroclastic flow deposits. From the Landsat imagery data, differences in appearance can be seen between pyroclastic flow deposits, forests, lakes and plantations. Landsat image data were used as an object for supervised neural network classification analysis. The supervised neural network classification divides the area based on the pixels that have been selected according to the class. The results of the supervised neural network analysis show that there are differences in the distribution of pyroclastic flow deposits between Mount Merapi and Mount Sinabung. The apparent differences in the Landsat imagery data that have been found by classification by the supervised neural network are the direction of flow, the area of deposit and the distribution of pyroclastic flow deposits. The differences in the distribution of pyroclastic flow deposits between Mount Sinabung and Mount Merapi were caused by several influential factors, such as the formation of volcanoes, regional geological conditions, volcano magmatic properties, volcanic type, lava domes and the Volcanic Explosivity Index.

The geological conditions of Sumatra affect the formation of Mt. Sinabung. The outline of Sumatra Island consists of three tectonic systems, namely the Sumatra Subduction System, the Mentawai Fault system and the Sumatra Fault System. Based on the geological reconstruction by Robert Hall (2000) [35], the initial formation of the Sumatra region began ~50 million years ago (early Eocene). There are at least 19 C-section segments, each with lengths of ±60–200 km, in the Sumatra Fault System, which has a total length of ±1900 km. Lake Toba on the island of Sumatra is evidence of a supervolcano and is the remains of the largest eruptions of Caldera Crescent (scale: 8 VEI).

The eruptions of Mount Sinabung eject many volcanic materials, one of which is pyroclastic flow. The pyroclastic flow deposits of Mount Sinabung spread to the southeast, tend to widen and are not transported far from the crater of Mount Sinabung. These characteristics result from the acid-intermediate (andesitic-rhyolitic) [17] magma of Mount Sinabung, which tends to be thick and prevents the pyroclastic flow deposit from being transported far. The other factor is due to the shape of the slopes of Mt. Sinabung, where the pyroclastic flows and the direction of the river in Sinabung valley are almost perpendicular. Mount Sinabung is a Type B mountain, which means there was no track record of eruption before it erupted in 2010 [14]. The eruption of 2010 was a phreatic-type eruption, as defined by the presence of volcanic dust. A phreatic eruption is the process of magma escaping to the surface because of the influence of steam caused by the direct or indirect interaction between water and magma [36]. The second eruption of Mount Sinabung was larger than that in 2010, which could have been caused by the formation of an old dome on Mount Sinabung after the first eruption. The material ejected in the next eruption mixed with the material on the surface or around the crater, destroying the already formed lava dome and forming a new lava dome. The eruptions of Mount Sinabung are generally rated as VEI 2, which means the eruption is relatively small [37]. However, the eruptions are continuous and interfere with the activity of the residents. A larger eruption is probable to occur since Sinabung is located in the Semangko Fault Zone, the largest fault on Sumatra Island.

Based on seismic data, Mount Merapi is estimated to have a magma pouch because no seismic data from inside the seismic zone exist, indicating that there is a soft material between harder materials in the seismic zone. This soft material is thought to be the magma pocket of Mount Merapi [19]. This hypothesis is supported by the fact that the temperatures around the top of the Woro and Gendol plates can reach 830 °C, which means there is a sufficiently shallow source of heat. It is estimated that these magma pockets arise because of this basic fault, where magma can accumulate. This magma pocket is thought to act as a valve that slows the upward magma migration of the magma chamber. Therefore, the scale of the eruption is reduced [18]. The absence of an earthquake below 5 km strengthens the case for the existence of a magma kitchen at a depth of ~8 km, as proposed by Beaducel (1998) based on tiltmeter and GPS data modeling [6].

Mount Merapi erupts almost every five years because of the existence of magma pockets in the magma chamber of Merapi Volcano, which are likely to have been formed by a major eruption in the past. The major eruptions recorded on Mount Merapi in the 21st Century were in 2006 and 2010, with the 2010 eruption causing more than 350 casualties [14].

Landsat imagery data show that the 2010 eruption issued pyroclastic material from inside Merapi's crater that spread around the Mt. Merapi area. Landsat 8 image data using Bands 4, 3 and 2 in the natural view show that the direction of the pyroclastic flow was to the south and southwest and that Merapi's pyroclastic flow sediment tends to elongate because it is transported far from the peak. The long range of the flow could be caused by the valleys that run in the same direction as the slope, as well as the basaltic-andesitic (intermediate-base) magma type. This type of magma tends to be dilute and can be transported far, along with the outflow of pyroclastic flows from the eruption. Merapi's eruption type can be categorized as a weak volcanic type that is not so explosive, but pyroclastic flow almost always occurs after an eruption. The eruptive activity of Merapi can be seen to undergo a long process, from the initial lava dome formation to incandescent lava and hot clouds (pyroclastic flow). The eruption in 2010 had a strong VEI value of four, and lahar flows occurred in addition to a pyroclastic flow [16]. Lahar flows damage buildings and bridges because of their high speed and because they carry material such as stone and sand. In addition, volcanic ash is widely distributed, the high silica content of which makes it dangerous to breathe. The volcanic ash sediment can be seen in Landsat 7 imagery data from 2010.

7. Conclusions

Mount Sinabung first erupted in 2010 after a long break of over 400 years, and this first eruption was followed by a series of six eruptions between 2013 until 2016. The biggest eruption occurred in 2014 and had a VEI of two, causing ~16 deaths and the destruction of farmland and residents' houses by pyroclastic flows. Mount Merapi erupts roughly every five years because of the high magma activity in its magma chamber, as well as the magma pocket that formed during a past eruption. We analyzed the differences in the pyroclastic deposit flows between Mt. Sinabung and Mt. Merapi, using Landsat 7 and 8 imagery with Bands 7, 4, 2 and 7, 5, 3, respectively, to see the pyroclastic precipitates and deposits of sand and rocks.

An artificial neural network method was used to analyze the pyroclastic precipitates and the pyroclastic precipitate area using Landsat imagery data taken annually from 2007 to 2017 for each volcano. The analysis showed that the pyroclastic sediment of Mount Sinabung spread to the east at the beginning of eruption and then widened to the southeast and south after the series of eruptions took place. The continuous eruption of Mount Sinabung caused the formation of pyroclastic deposits that accumulated around the body of the mountain. The pyroclastic sediment of Mount Sinabung tends to widen and is not transported far because of the direction of river flow on the mountain, which is not in the direction of the mountain's slope. In addition, the andesitic-rhyolitic (intermediate-acid) magma composition of Mount Sinabung prevents pyroclastic flow from being transported far, because the lava tends to be thick. In contrast, the pyroclastic flows of Mount Merapi flow to the south and to the southwest along the slope and river flow direction of Mount Merapi. Thus, the pyroclastic sediment of Mount Merapi does not accumulate much in the body of Merapi Volcano, but flows following the slope of the volcano. The composition of magma on Mount Merapi is also different from Mount Sinabung; Mount Merapi's magma is andesitic-basaltic (intermediate-base), so the flowing lava is thin and easily transported. Although the maximum VEI of Merapi is larger than that of Sinabung (VEIs of four and two, respectively), the pyroclastic precipitates of Mount Sinabung's eruption are wider than those of Mount Merapi (~18 and ~17 km^2, respectively).

Acknowledgments: This research was supported by the Basic Science Research Program through the National Research Foundation of Korea (NRF) funded by the Ministry of Science, ICT & Future Planning (NRF-2015M1A3A3A02013416) and the Korea Meteorological Administration Research and Development Program under Grant KMIPA (2015–3071).

Author Contributions: Prima Riza Kadavi applied the artificial neural networks model for detecting pyroclastic flow deposit area and wrote the paper. Won-Jin Lee designed the volcanic experimental equation corresponding to volcano eruption. Chang-Wook Lee suggested the idea, organized the paper work and interpreted the results of volcanic activities. All of the authors contributed to the writing of each part.

References

1.   Elhag, M. Quantification of Land Cover Changes Based on Temporal Remote Sensing Data for Ras Tarout Area, Saudi Arabia. *Indian J. Geo-Mar. Sci.* **2016**, in press.
2.   Elhag, M. Evaluation of Different Soil Salinity Mapping Using Remote Sensing Techniques in Arid Ecosystems, Saudi Arabia. *J. Sens.* **2016**, *2016*, 7596175. [CrossRef]
3.   Koomen, E.; Stillwell, J. *Modelling Land-Use Change*; Springer: New York, NY, USA, 2007.
4.   Anderson, J.R. *A Land Use and Land cover Classification System for Use with Remote Sensor Data Vol 964*; US Government Printing Office: Washington, DC, USA, 1976.
5.   Ullah, S.; Farooq, M.; Shafique, M.; Siyab, M.A.; Kareem, F.; Dees, M. Spatial assessment of forest cover and land-use changes in the Hindu-Kush mountain ranges of north ern Pakistan. *J. Mt. Sci.* **2016**, *13*, 1229–1237. [CrossRef]

6.   Beauducel, F. Structures Mechanical Behaviour of Merapi Volcano, Java: A Methodological Approach of the Deformation Field. Ph.D. Thesis, University Denis Diderot Paris VII, Paris, France, 1998.
7.   Lee, C.W.; Lu, Z.; Won, J.S.; Jung, H.S.; Dzurisin, D. Dynamic deformation of Seguam Volcano, Alaska, 1992–2008, from multi-interferogram InSAR processing. *J. Volcanol. Geotherm. Res.* **2013**, *260*, 43–51. [CrossRef]
8.   Lee, C.W.; Kim, J.W.; Jung, H.S.; Lee, J.K.; Degrandpe, K.; Lu, Z. Volcanic activity analysis of Mt. sinabung in Indonesia using remote sensing and GIS techniques. In Proceedings of the Near-Surface Asia Pacific Conference, Waikoloa, HI, USA, 7–10 July 2015; pp. 150–153.
9.   Lee, S.; Ryu, J.H.; Min, K.; Won, J.S. Landslide Susceptibility Analysis Using GIS and Artificial Neural Network. *Earth Surf. Process. Landf.* **2003**, *28*, 1361–1376. [CrossRef]
10.   Canziani, G.; Perrati, R.; Marinelli, C.; Dukatz, F. Artificial Neural Network and Remote Sensing in the Analysis of the Highly Variable Pampean Shallow Lakes. *Math. Biosci. Eng.* **2008**, *5*, 691–711. [CrossRef] [PubMed]
11.   Chen, K.; Yen, S.; Tsay, D. Neural classification of SPOT imagery through integration of intensity and fractal information. *Int. J. Remote Sens.* **1997**, *18*, 763–783. [CrossRef]
12.   Saravanan, K.; Sasithra, S. Review on Classification Based On Artificial Neural Networks. *Int. J. Ambient Syst. Appl. (IJASA)* **2014**, *2*. [CrossRef]
13.   Serpico, S.B.; Bruzzone, L.; Roli, F. An experimental comparison of neural and statistical non-parametric algorithms for supervised classification of remote sensing images. *Pattern Recognit. Lett.* **1996**, *17*, 1331–1341. [CrossRef]
14.   Ministry of Energy and Natural Resources of the Republic of Indonesia. Available online: http://www.vsi.esdm.go.id/ (accessed on 10 July 2017).
15.   Lupi, M.; Miller, S.A. Short-lived tectonic switch mechanism for long-term pulses of volcanic activity after mega-thrust earthquakes. *Solid Earth* **2014**, *5*, 13–24. [CrossRef]
16.   Hidayati, S.; Ishihara, K.; Iguchi, M.; Ratdomopurbo, A. Focal mechanism of volcano-tectonic earthquakes at merapi volcano, Indonesia. *Indones J. Phys.* **2008**, *19*, 75–82.
17.   Iguchi, M.; Ishira, K.; Hendrasto, M. Learned from 2010 Eruptions at Merapi and Sinabung Volcanoes in Indonesia. *Annu. Disaster Prev. Res. Inst. Kyoto Univ.* **2011**, *54*, 185–194.
18.   Ratdomopurbo, A.; Poupinet, G. An overview of the seismicity of merapi volcano (Java, Indonesia), 1983–1994. *J. Volcanol. Geotherm. Res.* **2000**, *100*, 193–214. [CrossRef]
19.   Surono, M.; Jousset, P.; Pallister, J.; Boichu, M.; Buongiorno, M.F. The 2010 explosive eruption of Java's Merapi volcano—A '100-year' event. *J. Volcanol. Geotherm. Res.* **2012**, *241–242*, 121–135. [CrossRef]
20.   Landsat 7 ETM+ and Landsat 8 Imagery. Available online: https://earthexplorer.usgs.gov/ (accessed on 10 July 2017).
21.   Richards, J.A. *Remote Sensing Digital Image Analysis*; Springer: Berlin, Germany, 1999; Volume 240.
22.   Eom, J.; Lee, C.W. Analysis on the Area of Deltaic Barrier Island and Suspended Sediments Concentration in Nakdong River Using Satellite Images. *Korean J. Remote Sens.* **2017**, *33*, 201–211.
23.   Lee, S.; Kim, G.; Lee, C.W. Preliminary Study for Tidal Flat Detection in Yeongjong-do according to Tide Level using Landsat Images. *Korean J. Remote Sens.* **2016**, *32*, 639–645. [CrossRef]
24.   Lee, C.W.; Cho, M.; Choi, Y.J. Method and System for Correction of Optical Satellite Image. U.S. Patent US9,117,276 B2, 25 Agusut 2015.
25.   Rosenblatt, F. *Principles of Neuro Dynamics*; Spartan: New York, NY, USA, 1962.
26.   Ahmed, E.A.; Adam, M.E.-N. Estimate of global solar radiation by using artificial neural network in Qena, Upper Egypt. *J. Clean Technol.* **2013**, *1*, 148–150. [CrossRef]
27.   Xiu, L.-N.; Liu, X.-N. Current status and future direction of the study on artificial neural network classification processing in remote sensing. *Remote Sens. Technol. Appl.* **2003**, *18*, 339–345.
28.   Kia, M.; Pirasteh, S.; Pradhan, B.; Mahmud, A.; Sulaiman, W.; Moradi, A. An artificial neural network model for flood simulation using GIS: Johor River Basin, Malaysia. *Environ. Earth Sci.* **2012**, *67*, 251–264. [CrossRef]
29.   Mishra, V.N.; Rai, P.K.; Mohan, K. Prediction of land use changes based on land change modeler (LCM) using remote sensing: A case study of Muzaffarpur (Bihar), India. *J. Geogr. Inst. Jovan Cvijic* **2014**, *64*, 111–127. [CrossRef]
30.   Rokni, K.; Ahmad, A.; Solaimani, K.; Hazini, S. A new approach for surface water change detection: Integration of pixel level image fusion and image classification techniques. *Int. J. Appl. Earth Obs. Geoinf.* **2015**, *34*, 226–234. [CrossRef]

31. McClelland, J.L.; Rumelhart, D.E. *Explorations in Parallel Distributed Processing*; MIT Press: Cambridge, UK, 1988.
32. Rumelhart, D.E.; Hinton, G.E.; Willams, R.J. Learning Internal Representations by Error Propagation. In *Parallel Distributed Processing, Explorations in the Microestruture of Cognition. Vol. 1: Foundations*; Rumelhart, D.E., McClelland, J.L., Eds.; MIT Press: Cambridge, UK, 1986.
33. Minsky, M.L.; Papert, S.A. *Perceptrons, Expanded Edition*; MIT Press: Cambridge, UK, 1969.
34. Lee, S.; Yang, M.; Lee, C.W. Time Series Analysis of Area of Deltaic Barrier Island in Nakdong River Using Landsat Satellite Image. *J. Korean Soc. Surv. Geod. Photogramm. Cartogr.* **2016**, *34*, 457–469. [CrossRef]
35. Robert, H. Cenozoic geological and plate tectonic evolution of SE Asia and the SW Pacific: Computer-based reconstruction, model and animations. *J. Asian Earth Sci.* **2002**, *20*, 353–431.
36. Mulyana, A.R. *Mapping of Disaster Prone Areas Mount Sinabung, Center for Volcanology and Geological Hazard Mitigation*; Ministry of Energy and Mineral Resources Indonesia: Jakarta, Indonesia, 2010.
37. Lee, C.W.; Lu, Z.; Jin, W.K. Monitoring Mount Sinabung in Indonesia Using Multi-Temporal InSAR. *Korean J. Remote Sens.* **2017**, *33*, 37–46. [CrossRef]

10

Optimized Neural Architecture for Automatic Landslide Detection from High-Resolution Airborne Laser Scanning Data

Mustafa Ridha Mezaal [1], **Biswajeet Pradhan** [1,2,*], **Maher Ibrahim Sameen** [1],
Helmi Zulhaidi Mohd Shafri [1] **and Zainuddin Md Yusoff** [1]

[1] Department of Civil Engineering, Faculty of Engineering, Universiti Putra Malaysia,
     Serdang 43400, Malaysia; gismustafa87@gmail.com (M.R.M.); maherrsgis@gmail.com (M.I.S.);
     helmi@eng.upm.edu.my (H.Z.M.S.); zmy@upm.edu.my (Z.M.Y.)
[2] School of Systems, Management and Leadership, Faculty of Engineering and Information Technology,
     University of Technology Sydney, Building 11, Level 06, 81 Broadway,
     Ultimo NSW 2007 (P.O. Box 123), Australia
*   Correspondence: Biswajeet24@gmail.com or biswajeet@lycos.com

Academic Editor: Saro Lee

Abstract: An accurate inventory map is a prerequisite for the analysis of landslide susceptibility, hazard, and risk. Field survey, optical remote sensing, and synthetic aperture radar techniques are traditional techniques for landslide detection in tropical regions. However, such techniques are time consuming and costly. In addition, the dense vegetation of tropical forests complicates the generation of an accurate landslide inventory map for these regions. Given its ability to penetrate vegetation cover, high-resolution airborne light detection and ranging (LiDAR) has been used to generate accurate landslide maps. This study proposes the use of recurrent neural networks (RNN) and multi-layer perceptron neural networks (MLP-NN) in landscape detection. These efficient neural architectures require little or no prior knowledge compared with traditional classification methods. The proposed methods were tested in the Cameron Highlands, Malaysia. Segmentation parameters and feature selection were respectively optimized using a supervised approach and correlation-based feature selection. The hyper-parameters of network architecture were defined based on a systematic grid search. The accuracies of the RNN and MLP-NN models in the analysis area were 83.33% and 78.38%, respectively. The accuracies of the RNN and MLP-NN models in the test area were 81.11%, and 74.56%, respectively. These results indicated that the proposed models with optimized hyper-parameters produced the most accurate classification results. LiDAR-derived data, orthophotos, and textural features significantly affected the classification results. Therefore, the results indicated that the proposed methods have the potential to produce accurate and appropriate landslide inventory in tropical regions such as Malaysia.

Keywords: landslide detection; LiDAR; recurrent neural networks (RNN); multi-layer perceptron neural networks (MLP-NN); GIS; remote sensing

1. Introduction

Landslides are dangerous geological disasters with catastrophic effects on human lives and properties. Landslides occur with high frequency in mountainous and hilly areas, such as the Cameron Highlands in Malaysia. Landslide incidence is related to a cluster of triggering factors, such as intense rainfall, volcanic eruptions, rapid snowmelt, elevated water levels, and earthquakes. Landslide inventory maps are crucial for measuring the magnitude and analyzing the susceptibility, hazard, and risk of earthquakes [1,2], as well as for examining distribution patterns and predicting the landscapes affected by landslide [3]. Mapping a landslide inventory in tropical areas is challenging because the dense vegetation cover in these regions obscures underlying landforms [4]. Moreover, the majority of available conventional landslide detection techniques are not rapid and accurate enough for inventory mapping given the rapid vegetation growth in tropical regions. Therefore, inventory mapping requires the use of more rapid and accurate techniques, such as light detection and ranging (LiDAR) [5], which uses active laser transmitters and receivers to acquire elevation data. In addition, LiDAR has the unique capability to penetrate densely vegetated areas [5] and provide detailed information on terrains with high point density. Moreover, it depicts ground surface features and provides useful information on topographical features in areas where landslide locations are obscured by vegetation cover [6,7].

Numerous studies have applied a multiresolution segmentation algorithm for the remote sensing of land features [8]. This algorithm requires the identification of three parameters (i.e., scale, shape, and compactness); the values of these parameters can be determined using the traditional trial-and-error method, which is very time consuming and laborious [5]. Moreover, using the algorithm to delineate the boundary of an object at different scales remains challenging [9]. Thus, optimal parameters for segmentation should be identified via semiautomatic and automatic approaches [10–12]. The automatic selection of segmentation parameters requires the use of the advanced supervised approach presented in [13].

Processing a large number of irrelevant features causes overfitting [14]. By contrast, the best classification results are obtained by selecting the most relevant feature [15]. Landslide identifcation in a particular area can be improved by selecting the most significant feature [15,16]. As shown in [2], selecting the most significant feature facilitates the differentiation of landslides from non-landslides. Accuracy can be improved by decreasing the number of features, as recommended in [17]. The efficiency of feature selection techniques for landslide detection has been proven in [18–20].

The neural network (NN) is effective in remote sensing applications [21], particularly in solving different image classification problems [22] specified by nonlinear mathematical fitting for function approximation. NN architectures are classified into the recurrent neural network (RNN), back-propagation neural network, probability neural network, and multilayer perceptron neural network (MLP-NN). NN-based classifiers can adapt to different types of data and inputs, and can overcome the issue of mixed pixels by providing fuzzy output and fit with multiple images [23,24]. These classifiers include parallel computation, which is superior to statistical classification approaches because it is non-parametric and does not require the prior knowledge of a distribution model for input data [25]. Moreover, NN-based classifiers can evaluate non-linear relationships between the input data and desired outputs and are distinguished by their fast generalization capability [26]. NN-based classifiers have been successfully in function approximation, prediction, pattern recognition, landslide detection, image classification, automatic control, and landslide susceptibility [27–32]. Authors of [33] found that MLP-NN can be effectively applied in landslide detection using multi-source data. The RNN model can effectively predict landslide displacement [34]. The above neural architecture techniques have not been extensively used for landslide detection using only LiDAR data. This research gap urged us to apply the RNN and MLP-NN models in landslide detection based on very high-resolution LiDAR data. To achieve this objective, we optimized multiresolution segmentation parameters via a supervised approach. Using the correlation-based feature selection (CFS) algorithm, we selected the most significant feature from high-resolution airborne laser scanning data.

2. Study Area

This study was performed in a small section of the Cameron Highlands, which is notorious for its frequent occurrence of landslides. The study area covers an area of 26.7 km^2. It is located on northern peninsular Malaysia within the zone comprising latitudes 4°26'3'' to 4°26'18'' and longitudes 101°23'48'' to 101°24'4'' (Figure 1). The annual average rainfall and temperature in this region are approximately 2660 mm and 24/14 °C (daytime/nighttime temperatures), respectively. Approximately 80% of its area is forested with a flat (0°) to hilly (80°) land form.

Figure 1. Location of the study area. The red boundary represents the analysis area and the yellow boundary represents the test area.

Two sites were selected to implement and test the proposed models (Figure 1). All the prerequisite considerations were taken in to account during test site selection to avoid missing any land cover classes. To obtain an accurate map of the analysis and test sites, the training sample size was measured via the stratified random sample method.

3. Methodology

3.1. Overall Methodology and Pre-Processing

LiDAR data and landslide inventories were first pre-processed to eliminate noise and outliers. A high-resolution digital elevation model (DEM) at 0.5 m was then derived from LiDAR point clouds to generate other LiDAR-derived products (i.e., slope, aspect, height or (normalized digital surface model (nDSM)), and intensity. LiDAR-derived products and orthophotos were then composited by rectifying their geometric distortions to generate one coordinate system and were finally prepared in geographic information system (GIS) for feature extraction. Suitable parameters (scale, shape, and compactness) at various levels of segmentation were obtained via a supervised approach, i.e., a fuzzy-based segmentation parameter optimizer (FbSP optimizer) [13]. The stratified random method was used to evaluate the training dataset in accordance with the procedure in [35]. The correlation-based selection algorithm (CFS) [36] was used to rank features from the most to least important. RNN and MLR-NN models were applied to detect landslide locations. The results of the models were validated using a 10-fold cross validation method. In addition, the models were evaluated in another part of the study area (i.e., the test site). Slope and aspect layers were overlaid with the results to identify other landslide characteristics (i.e., direction and run off). The study flow is illustrated in Figure 2.

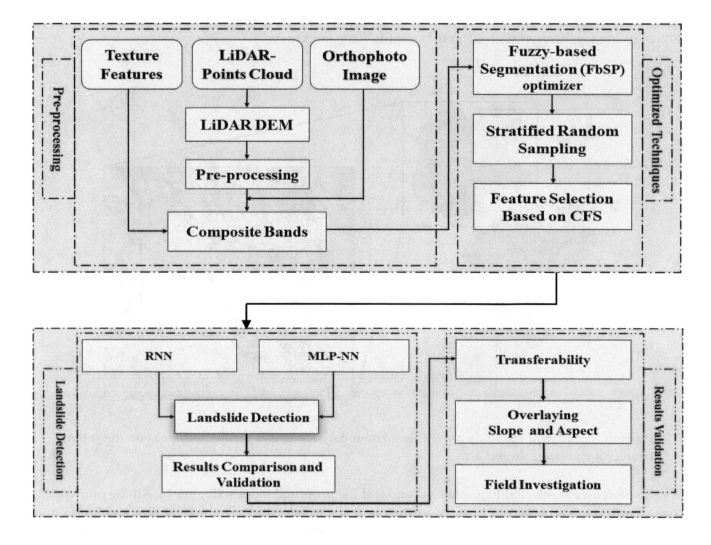

Figure 2. Overview of the proposed method. LiDAR: light detection and ranging; RNN: recurrent neural networks; MLP-NN: multi-layer perceptron neural networks; CFS: correlation-based feature selection; DEM: digital elevation model.

3.2. Landslide Inventory

The landslide inventory; produced previously by Pradhan and Lee, [39] was used to develop the proposed detection method and the total number of landslides is 21 in the study area covering 3781 m^2 (Figure 3).

3.3. Data

LiDAR point-cloud data were collected on 15 January 2015 at a point density of 8 points/m^2 and frequency pulse rate of 25,000 Hz. The absolute accuracy of the data (root-mean square errors) was restricted to 0.15 m and 0.3 m in the vertical and horizontal axes, respectively. Orthophotos were obtained using the same acquisition system that relied on the abovementioned cloud data. A DEM was derived from LiDAR point clouds with a spatial resolution of 0.5 m after non-ground points were removed using inverse distance weighting with a spatial reference of GDM2000/Peninsula RSO. Subsequently, LiDAR-based DEM was used to generate derived layers to facilitate the identification and characterization of landslide locations [37].

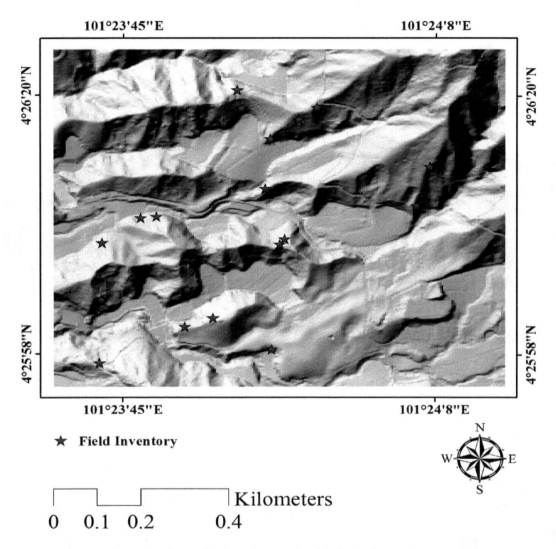

Figure 3. Shows the locations of landslide in the study area.

According to the authors of [38], slope directly and highly affects landslide phenomenology. The authors of [39] also inferred that slope is the principal factor that affects landslide occurrence. The author of [40] indicated that a hillshade map provides a good image of terrain movements, thus facilitating the development of landslide maps. Texture and geometric features are crucial for improving the classification accuracy of landslide mapping [14]. Landslide intensity and texture derived from LiDAR data are affected by the accuracy of landslide detection [9]. The accuracy and capacity of DEM to represent surface features are determined by terrain morphology, sampling density, and the interpolation algorithm [41]. In this study, hillshade, height (nDSM), slope, and aspect were generated from LiDAR-based DEM. As shown in Figure 4, landslide locations were detected using visible bands and texture features.

3.4. Image Segmentation

The sizes and shapes of image objects [42] are determined via image segmentation, the preliminary step in object-based classification. Optimal segmentation parameters depend on the environment under analysis, the selected application, and the underlying input data [8]. Previous studies have used the multiresolution segmentation algorithm with eCognition software for image segmentation [8,9]. Three parameters (scale, shape, and compactness) are defined in this algorithm. According to [5], these parameters can be obtained via the traditional trial-and-error method, which is time consuming and laborious. Therefore, the fuzzy logic supervised approach presented by [13] was adopted in this study.

Figure 4. *Cont.*

Figure 4. LiDAR-derived data; (**A**) Orthophotos; (**B**) digital terrain model (DTM); (**C**) digital surface model (DSM); (**D**) Intensity; (**E**) Height; (**F**) Slope; and (**G**) Aspect.

3.5. Training Sets

The authors of [35] suggested the use of stratified random sampling method to obtain an adequately sized training dataset for every class without any bias during sample selection. Accordingly, the present study adopted stratified random sampling to evaluate training samples and achieve high performance without strong bias. Four classes with different numbers of objects were set as shown in Table 1.

Stratified random sampling is a prerequisite to obtain prior knowledge of the two sites considered for landslide inventory. Hence, segmentation parameters were first optimized. Then, the landslide inventory was overlapped with the segmented layer for object labeling. ArcGIS 10.3 was used to construct sample sets automatically at each optimal scale. Subsequently, stratified random sampling was applied on the labeled objects. This process was performed 20 times at each optimal scale.

Table 1. Number of selected training objects in four classes.

Class Name	Number of the Object for Each Class
Landslide	52
Cut slope	67
Bare soil	80
Vegetation	150

3.6. Correlation-Based Feature Selection

The authors of [15] reported that the selection of only the most relevant features improves the quality of landslide identification and classification. Working with large numbers of features causes numerous problems. As reported in [43] and [14], some of these problems include the slow run time of algorithms due to the consideration of numerous resources, low accuracy when the number of features exceed the number of observation features, and overfitting when irrelevant features are used as inputs. Therefore, the most significant features should be selected to enhance the accuracy of feature extraction. In this study, relevant features were extracted using the CFS algorithm with Weka 3.7 software. Furthermore, the CFS algorithm was applied to all LiDAR-derived data, visible bands, and textural features, and was used to determine the feature subsets required to develop models for landslide identification. The CFS algorithm comprises two basic steps: the ranking of initial features and the elimination of the least important features through an iterative process.

3.7. MLP-NN

NNs are a family of biological learning models in machine learning. The NN model comprises interconnected neurons or nodes, which are structured into layers with random or full interconnections among successive layers [44]. The NN model comprises input, hidden, and output layers that are responsible for receiving, processing, and presenting results, respectively [44]. Each layer contains nodes connected by numeric weights and output signals. The weights are the functions of the sum of the inputs to the node modified by a simple activation function [45]. The possibility of learning is the most important feature that attracts researchers to use NNs.

Back-propagation, which was first proposed by Paul Werbos in 1974 and independently rediscovered by Rumelhart and Parker, is the most common learning algorithm used in NN. It aims to minimize the error function via the iterative approach as shown in Equation (1). NNs have been successfully used in remote sensing applications. However, this model has some limitations, specifically, high computational complexity and overlearning [46,47].

$$E = \frac{1}{2} \sum_{i=1}^{L} (d_i - o_i)^2 \tag{1}$$

where d_i and o_i represent the desired output and the current response of node $"i"$ in the output layer, respectively. $"L"$ is the number of nodes in the output layer. Corrections to weight parameters were calculated and effected with the previous values in the iterative method, as demonstrated in Equation (2):

$$\begin{cases} \Delta w_{i,j} = -\mu \frac{\partial E}{\partial w_{i,j}} \\ \Delta w_{i,j}(t+1) = \Delta w_{i,j} + \alpha \Delta w_{i,j}(t) \end{cases} \tag{2}$$

where delta rule $\Delta w_{i,j}$ is the weight parameter between nodes i and j; μ is a positive constant that controls the amount of adjustment and is referred to as learning rate; α is the momentum factor, which takes a value between 0 and 1; and t is the iteration number. α is referred to as the stabilizing factor because it smoothens quick changes between weights [48].

3.8. RNN

RNNs are designed to model sequences in NNs with feedback connections. They are very powerful in computational analysis and are biologically more reliable than other NN techniques given their lack of internal states. The memory of past activations in RNN is very effective with feedback connections, making them suitable for learning the temporal dynamics of sequential data. RNN is very powerful when used to map input and output sequences because it uses contextual information. However, traditional RNNs face the challenge of exploding or vanishing gradients. Hochreiter and Schmidhuber [49] proposed long short-term memory (LSTM) to tackle this issue.

Hidden units in LSTM are replaced with memory blocks that contain three multiplicative units (input, output, forget gates) and self-connected memory cells to allow for reading, writing, and resetting through a memory block and behavioral control. A single LSTM unit is shown in Figure 5. c_t is the sum of inputs at time step t and its previous time step activations. LSTM updates time step i given inputs x_t, h_{t-1}, and c_{t-1} as reported in [50].

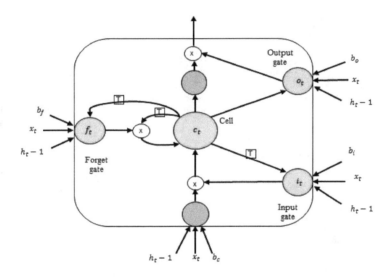

Figure 5. Structure of a memory cell in long short-term memory (LSTM)-RNN.

Input gates:
$$i_t = \sigma(W_{xi} \cdot x_t + W_{hi} \cdot h_{t-1} + W_{ci} \cdot c_{t-1} + b_i) \tag{3}$$

Forget gates:
$$f_t = \sigma\left(W_{xf} \cdot x_t + W_{hf} \cdot h_{t-1} + W_{xf} \cdot x_t + W_{cf} \cdot c_{t-1} + b_f\right) \tag{4}$$

Cell units:
$$c_t = i_t \cdot tanh\left(W_{xc} \cdot x_t + W_{hc} \cdot h_{t-1} + b_c + b_f \cdot c_{t-1}\right) \tag{5}$$

Output gates:
$$o_t = \sigma(W_{xo} \cdot x_t + W_{ho} \cdot h_{t-1} + W_{co} \cdot x_t + b_o) \tag{6}$$

The hidden activation (output of the cell) is also given by a product of the two terms:

$$c_t = o_t \cdot \tanh(c_t) \tag{7}$$

where σ and tanh are an element-wise non-linearity, such as a sigmoid function and hyperbolic tangent function, respectively; W is the weight matrix; x_t refers to input at time step t; t, h_{t-1} represents the hidden state vector of the previous time step; and b_c denotes the input bias vector. The memory cell unit c_t is a sum of two terms: the previous memory cell unit c_{t-1}, which is modulated by f_t and c_t, a function of the current input, and previous hidden state, modulated by the input gate i_t due to i_t

and f_t being sigmoidal. Their values range within [0, 1], and i_t and f_t can be considered as knobs that the LSTM learns to selectively forget its previous memory or consider its current input, whilst o_t is an output gate that learns how much of the memory cell to transfer to the hidden layers.

3.9. Neural Network Models

3.9.1. MLP-NN

This study proposed the network architectures RNN and MLP-NN. Figure 6 depicts the MLP-NN model architecture, which has two hidden layers of 50 hidden units. Ten features were taken as inputs in the model to detect different types of objects, such as landslide, cut slope, bare soil, and vegetation. The MLP-NN model was trained through a back-propagation technique with the Adam optimizer and a batch size of 64. The hyper-parameters used in this NN were carefully selected through grid search and a 10-fold cross validation process.

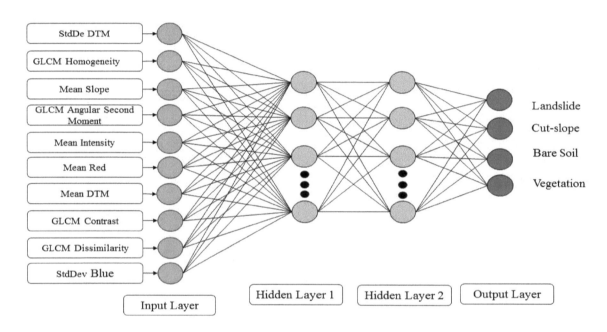

Figure 6. Architecture of the MLR-NN model; GLCM: gray level co-occurrence matrix, StdDev: standard deviation.

3.9.2. RNN

RNN is a sequence problem considered as the addition of loops to architecture. For example, in any layer under consideration, signals can be passed to each neuron and are subsequently forwarded to the next layer. The network output can be input to the network in the next input feature, and so on, as shown in Figure 7. In this study, RNN received 10 features as inputs to differentiate landslides from other objects (cut slope, bare soil, and vegetation). RNN consisted of an LSTM layer with 50 hidden units, two fully connected layers, a dropout layer, and a softmax layer. The back-propagation technique was used in trained the RNN model with Adam optimizer and a batch size of 128.

To avoid overfitting, a dropout layer was used in the RNN model and the NN learned weights from the training dataset. However, overfitting may occur when new data are inputted. The dropout layer randomly set some selected activations to zero, thus alleviating overfitting. The selected activations were used only during training and not during testing. The parameter was controlled by the number of activations that the dropout layer referred to as keep probability.

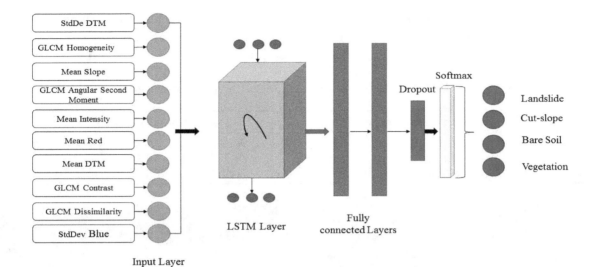

Figure 7. Architecture of the RNN model.

3.9.3. Optimization of Model Hyper-Parameters

The hyper-parameters of the RNN and MLP-NN models were optimized via a systematic grid search in scikit-learn [51] for 100 epochs. Despite its high computational cost, the systematic grid search provides better results because it systematically tunes the hyper-parameter values. Parameter combinations were selected for the models. The models were evaluated using a 10-fold cross-validation method. Among the evaluated parameters, the model with the highest validation accuracy was selected. Table 2 presents the most optimized parameters obtained for the models.

Table 2. Optimized model hyper-parameters; RNN: recurrent neural networks, MLP-NN: multi-layer perceptron neural networks.

Optimized Parameter	Suitable Value	Description
Minibatch size	126 (RNN) 64 (MLP-NN)	Number of training cases over which the Adam update is computed.
Loss function	categorical cross-entropy	The objective function or optimization score function is also called as multiclass legless, which is appropriate for categorical targets.
Optimizer	Adam	Adaptive moment estimation
dropout rates	0.6	Dropping out units (hidden and visible)

4. Results and Discussion

4.1. Supervised Approach for Optimizing Segmentation

The supervised approach was employed to optimize the parameters (i.e., scale, shape, and compactness) of the multiresolution segmentation algorithm for landslide identification and for differentiation from non-landslides (bare soil, cut slope, and vegetation). The optimized parameters rapidly increased the accuracy of classification to the optimum level by delineating the segmentation boundaries of the landslide. The application of optimized segmentation parameters allowed for the spatial and textural identification of features (landslide and non-slides). In our proposed method, accurate segmentation results should be first obtained prior to performing subsequent steps.

The optimal parameters of the multiresolution segmentation algorithm were obtained. The selected values for the three parameters are shown in Table 3. The initial segmentation parameters set in the supervised approach were 50, 0.1, and 0.1 for scale, shape, and compactness, respectively. After 100 iterations with these initial values, the optimal values obtained for scale, shape, and compactness were 75.52, 0.4, and 0.5, respectively in the analysis area. Meanwhile, the test area values were 100, 0.45 and 0.74, respectively. Figure 8a,b show the initial and optimal segmentation processes. The results of optimized segmentation accurately delineated landslide objects in the analysis and test areas.

Table 3. Multi-resolution segmentation parameters.

	Initial Parameters			Optimal Parameters		
Number	Scale	Shape	Compactness	Scale	Shape	Compactness
1	50	0.1	0.1	75.52	0.4	0.5
2	80	0.1	0.1	100	0.45	0.74

(**a**) Initial Segmentation   (**b**) Optimized Segmentation

Figure 8. Parameter optimization of the multiresolution segmentation algorithm: (**a**) initial segmentation and (**b**) optimized segmentation.

4.2. Relevant Feature Subset Based on a CFS Algorithm

In this study, the feature input consisted of 39 items of LiDAR-derived data (i.e., slope, height, and intensity), texture features (i.e., GLCM StdDev and GLCM homogeneity), and visible band. The optimal combination of features was selected via ten experiments using a CFS algorithm. Selection began from (1, 2, 4, 6, 8, 10, 12, 14, 16, 18, 20, 22, 24, 26, 28, 30, 32, 34, 36, 39) of the features. The most relevant feature subsets were obtained after 100 iterations in every experiment; this result is in line with the procedure proposed by Sameen et al. [52]. High classification accuracy was achieved when 10 of the features were applied, indicating that LiDAR-derived data, visible bands, and textural features were more effective in detecting the landslide location. Table 3 shows the most significant results of feature selection based on the CFS algorithm.

4.3. Results of Landslide Detection

Classification techniques affect the quality of the classification maps. Many classification algorithms have been established for each category, and each has its merits and demerits. In the present work, the RNN and MLP-NN models with optimized parameters were used for landslide detection with good accuracy. Figure 9 shows the classification results of the RNN and MLP-NN models in the analysis area. The qualitative assessment of the RNN model yielded high-quality results, as shown in Figure 9A. Well-defined landslide boundaries were detected and correctly differentiated from other objects (cut-slope, bare soil and vegetation). On the other hand, the qualitative assessment of MLP-NN produced low-quality results, as shown in Figure 9B.

Figure 9. Results of the qualitative assessment of (**A**) RNN and (**B**) MLP-NN for the analysis area.

The proposed models were evaluated using another LiDAR dataset (test site) from the Cameron Highlands. All features (all existing objects) of the test area were carefully considered. Segmentation parameters were optimized using the FbSP optimizer. A 10-fold cross-validation approach introduced by Bartels et al. [53] was used to resolve this issue with high accuracy. Environmental conditions and differences in landslide characteristics resulted in misclassification [9]. Differences in the sensors used, illumination conditions, and the spatial resolutions of images are some of the challenges faced by the proposed NN models [54]. The results of qualitative assessment indicated that the proposed NNs with optimized techniques correctly detected landslide locations in the test site, as shown in Figure 10. The qualitative assessment of the RNN model yielded high-quality results, as shown in Figure 10A,B. On the other hand, the qualitative assessment of the MLP-NN model produced low-quality results, as shown in Figure 10.

Figure 10. Results of the qualitative assessment of (**A**) RNN and (**B**) MLP-NN for the test area.

It is crucial to take the required measures to avoid the issue of the landslide separation from the bare land. The morphology characteristics of the landslide map is different from other types of land cover. For example, the shape, slope and other characteristics (i.e., dip direction, width and length) of the surface terrain may be changed after landslide occurs. Therefore, by using relevant features derived from very high resolution LiDAR, data such as texture and geometric features can be used to separate between landslides and bare land. In addition, applying different optimization techniques helped us to improve the classification accuracy in landslide detection over other landcover classes, such as bare land, man-made, etc., as described previously by Pradhan and Mezaal [9]. Their results demonstrated that using optimized techniques with very high resolution LiDAR data (0.5) enabled them to separate landslide and other types of land cover. In addition, the most relevant features in Table 4 were optimized during this study. Furthermore, authors of [16] suggested that using the object feature from LiDAR data is a suitable solution for landslide identification.

Table 4. Correlation-based feature selection (CFS) results for the most relevant feature subset at a scale of 75.52; StdDe: Standard deviation, DTM: Digital terrain model, GLCM: Gray level co-occurrence matrix.

Feature	Iteration	Rank
StdDe DTM	20	1
GLCM homogeneity	18	2
Mean slope	20	3
GLCM angular second moment	20	4
Mean intensity	17	5
Mean red	20	6
Mean DTM	20	7
GLCM contrast	18	8
GLCM dissimilarity	15	9
StdDev blue	20	10

The landslide detection results showed that the proposed model is robust. Optimizing the segmentation parameters, namely, scale, shape, and compactness, using the fuzzy logic supervised approach resulted in the effective differentiation of landslide from non-landslide (bare soil, cut slope and vegetation) objects. Creating accurate objects through the optimized segmentation process allowed the use of spatial, orthophoto, and textural features for feature detection. Landslides should be differentiated from non-landslides based on the accurate segmentation of spatial and textural features. The selection of relevant features in landslide detection relies on the experience of the analysts. Thus, a feature selection method is crucial for accurate and reliable landslide detection. The optimal features selected via the CFS method simplified landslide detection by the NN model. Computation time and reliance on the expert knowledge of the analyst were reduced. Moreover, the optimized parameters of the NN models improved the performance of the models, reduced the complexity of the models, and decreased overfitting in the training sample.

4.4. Performance of the MLP-NN and RNN

The models were implemented in Python using the open source TensorFlow deep learning framework developed by Google [26]. Meanwhile, the accuracy of the proposed NN models was tested using a 10-fold cross-validation method. The results are presented in Table 5. The best accuracy of 83.33% in the analysis area was achieved by the RNN model. The MLR-NN model achieved an accuracy of 78.38% in the same area. Furthermore, the RNN model outperformed the MLR-NN model in terms of stability of accuracy across different folds of the tested dataset. In the test area, accuracies of 81.11% and 74.56% were achieved with the RNN and MLP-NN models, respectively. These results indicated that the RNN model has better accuracy than the MLR-NN model in the analysis and test areas and indicated the high stability of the RNN model in detecting the spatial distribution of landslides.

Table 5. Cross-validation accuracy results of the proposed models.

Neural Network Model	Analysis Area	Test Area
RNN model	83.33%	81.11%
MLP-NN model	78.38%	74.56%

However, producing neural network models such as LSTM and convolution layers with fully connected networks is a crucial task. Complex networks with more hidden units and many modules often tend to have a better overfit due to the detection ability with respect to any possible interaction so the model becomes too specific to the training dataset. Thus, optimizing the network structures is very crucial for avoiding over-fitting. This study indicated that the hyperparameters in both models have a significant effect on their results. For example, the effect of learning rate varied from 0.1, 0.01, 0.05 and 0.001 in landslide detection. The highest accuracy was obtained when a learning rate reached 0.001. In contrast, increasing the learning rate to 0.1 significantly reduced the accuracy in both models. The batch size parameter in both models had significant effects on the result accuracy. The results of MLR-NN and RNN models showed high accuracies with batch sizes of 64 and 128, respectively. This indicates that RNN model achieved high accuracy with the increase of the batch size, whereas the accuracy of MLB-NN model was decreased.

Furthermore, it was revealed that the dropout rate had a substantial influence on the results of the RNN model. The RNN model showed higher accuracy when the dropout rate reached 0.6. The results of the RNN model indicated that the accuracy increased when the dropout rate parameter was increased.

The results of two models (Table 5), show that the accuracies of the RNN model outperformed the MLP-NN model in both study areas. This is due to several reasons, for example the fact that the MLP-NN model uses only local contexts and therefore it does not capture the temporal and spatial correlation in the dataset. Meanwhile, the hidden units of the RNN model contain historical information from the previous step. This indicates RNN model has more information about the data structure and accurate as compared to the MLP-NN model.

4.5. Sensitivity Analysis

The optimization of network architecture is necessary and should be considered over the use of standard parameters [28] because network architecture models are principally influenced by the analytical task and data type. Data could differ in size, relationships between independent and dependent variables, and complexity. Therefore, the neural architecture of the RNN and MLP-NN networks was enhanced using a grid search implemented in SciPy-python. The combinations of 10 parameters that can best identify landslide locations in densely vegetated areas were optimized.

The Adam optimizer is the most suitable algorithm for the optimization of the two NN models. Using the Adam optimizer with default parameters (learning rate $\mu = 0.001$, beta $\beta_1 = 0.9$, epsilon $\epsilon = $ 1e-08 and weight decay = 0.0) yielded an accuracy of 0.77 and 0.825 for MLP-NN and RNN models, respectively, as shown in Figure 11. Rmsprop and Nadam optimizers also achieved excellent results for the two models. Overall, the Adam algorithm is more suitable for analyzing landslide data. However, better accuracy was obtained when Adadelta was used with the RNN model. Meanwhile, adding the weight decay in the neural network did not affect the results.

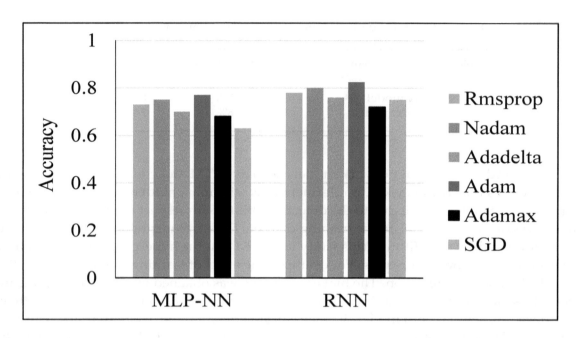

Figure 11. Impact of the optimization algorithm on the performance of MLP-NN and RNN models; SGD: Stochastic Gradient Descent.

Batch size, which refers to the number of training examples computed during optimization, has substantial effects on model accuracy. The results of the RNN and MLR-NN models are shown in Figure 12 and depict how increasing batch size from 2 to 128 (by ×2) affected model accuracy. The MLR-NN and RNN models exhibited the best accuracies with batch sizes of 64 and 128, respectively.

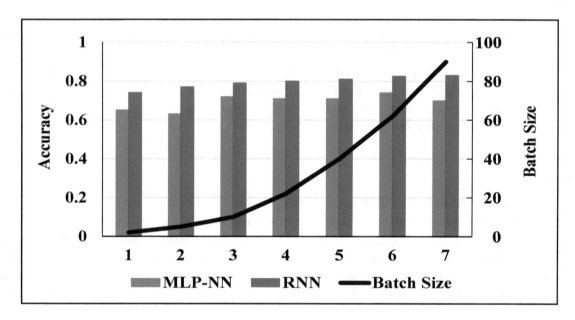

Figure 12. Impact of batch size on the performance of the MLP-NN and RNN models.

Overfitting can be avoided when dropouts are controlled through the number of parameters in the RNN model. Figure 13 illustrates the sensitivity analysis of the effects of dropout rate with various keep probability parameters on the RNN model. The results showed that the appropriate dropout rate is 0.6 for the RNN model. The selected dropout rate considerably affects the performance of NN models. The keep probability was selected in each dataset and analysis was conducted via a systematic grid search.

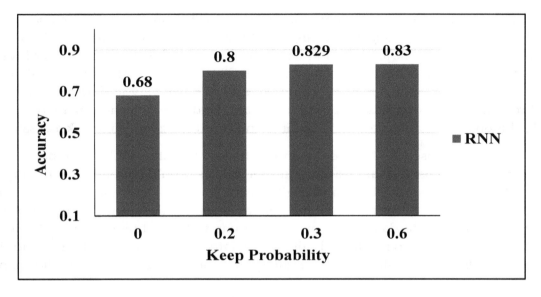

Figure 13. Influence of dropout rate on the performance of the RNN model.

4.6. Field Investigation

The reliability of the proposed methods was validated via field investigation using a handheld Global Position System (GPS) device (GeoExplorer 6000) to locate landslides (Figure 14) and to produce a precise and reliable inventory map of the Cameron Highlands. The more detailed information (landslide extent, source area, deposition, and volume) was obtained from in situ measurements which ultimately demonstrate the reliability of the produced inventory map in the field through use of a GeoExplorer 6000 handheld GPS. The results illustrated that the neural network techniques were able to detect true landslide locations which occurred in past years. Therefore, the results of this study verified that the proposed models can detect landslide locations and generate a reliable landslide inventory map.

Figure 14. Field photographs showing landslide locations during field investigation in (**A**) Tanah Rata and (**B**) Tanah Runtuh.

5. Conclusions

The Cameron Highlands, Malaysia form an ideal site for testing the feasibility of RNN and MLP-NN models for landslide detection based on high-resolution LiDAR data. The optimization of segmentation parameters is crucial for improving model performance and computational efficiency with different spatial subsets in the Cameron Highlands. Furthermore, optimization is essential for feature selection to improve the classification accuracy and the computational efficiency of the proposed methodology. The optimization of NN model parameters helped improve the performance of the model by reducing model complexity and preventing overfitting in the training sample. The RNN model exhibited better accuracy in the analysis and test areas than the MLR-NN model. This investigation showed that network architectures based on optimized techniques, very high resolution (VHR) airborne LiDAR-derived data, and spatial features could be used to effectively identify landslide locations in tropical regions. Therefore, this proposed automatic landslide detection method is a potential geospatial solution for managing landslide hazards and conducting landslide risk assessments.

Given that the proposed RNN model is more efficient than the MLP-NN model and has the potential to process the most relevant features, further studies should be conducted to fully optimize network structures for higher flexibility and eligibility for landslide detection. More theoretical tasks are recommended to enhance the representation of variables and data structure by the RNN model and the storage capacity of the data. Faster and more accurate NN techniques for landslide detection should be developed to overcome all the limitations related to accuracy and time. In addition, the RNN model can be integrated with other NN techniques to help improve other landslide applications.

Author Contributions: Mustafa Ridha Mezaal and Biswajeet Pradhan conceived and designed the experiments; Mustafa Ridha Mezaal, Biswajeet Pradhan and Maher Ibrahim Sameen performed the experiment and analyzed the data; Mustafa Ridha Mezaal and Biswajeet Pradhan contributed reagents/materials/analysis tools; Mustafa Ridha Mezaal and Biswajeet Pradhan wrote the paper. Biswajeet Pradhan is the main supervisor of PhD candidate Mustafa Ridha Mezaal; Helmi Zulhaidi Mohd Shafri and Zainuddin Md Yusoff supported as part of the co-supervisory committee of the first author.

References

1. Guzzetti, F.; Mondini, A.C.; Cardinali, M.; Fiorucci, F.; Santangelo, M.; Chang, K.-T. Landslide inventory maps: New tools for an old problem. *Earth-Sci. Rev.* **2012**, *112*, 42–66. [CrossRef]

2. Van Westen, C.J.; Castellanos, E.; Kuriakose, S.L. Spatial data for landslide susceptibility, hazard, and vulnerability assessment: An overview. *Eng. Geol.* **2008**, *102*, 112–131. [CrossRef]

3. Parker, R.N.; Densmore, A.L.; Rosser, N.J.; De Michele, M.; Li, Y.; Huang, R.; Petley, D.N. Mass wasting triggered by the 2008 Wenchuan earthquake is greater than orogenic growth. *Nat. Geosci.* **2011**, *4*, 449–452. [CrossRef]

4. Chen, R.-F.; Lin, C.-W.; Chen, Y.-H.; He, T.-C.; Fei, L.-Y. Detecting and Characterizing Active Thrust Fault and Deep-Seated Landslides in Dense Forest Areas of Southern Taiwan Using Airborne LiDAR DEM. *Remote Sens.* **2015**, *7*, 15443–15466. [CrossRef]

5. Pradhan, B.; Jebur, M.N.; Shafri, H.Z.M.; Tehrany, M.S. Data Fusion Technique Using Wavelet Transform and Taguchi Methods for Automatic Landslide Detection From Airborne Laser Scanning Data and QuickBird Satellite Imagery. *IEEE Trans. Geosci. Remote Sens.* **2016**, *54*, 1610–1622. [CrossRef]

6. McKean, J.; Roering, J. Objective landslide detection and surface morphology mapping using high-resolution airborne laser altimetry. *Geomorphology.* **2004**, *57*, 331–351. [CrossRef]

7. Whitworth, M.; Giles, D.; Murphy, W. Airborne remote sensing for landslide hazard assessment: A case study on the Jurassic escarpment slopes of Worcestershire, UK. *Q. J. Eng. Geol. Hydrogeol.* **2005**, *38*, 285–300. [CrossRef]

8. Blaschke, T. Object based image analysis for remote sensing. *ISPRS J. Photogramm. Remote Sens.* **2010**, *65*, 2–16. [CrossRef]

9. Pradhan, B.; Mezaal, M.R. Optimized Rule Sets for Automatic Landslide Characteristic Detection in a Highly Vegetated Forests. In *Laser Scanning Applications in Landslide Assessment*; Springer: New York, NY, USA, 2017; pp. 51–68.

10. Anders, N.S.; Seijmonsbergen, A.C.; Bouten, W. Segmentation optimization and stratified object-based analysis for semi-automated geomorphological mapping. *Remote Sens. Environ.* **2011**, *115*, 2976–2985. [CrossRef]

11. Belgiu, M.; Drăguţ, L. Comparing supervised and unsupervised multiresolution segmentation approaches for extracting buildings from very high resolution imagery. *ISPRS J. Photogramm. Remote Sens.* **2014**, *96*, 67–75. [CrossRef] [PubMed]

12. Drăguţ, L.; Tiede, D.; Levick, S.R. ESP: A tool to estimate scale parameter for multiresolution image segmentation of remotely sensed data. *Int. J. Geogr. Inf. Sci.* **2010**, *24*, 859–871. [CrossRef]

13. Zhang, Y.; Maxwell, T.; Tong, H.; Dey, V. Development of a supervised software tool for automated determination of optimal segmentation parameters for ecognition. In Proceedings of the ISPRS TC VII Symposium–100 Years, ISPRS, Vienna, Austria, 5–7 July 2010.

14. Chen, W.; Li, X.; Wang, Y.; Chen, G.; Liu, S. Forested landslide detection using LiDAR data and the random forest algorithm: A case study of the Three Gorges, China. *Remote Sens. Environ.* **2014**, *152*, 291–301. [CrossRef]

15. Kursa, M.B.; Rudnicki, W.R. Feature selection with the Boruta package. *J. Stat. Softw.* **2010**, *36*, 1–13. [CrossRef]

16. Li, X.; Cheng, X.; Chen, W.; Chen, G.; Liu, S. Identification of forested landslides using LiDar data, object-based image analysis, and machine learning algorithms. *Remote Sens.* **2015**, *7*, 9705–9726. [CrossRef]

17. Stumpf, A.; Kerle, N. Object-oriented mapping of landslides using Random Forests. *Remote Sens. Environ.* **2011**, *115*, 2564–2577. [CrossRef]

18. Borghuis, A.; Chang, K.; Lee, H. Comparison between automated and manual mapping of typhoon-triggered landslides from SPOT-5 imagery. *Int. J. Remote Sens.* **2007**, *28*, 1843–1856. [CrossRef]

19. Danneels, G.; Pirard, E.; Havenith, H.B. Automatic landslide detection from remote sensing images using supervised classification methods. In *Geoscience and Remote Sensing Symposium*; IEEE: Hoboken, NJ, USA, July 2007; pp. 3014–3017.

20. Moine, M.; Puissant, A.; Malet, J.-P. Detection of Landslides From Aerial and Satellite Images With a Semi-Automatic Method. Application to the Barcelonnette Basin (Alpes-de-Hautes-Provence, France). February 2009, pp. 63–68. Available online: https://halshs.archives-ouvertes.fr/halshs-00467545/document (accessed on 16 July 2017).

21. Pratola, C.; Del Frate, F.; Schiavon, G.; Solimini, D.; Licciardi, G. Characterizing land cover from X-band COSMO-SkyMed images by neural networks. In Proceedings of the Urban Remote Sensing Event (JURSE), Munich, Germany, 11–13 April 2011; pp. 49–52.

22. Singh, A.; Singh, K.K. Satellite image classification using Genetic Algorithm trained radial basis function neural network, application to the detection of flooded areas. *J. Vis. Commun. Image Represent.* **2017**, *42*, 173–182. [CrossRef]

23. Singh, K.K.; Singh, A. Detection of 2011 Sikkim earthquake-induced landslides using neuro-fuzzy classifier and digital elevation model. *Nat. Hazards* **2016**, *83*, 1027–1044. [CrossRef]

24. Mehrotra, A.; Singh, K.K.; Nigam, M.J.; Pal, K. Detection of tsunami-induced changes using generalized improved fuzzy radial basis function neural network. *Nat. Hazards* **2015**, *77*, 367–381. [CrossRef]

25. Benediktsson, J.A.; Sveinsson, J.R. Feature extraction for multisource data classification with artificial neural networks. *Int J. Remote Sens.* **1997**, *18*, 727–740. [CrossRef]

26. Yuan, H.; Van Der Wiele, C.F.; Khorram, S. An automated artificial neural network system for land use/land cover classification from Landsat TM imagery. *Remote Sens.* **2009**, *1*, 243–265. [CrossRef]

27. Fu, G.; Liu, C.; Zhou, R.; Sun, T.; Zhang, Q. Classification for High Resolution Remote Sensing Imagery Using a Fully Convolutional Network. *Remote Sens.* **2017**, *9*, 498. [CrossRef]

28. Sameen, M.I.; Pradhan, B. Severity Prediction of Traffic Accidents with Recurrent Neural Networks. *Appl. Sci.* **2017**, *7*, 476. [CrossRef]

29. Gorsevski, P.V.; Brown, M.K.; Panter, K.; Onasch, C.M.; Simic, A.; Snyder, J. Landslide detection and susceptibility mapping using LiDAR and an artificial neural network approach: A case study in the Cuyahoga Valley National Park, Ohio. *Landslides* **2016**, *13*, 467–484. [CrossRef]

30. Chang, K.T.; Liu, J.K.; Chang, Y.M.; Kao, C.S. An Accuracy Comparison for the Landslide Inventory with the BPNN and SVM Methods. *Gi4DM 2010*; Turino, Italy, 2010. Available online: https://www.researchgate.net/profile/Jin_King_Liu/publication/267709454_An_Accuracy_Comparison_for_the_Landslide_Inventory_with_the_BPNN_and_SVM_Methods/links/5511773f0cf29a3bb71de12c.pdf (accessed on 20 March 2017).

31. Robert, H.N. *Neurocomputing*; Addison-Wesley Pub. Co.: Boston, CA, USA, 1990; pp. 21–42.

32. Zurada, J.M. *Introduction to Artificial Neural Systems*; West Pub. Co.: Eagan, MN, USA, 1992; pp. 163–248.

33. Chang, K.T.; Liu, J.K. Landslide features interpreted by neural network method using a high-resolution satellite image and digital topographic data. In Proceedings of the XXth ISPRS Congress, Istanbul, Turkey, 12–23 July 2004.

34. Chen, H.; Zeng, Z.; Tang, H. Landslide deformation prediction based on recurrent neural network. *Neural Process. Lett.* **2015**, *41*, 169–178. [CrossRef]

35. Ma, H.-R.; Cheng, X.; Chen, L.; Zhang, H.; Xiong, H. Automatic identification of shallow landslides based on Worldview2 remote sensing images. *J. Appl. Remote Sens.* **2016**, *10*, 016008. [CrossRef]

36. Hall, M.A. Correlation-based Feature Selection for Machine Learning. Doctoral dissertation, The University of Waikato, Hamilton, New Zealand, April 1999.

37. Miner, A.; Flentje, P.; Mazengarb, C.; Windle, D. Landslide Recognition Using LiDAR Derived Digital Elevation Classifiers-Lessons Learnt from Selected Australian Examples. Available online: http://ro.uow. edu.au/cgi/viewcontent.cgi?article=1590&context=engpapers (accessed on 10 June 2017).

38. Martha, T.R.; Kerle, N.; van Westen, C.J.; Jetten, V.; Kumar, K.V. Segment optimization and data-driven thresholding for knowledge-based landslide detection by object-based image analysis. *IEEE Trans. Geosci. Remote Sens.* **2011**, *49*, 4928–4943. [CrossRef]

39. Pradhan, B.; Lee, S. Regional landslide susceptibility analysis using back-propagation neural network classifier at Cameron Highland, Malaysia. *Landslides* **2010**, *7*, 13–30. [CrossRef]

40. Olaya, V. Basic land-surface parameters. *Dev. Soil Sci.* **2009**, *33*, 141–169.

41. Barbarella, M.; Fiani, M.; Lugli, A. Application of LiDAR-derived DEM for detection of mass movements on a landslide. *Int. Arch. Photogramm. Remote Sens. Spat. Inf. Sci.* **2013**, *1*, 89–98. [CrossRef]

42. Duro, D.C.; Franklin, S.E.; Dubé, M.G. Multi-scale object-based image analysis and feature selection of multi-sensor earth observation imagery using random forests. *Int. J. Remote Sens.* **2012**, *33*, 4502–4526. [CrossRef]

43. Kohavi, R.; John, G.H. Wrappers for feature subset selection. *Artif. Intell.* **1997**, *97*, 273–324. [CrossRef]

44. Mokhtarzade, M.; Zoej, M.V. Road detection from high-resolution satellite images using artificial neural networks. *Int. J. Appl. Earth Obs. Geoinf.* **2007**, *9*, 32–40. [CrossRef]

45. Gardner, M.W.; Dorling, S.R. Artificial neural networks (the multilayer perceptron)—A review of applications in the atmospheric sciences. *Atmos. Environ.* **1998**, *32*, 2627–2636. [CrossRef]

46. Baczyński, D.; Parol, M. Influence of artificial neural network structure on quality of short-term electric energy consumption forecast. *IEE Proc.-Gener., Transm. Distrib.* **2004**, *151*, 241–245. [CrossRef]

47. Mia, M.M.A.; Biswas, S.K.; Urmi, M.C.; Siddique, A. An algorithm for training multilayer perceptron (MLP) for Image reconstruction using neural network without overfitting. *Int. J. Sci. Techno. Res.* **2015**, *4*, 271–275.

48. Yang, G.Y.C. *Geological Mapping from Multi-Source Data Using Neural Networks*; Geomatics Engineering, University of Calgary: Calgary, AB, Canada, 1995.

49. Hochreiter, S.; Schmidhuber, J. Long short-term memory. *Neural Comput.* **1997**, *9*, 1735–1780. [CrossRef] [PubMed]

50. Donahue, J.; Anne Hendricks, L.; Guadarrama, S.; Rohrbach, M.; Venugopalan, S.; Saenko, K.; Darrell, T. Long-term recurrent convolutional networks for visual recognition and description. In Proceedings of the IEEE Conference on Computer Vision and Pattern Recognition, Boston, MA, USA, 7–12 June 2015; pp. 2625–2634.

51. Pedregosa, F.; Varoquaux, G.; Gramfort, A.; Michel, V.; Thirion, B.; Grisel, O.; Vanderplas, J. Scikit-learn: Machine learning in Python. *J. Mach. Learn. Res.* **2011**, *12*, 2825–2830.

52. Sameen, M.I.; Pradhan, B.; Shafri, H.Z.; Mezaal, M.R.; bin Hamid, H. Integration of Ant Colony Optimization and Object-Based Analysis for LiDAR Data Classification. *IEEE J. Sel. Top. Appl. Earth Obs. Remote Sens.* **2017**, *10*, 2055–2066. [CrossRef]

53. Bartels, M.; Wei, H. Threshold-free object and ground point separation in LIDAR data. *Pattern Recognit. Lett.* **2010**, *31*, 1089–1099. [CrossRef]

54. Rau, J.Y.; Jhan, J.P.; Rau, R.J. Semiautomatic object-oriented landslide recognition scheme from multisensor optical imagery and DEM. *IEEE Trans. Geosci. Remote Sens.* **2011**, *52*, 1336–1349. [CrossRef]

11

Severity Prediction of Traffic Accidents with Recurrent Neural Networks

Maher Ibrahim Sameen [†] and Biswajeet Pradhan [†,*]

Department of Civil Engineering, Geospatial Information Science Research Center (GISRC),
Faculty of Engineering, University Putra Malaysia, UPM, Serdang 43400, Malaysia; maherrsgis@gmail.com
* Correspondence: biswajeet24@gmail.com or biswajeet@upm.edu.my
† These authors contributed equally to this work.

Academic Editor: Saro Lee

Abstract: In this paper, a deep learning model using a Recurrent Neural Network (RNN) was developed and employed to predict the injury severity of traffic accidents based on 1130 accident records that have occurred on the North-South Expressway (NSE), Malaysia over a six-year period from 2009 to 2015. Compared to traditional Neural Networks (NNs), the RNN method is more effective for sequential data, and is expected to capture temporal correlations among the traffic accident records. Several network architectures and configurations were tested through a systematic grid search to determine an optimal network for predicting the injury severity of traffic accidents. The selected network architecture comprised of a Long-Short Term Memory (LSTM) layer, two fully-connected (dense) layers and a Softmax layer. Next, to avoid over-fitting, the dropout technique with a probability of 0.3 was applied. Further, the network was trained with a Stochastic Gradient Descent (SGD) algorithm (learning rate = 0.01) in the Tensorflow framework. A sensitivity analysis of the RNN model was further conducted to determine these factors' impact on injury severity outcomes. Also, the proposed RNN model was compared with Multilayer Perceptron (MLP) and Bayesian Logistic Regression (BLR) models to understand its advantages and limitations. The results of the comparative analyses showed that the RNN model outperformed the MLP and BLR models. The validation accuracy of the RNN model was 71.77%, whereas the MLP and BLR models achieved 65.48% and 58.30% respectively. The findings of this study indicate that the RNN model, in deep learning frameworks, can be a promising tool for predicting the injury severity of traffic accidents.

Keywords: severity prediction; GIS; traffic accidents; deep learning; recurrent neural networks

1. Introduction

Traffic accidents are a primary concern due to many fatalities and economic losses every year worldwide. In Malaysia, recent statistics show that there are nearly 24 deaths per 100,000 people for all road users [1]. Expressways are potential sites of fatal highway accidents in Malaysia. Better accident severity prediction models are critical to enhancing the safety performance of road traffic systems. According to recent literature, driver injury severity can be classified into a few categories such as property damage, possible/evident injury, or disabling injury/fatality [2]. Therefore, modeling accident severity can be addressed as a pattern recognition problem [3], which can be solved by deep learning, statistical techniques and sometimes by physical modelling approaches [4–7]. In a deep learning model, an input vector is often mapped into an output vector through a set of nonlinear functions. In the case of accident severity, the input vectors are the characteristics of the accident, such as driver behavior and highway, vehicle and environment characteristics. The output vector is the corresponding classes of accident severity. Deep learning allows computational models to learn hierarchal representations of data with multiple levels of abstraction at different processing layers. The advantage of deep learning neural networks over statistical techniques is that they involve a more general mapping procedure i.e., a specific function is not required in model building [8]. However, these techniques can be treated as black box methods, if the network architecture is not carefully designed and its parameters are not optimized.

Several studies have investigated Neural Network (NN) and deep learning methods in transportation related applications [9–12]. For example, Abdelwahab and Abdel-Aty [8] used NN to predict driver injury severity from various accident factors (i.e., driver, vehicle, roadway and environment characteristics). In their study, the NN model developed performed better than the ordered probit model. In another paper, Delen et al. [10,13] applied NN to model injury severity of road accidents using 17 significant parameters. The NN model was used to predict the injury severity levels, resulting in a low overall accuracy of 40.71%. More recently, Hashmienejad and Hasheminejad [14] proposed a novel rule-based technique to predict traffic accident severity based on users' preferences instead of conventional data mining methods. The proposed method outperformed the NN and support vector machine methods. In a recent paper, Alkheder et al. [11] used NN methods to predict the injury severity (minor, moderate, severe, death) of traffic accidents in Abu Dhabi. Their analysis was based on 5973 traffic accident records that had occurred over a 6-years period. The overall accuracy of the model for the training and testing data were 81.6% and 74.6%, respectively. In addition, Zeng and Huang [15] proposed a training algorithm and network structure optimization method for crash injury severity prediction. Their results indicated that the proposed training algorithm performed better than the traditional back-propagation algorithm. The optimized NN, which contained less nodes than the fully connected NN, achieved reasonable prediction accuracy. Also, the fully connected and optimized NN models outperformed the ordered logit model. Also, their results showed that optimization of the NN structure could improve the overall performance of model prediction.

In recent years, deep learning has become a popular technique in image [16] and natural language processing applications [17]. Many researchers have applied deep learning based techniques in transportation related applications such as traffic flow [18] and accident hotspots prediction [19]. A detailed literature review showed that optimization of network structures is critical for crash severity prediction. Generally speaking, deep learning allows compositionality and the design of flexible network structures with multilayer modules, and therefore it is expected that the performance of deep learning models will be better than the traditional NN models. A deep learning architecture is a multilayer stack of simple modules, most of which are subjected to learning computing nonlinear input–output mappings [20]. Each module in the stack transforms its input to increase both selectivity and invariance of the representation. A Recurrent Neural Network (RNN) is a type of deep learning module, which is more appropriate for sequential data (e.g., traffic accident data with temporal correlations). It is effective for a wide range of applications including text generation [21] and speech recognition [22]. The application of RNNs in various fields motivated the authors to evaluate these models in deep learning frameworks for severity prediction of traffic accidents.

The main novelty of this work is the development of a RNN model that accurately predicts injury severity of traffic accidents utilizing the temporal structure of accident data. The specific objective is to design a deep learning model using a RNN for severity prediction of traffic accidents. The hyperparameters of the model will be selected through a systematic grid search technique. In addition, the sensitivity of model parameters and configurations will be assessed, and then will be compared with the well-known Multilayer Perceptron (MLP) and Bayesian Logistic Regression (BLR) models to understand its advatange. Finally, the impacts of accident-related factors on injury severity outcome will be determined using the profile-based method.

2. Recurrent Neural Networks (RNNs)

Recurrent Neural Networks (RNNs) are neural networks with feedback connections specifically designed to model sequences. They are computationally more powerful and biologically more reasonable than feed-forward networks (no internal states). The feedback connections provide a RNN the memory of past activations, which allows it to learn the temporal dynamics of sequential data. A RNN is powerful because it uses contextual information when mapping between input and output sequences. However, the traditional RNNs have a problem called vanishing or exploding gradient. To handle this problem, Hochreiter and Schmidhuber [23] proposed the Long Short-Term Memory (LSTM) algorithm.

In LSTM, the hidden units are replaced by memory blocks, which contain one or more self-connected memory cells and three multiplicative units (input, output, forget gates). These gates allow writing, reading, and resetting operations within a memory block, and they control the overall behavior internally. A representation of a single LSTM unit is shown in Figure 1. Let c_t be the sum of inputs at time step t, then LSTM updates for time step i at given inputs x_t, h_{t-1}, and c_{t-1} are [24]:

$$i_t = \sigma(W_{xi}.x_t + W_{hi}.h_{t-1} + W_{ci}.c_{t-1} + b_i) \tag{1}$$

$$f_t = \sigma\left(W_{xf}.x_t + W_{hf}.h_{t-1} + W_{cf}.c_{t-1} + b_f\right) \tag{2}$$

$$c_t = i_t.tanh(W_{xc}.x_t + W_{hc}.h_{t-1} + b_c) + f_t.c_{t-1} \tag{3}$$

$$o_t = \sigma(W_{xo}.x_t + W_{ho}.h_{t-1} + W_{co}.c_t + b_o) \tag{4}$$

$$h_t = o_t.tanh(c_t) \tag{5}$$

where σ is an element-wised non-linearity such as a sigmoid function, W is the weight matrix, x_t is the input at time step t, h_{t-1} is the hidden state vector of the previous time step and b_i denotes the input bias vector.

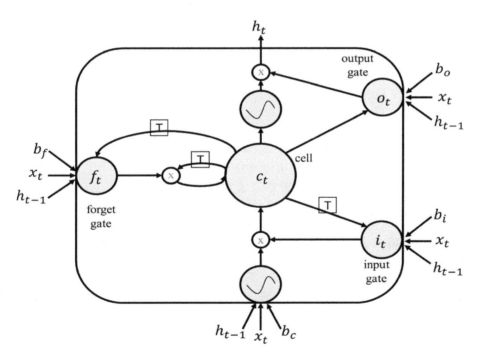

Figure 1. The structure of a memory cell in the Long Short-Term Memory–Recurrent Neural Network (LSTM–RNN).

3. The Proposed Network Framework

3.1. Network Architecture

Figure 2 shows the high-level architecture of the proposed severity prediction model based on deep learning. The advantage of LSTM-RNN over the traditional neural networks is that the traffic accidents are correlated to the historical and future incidents [25]. The RNN model comprises of five main layers which include input, LSTM, two dense layers and a Softmax layer.

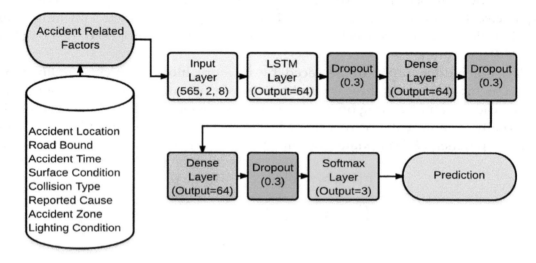

Figure 2. The high-level architecture of the proposed RNN model used in this study.

The main input of the network is a set of traffic accident-related factors, and the output is the corresponding accident severity classes (i.e., property damage only, possible/evident injury, disabling/fatality). The input dimension of the LSTM layer is equal to eight (number of input factors) with 64 nodes. The output of each node in the LSTM layer is results from the Rectified Linear Unit (ReLU) activation function applied to a weighted sum of both inputs from the previous layer and the previous outputs of the layer. The ReLU computes the function $f(x) = \max(0, x)$. This activation function accelerates (e.g., a factor of 6 in [26]) the convergence of Stochastic Gradient Descent (SGD) compared to the sigmoid/tanh functions. In addition, it can be implemented without expensive operations as in the case of sigmoid/tanh functions. Then, two fully connected layers were trained on top of the LSTM layers. They were added to match the output of the LSTM layer and the size of the accident severity classes. The output layer was a fully connected feed-forward layer and used to directly map the learned features to the three accident severity classes. A Softmax function is used to activate the output layer. In addition, three dropout layers with probability of 0.3 were used to reduce the complexity of the model to prevent over-fitting [27].

3.2. Training Methodology

The network was trained with Backpropagation Through Time (BPTT) [28] and Stochastic Gradient Descent (SGD) algorithms in Tensorflow on a personal CPU system (Core i7 with 16 GB RAM). As a RNN can be seen as a normal feedforward NN with shared weights, the BPTT begins by unfolding the RNN through several steps in time. The network training then proceeds in a manner similar to training a feed-forward neural network with backpropagation, except that the training patterns were visited in a sequential order. It determines the gradients of the objective function (Equation (6)) with respect to a parameter at each time step. Finally, the BPTT calculates the gradient outputs by taking the average of the individual step-dependent gradients.

$$H_{\hat{y}}(y) = -\sum_{i} y_l \log(y_l) \tag{6}$$

where y is the predicted probability distribution and $\hat{y}$ is the actual distribution (the one-hot vector with injury severity outcomes).

In addition, the SGD algorithm uses few examples from the input training vector and computes the outputs and the errors, and then adjusts the weights. The process is recurrent for many small sets of examples until the average of the objective function stops. The SGD method could determine a good set of weights when compared with other optimization techniques [29,30]. The training was run on a batch size 32 with 100 epochs. The best model was achieved using SGD with the decay of 0.9 and a momentum = 0.8. The best learning rate was 0.01. In addition, gradient clipping [31] with threshold 2.0 was useful to stabilize the training and to avoid gradient explosion.

3.3. Mitigating Overfitting

Networks with more complicated functions can perform better generalization performance as they learn different representations in each of their layers. However, complicated networks can easily over-fit the training data, producing poor generalization capacity and testing performance. Over-fitting occurs when a network model with high capacity fits the noise in the data instead of the underlying relationship.

Therefore, to avoid over-fitting in the proposed RNN model, three standard techniques were used. These were Gaussian noise injection into the training data [32], using a ReLU activation function in the hidden layers [20], and subsequently applying the dropout technique [29]. Dropout leads to big improvements in the prediction performance of the model. When a dropout technique is applied, the generalization capacity of the model is improved because the network is forced to learn multiple independent representations of the data. A probability of 30% was used in the dropout technique. This is because low probability has minimal effect and a high probability results in under-learning by the network.

3.4. Hyperparameter Tuning

The hyper-parameters of the RNN model were selected by performing a systematic grid search implemented in scikit-learn [33] using 100 epochs. Even though the systematic grid search requires high computational cost, better results could be obtained by systematically tuning the hyper-parameter values. Models with various combination of parameters were constructed, and a 3-fold cross-validation was used to evaluate each model. The parameters of the model with the highest validation accuracy were found to be the best parameters among the evaluated ones. Table 1 shows the optimized parameters used in the network.

Table 1. The optimized hyperparameters of the proposed RNN model.

Hyper-Parameter	Best Value	Description
Minibatch size	32	Number of training cases over which SGD update is computed.
Loss function	Categorical crossentropy	The objective function or optimization score function is also called as multiclass logloss which is appropriate for categorical targets.
Optimizer	SGD	Stochastic gradient descent optimizer.
Learning rate	0.01	The learning rate used by SGD optimizer
Gradient momentum	0.80	Gradient momentum used by SGD optimizer.
Weight decay	0.9	Learning rate decay over each update.

4. Experimental Results and Discussion

The proposed RNN model was implemented in Python using the open source TensorFlow deep learning framework developed by Google [34]. TensorFlow has automatic differentiation and parameter sharing capabilities, which allows a wide range of architectures to be easily defined and executed [34]. The proposed network was trained with 791 samples and validated with 339 samples using the Tensorflow framework. The SGD optimization algorithm was applied, with a batch size of 32 and a learning rate of 0.01.

4.1. Data

The traffic accident data for the period 2009–2015 from the North-South Expressway (NSE, Petaling Jaya, Malaysia), Malaysia were used in this study. The NSE is the longest expressway

(772 km) operated by Projek Lebuhraya Usaha Sama (PLUS) Berhad (the largest expressway operator in Malaysia, Petaling Jaya, Malaysia) and links many major cities and towns in Peninsular Malaysia. The data were obtained from the PLUS accident databases. The files used in this study were accident frequency and accident severity files in the form of an Excel spreadsheet. The accident frequency file contains the positional and descriptive accident location and the number of accidents in each road segment of 100 m. The accident records were separated according to the road bound (south, north). In contrast, the accident severity file contains the general accident characteristics such as accident time, road surface and lighting conditions, collision type, and the reported accident cause. To link the two files, the unique identity field (accident number) was used.

According to the vehicle type information in the accident data, three scenarios were found: (1) single-vehicle with object accidents, (2) two-vehicle accidents, (3) and multiple vehicle accidents (mostly three vehicles). Training a deep learning model requires many samples to capture the data structure and to avoid model overfitting. Therefore, the analysis in this study did not focus just on a single scenario, but instead, all scenarios were included in training the proposed RNN model. In total, 1130 accident records were reported during 2009–2015. Of these, 740 (approximately 65.4%) resulted in drivers damaging property only. On the other hand, 172 (15.2%) drivers were involved in possible/evident injury, and 218 (19.4%) in disabling injury.

The section of the NSE used in this study has a length of 15 km running from Ayer Keroh (210 km) to Pedas Linggi (225 km) (Figure 3). The accident severity data showed that the last section (220–225) of the NSE experienced several accidents resulting in serious injury (82) than the other sections (Table 2). Most accidents have occurred on the main route and southbound of the expressway. During the accident events, the actual accident causes were documented. The data showed that lost control, brake failure, and obstacles were the main accident causes on the NSE. With respect to lighting and surface conditions, most accidents occurred in daylight conditions and with a dry road surface. The main collision types in the accident records were out of control and rear collision. In addition, the accident time factor showed that 91.68% of the accidents occurred during the daytime. Additionally, the data also revealed that two-car accidents, single heavy car with an object and motorcycle with an object were the most recorded crashes on the NSE.

Figure 3. Location of the North-South Expressway (NSE) section analyzed in this study.

Table 2. Driver injury severity distribution according to accident related factors.

Factor	Property Damage Only	Evident Injury	Disabling Injury	Total
Location				
210–214	185	172	58	415
215–219	234	47	56	337
220–225	238	58	82	378
Road-bound				
South	453	99	139	691
North	287	73	79	439
Accident zone				
Interchange	14	3	0	17
Junction				
Lay-by	2	0	1	3
Main Route	666	155	209	1030
North Bound Entry Ramp	8	2	0	10
North Bound Exit Ramp	4	2	0	6
Rest and Service Area	21	4	2	27
South Bound Entry Ramp	2	0	1	3
South Bound Exit Ramp	7	1	3	11
Toll Plaza	16	5	2	23
Accident reported cause				
Bad Pavement Condition	0	1	0	1
Brake Failure	6	2	1	9
Bump-bump	37	12	27	76
Dangerous Pedestrian Behaviour	0	0	1	1
Drunk	0	0	1	1
Loss of Wheel	1	0	2	3
Lost control	75	18	22	115
Mechanical	5	1	0	6
Mechanical/Electrical Failure	11	0	1	12
Obstacle	43	12	6	61
Other Bad Driving	15	1	4	20
Other Human Factor/Over Load/Over Height	3	0	0	3
Over speeding	345	61	91	497
Parked Vehicle	4	4	10	18
Skidding	1	0	0	1
Sleepy Driver	134	44	42	220
Stray Animal	13	1	2	16
Tire burst	47	15	8	70
Lighting condition				
Dark with Street Light	47	6	8	61
Dark without Street Light	225	74	89	388
Dawn/Dusk	35	9	9	53
Day Light	433	83	112	628
Surface condition				
Dry	460	146	190	796
Wet	280	26	28	334
Collision type				
Angular Collision	9	2	0	11
Broken Windscreen	2	0	0	2
Cross direction	2	0	1	3
Head-on Collision	0	1	4	5
Hitting Animal	12	1	2	15
Hitting Object On Road	44	12	7	63
Others	20	0	6	26
Out of Control	457	92	107	656
Overturned	33	11	7	51
Rear Collision	137	48	81	266
Right Angle Side Collision	11	1	1	13
Side Swipe	13	4	2	19
Accident time				
Day time	677	156	203	1036
Night time	63	16	15	94

Table 2. *Cont.*

Factor	Property Damage Only	Evident Injury	Disabling Injury	Total
Vehicle type				
Car-Bus	7	3	6	16
Car-Car	499	68	60	627
Car-Heavy Car	51	11	14	76
Car-Motorcycle	4	7	22	33
Heavy Car	131	23	25	179
Heavy Car-Bus	2	3	3	8
Heavy Car-Heavy Car	24	9	15	48
Heavy Car-Motorcycle	0	1	6	7
Heavy Car-Taxi	2	0	0	2
Motorcycle	11	42	60	113
Motorcycle-Taxi	0	1	1	2
Motorcycle-Van	0	0	2	2
Taxi	1	0	1	2
Van	8	4	3	15

Before feeding the accident data into the recurrent neural network, the data must be preprocessed. The steps included were: removing missing data, data transformation, and detection of highly correlated factors. Some data were missing in the accident records; therefore, the complete raw data in which a missing data is found was removed. The data transformation included one-hot encoding for the categorical factors. In addition, correlation between predictors was assessed to detect any multicollinearity problem. First, the multiple R^2 was calculated for each factor. Second, the Variance Inflation Factor (*VIF*) was calculated from the multiple R^2 for each factor (Table 3). The highest correlation of 0.27 was found between lighting condition factor and surface condition factor. However, the highest multiple R^2 and *VIF* were found to be 0.135 and 1.156 for surface condition factor. There was no multicollinearity found among the factors if *VIF* = 1.0, however when the value exceeded 1.0, then moderate multicollinearity was found. In both cases, no high correlation was found during the model training and testing phase. Therefore, none of the factors was removed.

Table 3. Multicollinearity assessment among the accident related factors.

Factor	Multiple R^2	VIF
Accident location	0.062	1.066
Road bound	0.012	1.012
Accident zone	0.040	1.042
Accident reported cause	0.060	1.063
Lighting condition	0.095	1.105
Surface condition	0.135	1.156
Collision type	0.033	1.034
Accident time	0.009	1.009
Vehicle type	0.090	1.099

4.2. Results of RNN Model Performance

Figure 4 shows the accuracy performance and loss of the RNN model calculated for 100 epochs (iterations) using the training (80%) and validation (20%) datasets. In general, model accuracy on the training and validation datasets increases after each iteration with fluctuations. The fluctuations in accuracy are due to the use of the dropout technique in the model, which results in different training during every iteration. The dropout technique, which was used to prevent over-fitting, introduces some of the randomnesses to the network. In the first iteration, the accuracy was 61.28% and 66.37% for training and testing data, respectively. As the model trains during the first pass through the data, both training and validation losses decline, indicating that the model is learning the structure of the traffic accident data and possibly its temporal correlations. In the first and consecutive iterations the validation loss did not increase significantly and was always less than the training loss, indicating that the network did not overfit the training data and accurately generalized to the unseen validation data. After 100 epochs, the validation accuracy of the model was 71.77%.

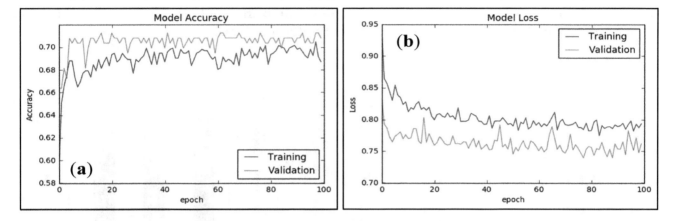

Figure 4. Accuracy performance and loss of the RNN model calculated for 100 epochs; (**a**) model accuracy; and (**b**) model loss.

4.3. Sensitivity Analysis of Optimization Algorithm

Several optimization algorithms such as SGD, Adagrad [35], RMSprop [36] and Adam [37] are available to adjust the weights and biases of a network by iterating through the training data. Each algorithm has its own advantages and disadvantages, and there are no clear guidelines for selecting an optimizer for a particular problem. Therefore, in this study, several optimization algorithms were evaluated as we searched for the best to train the proposed RNN model. The evaluated algorithms are SGD, RMSprop, Adagrad, Adadelta, Adam, Adamax, and Nadam [38]. The parameters of each algorithm were obtained from the aforementioned literature. Table 4 shows the performance of each optimization technique in predicting the severity of traffic accidents. The best validation accuracy (71.77) was achieved by the SGD method with a learning rate of 0.01. The SGD uses a small batch of the total data set to calculate the gradient of the loss function with respect to the weights and biases, using a backpropagation algorithm. The algorithm moves down the gradient by an amount proportional to its learning rate. Optimization methods such as Adam and Nadam have performed reasonably well.

Table 4. The performance of different optimization methods evaluated in this study.

Optimizer	Parameters	Training Accuracy (%)	Validation Accuracy (%)
SGD	lr = 0.01, momentum = 0.0, decay = 0.0, nesterov = False	71.79	71.77
RMSprop	lr = 0.001, rho = 0.9, epsilon = 1×10^{-8}, decay = 0.0	71.90	70.80
Adagrad	lr = 0.01, epsilon = 1×10^{-8}, decay = 0.0	71.35	71.24
Adadelta	lr = 1.0, rho = 0.95, epsilon = 1×10^{-8}, decay = 0.0	70.24	71.24
Adam	lr = 0.001, beta_1 = 0.9, beta_2 = 0.999, epsilon = 1×10^{-8}, decay = 0.0	73.12	71.68
Adamax	Lr = 0.002, beta_1 = 0.9, beta_2 = 0.999, epsilon = 1×10^{-8}, decay = 0.0	70.46	71.24
Nadam	Lr = 0.002, beta_1 = 0.9, beta_2 = 0.999, epsilon = 1×10^{-8}, schedule_decay = 0.004	74.67	71.68

4.4. Sensitivity Analysis of Learning Rate and RNN Sequence Length

Furthermore, the effect of the learning rate and RNN sequence length on the generalization ability of the proposed network in predicting the severity of traffic accidents was investigated. The learning rates varied by 0.5, 0.1, 0.05, 0.01, and 0.001. The sequence length varied by 2, 5, and 10. The choices of learning rate and sequence length were purely on an arbitrary basis. Figure 5 shows the results of these experiments. The highest validation accuracy was achieved when a learning rate of 0.01 was used. Reducing the learning rate up to 0.001 did not improve the validation accuracy, but reduced it to 70.0%. Similarly, large learning rates (e.g., 0.5) significantly reduced the performance of the network with a validation accuracy of 66.37%. On the other hand, the best validation accuracy (71.77%) was achieved with a sequence length of 2. In addition, the model performed better with a sequence length of 10 (70.87%) than 5 (70.43%). The unseen structured pattern in the model performance with different learning rates and sequence lengths indicated that the optimization of these parameters is critical to design a network that can best predict the severity of traffic accidents.

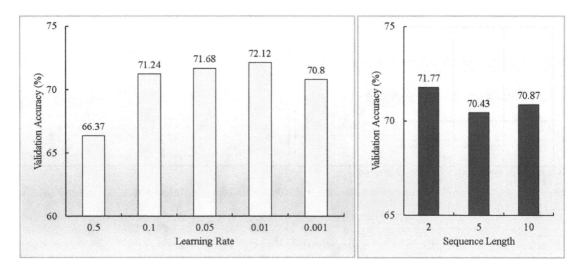

Figure 5. The sensitivity of the RNN model for different learning rate and sequence length configurations.

4.5. Network Depth Analysis

Generally, in a deep learning model, several modules and multilayers can be stacked on top of each other so that it is significant to analyze the network depth (number of hidden layers) to understand the network behavior. Table 5 shows the accuracy of the RNN model with various numbers of dense layers used on top of the LSTM layer. The best accuracy was achieved by using two dense (fully-connected) layers with 64 hidden units. This model achieved training and validation accuracies of 71.79% and 71.77% respectively. When another dense layer was added to the network, the training and validation accuracies were reduced to 70.09% and 71.24 respectively. When using five dense layers, the validation accuracy (70.35%) was found to be better than the training accuracy (67.26%); however, it could not perform better than using only two dense layers. The model started over-fitting when more than five dense layers were used. The validation accuracy of the network with eight dense layers was 56.37%, i.e., less by almost 10% than the training accuracy.

Table 5. The training and validation accuracy of the proposed RNN model with a different number of dense layers.

Number of Dense Layers	Training Accuracy (%)	Validation Accuracy (%)
2	71.79	71.77
3	70.09	71.24
5	67.26	70.35
8	65.27	56.37

In addition, the effect of the number of LSTM layers on model accuracy was assessed. Table 6 shows that the best training (71.79%) and validation (71.77%) accuracy could be achieved with a network of one LSTM layer. The accuracy was slightly reduced with another LSTM layer which was added to the network. Finally, when three LSTM layers were used, the accuracy of the model was gradually decreased and overfitted the training data.

Table 6. The training and validation accuracy of the proposed RNN model with a different number of LSTM layers.

Number of LSTM Layers	Training Accuracy (%)	Validation Accuracy (%)
1	71.79	71.77
2	71.30	71.58
3	67.11	65.34

4.6. Extraction Factor Contribution in the RNN Model

To compute the contribution of each factor in predicting the severity of traffic accidents, the profile method [39,40] was used. This technique examines the evolution of each factor with a scale of values, while the remaining factors keep their values fixed. Each factor x_i takes 11 values resulting from the division of the range, between its minimum and maximum value, into 10 equal intervals. All factors except one were initially fixed at their minimum value and then were fixed successively at their first quartile, median, third quartile and maximum values. Five values of the response variable were obtained for each 11 value and adopted by x_i, and the median of those five values was calculated. Finally, a curve with the profile of variation was obtained for every factor.

Table 7 shows the calculated weight of each factor by the profile method implemented in Python. The results indicate that the road bound and accident time have a significant effect on injury severity; the drivers had a greater risk of injuries during daytime along the southbound lanes of the NSE. Also, dry surface conditions were found to be more dangerous than a wet surface, as far as driver injury severity is concerned. During rainy times (surface is wet), drivers decrease their speed due to traffic jams, hence decreasing the severity level of driver injury. Lighting conditions, such as dark with and without street lights, increase the potential of possible injury and fatality, whereas daylight reduces the severity level of driver injury. The negative weight of the accident location factor indicates that severe accidents are most likely to happen in the section 220–225 km of the expressway. When vehicle types such as cars and motorcycles are involved in crashes, drivers are more prone to possible injury and fatality than for heavy vehicles and buses. Fatigue and speeding can give the driver a greater chance of experiencing a severe injury than other causes reported in the dataset. Accidents at the entry and exit ramps, toll plaza, and main route are more dangerous than accidents in other zones. Collision type also plays a significant role in increasing or decreasing the severity level of an accident. The analysis in this paper shows that collisions such as out of control and rear collisions increase the injury severity, whereas right angle side collision and sideswipe decrease the injury severity on the NSE.

Table 7. The calculated weights of accident related factors.

Factor	Weight
Accident location	−0.0074
Road bound	0.2899
Accident zone	0.0779
Accident reported cause	0.0469
Lighting condition	−0.0892
Surface condition	0.1785
Collision type	−0.0603
Accident time	0.3612
Vehicle type	−0.0468

4.7. Comparative Experiment

In this experiment, the proposed RNN model was compared with the traditional MLP model and the BLR approach. The advantages of MLP networks over Radial Basis Function (RBF) networks are that MLP networks are compact, less sensitive to including unnecessary inputs, and more effective for modeling data with categorical inputs.

The Grid Search algorithm was used to search the space of MLP network architectures and configurations. The number of units in the hidden layer, activation functions and the learning rate were optimized. The search space of the number of hidden units was 4 to 64 units. Five activation functions were tested such as Gaussian, Identity, Sigmoid, Softmax, and Exponential for the hidden and output layers. In addition, five values of learning rate (0.5, 0.1, 0.05, 0.01, and 0.001) were evaluated. The suboptimal network was then determined according to the best validation accuracy. The suboptimal network is the one that best represents the relationship between the input and output variables. In total, 100 network architectures with different combinations of selected parameters and network configurations were executed, and the validation accuracy on 20% of the data was computed for each created model. Then, the best network was retained. The best network had eight (8) hidden units with identity activation function, a Softmax activation in the output layer, and a learning rate of 0.01. The training and validation performance of the network was 68.90% and 65.48% respectively.

On the other hand, the BLR [41] and the computation of Markov chain Monte Carlo (MCMC) simulations were implemented in OpenBUGS software. BUGS modeling language (Bayesian Inference using Gibbs Sampling) is an effective and simplified platform to allow the computation using MCMC algorithms for all sorts of Bayesian models including BLR applied. The simulation of the posterior distribution of beta (β) allowed estimating the mean, standard deviation, and quartiles of the parameters of each explanatory variable. In the simulation stage, two MCMC chains were used to ensure convergence. The initial 100,000 iterations were discarded as burn-ins to achieve convergence and a further 20,000 iterations for each chain were performed and kept to calculate the posterior estimates of interested parameters.

In developing the Bayesian model, monitoring convergence is important because it ensures that the posterior distribution was achieved at the beginning of sampling of parameters. Convergence of multiple chains is assessed using the Brooks-Gelman-Rubin (BGR) statistic [42]. A value of less than 1.2 BGR statistic indicates convergence [42]. Convergence is also assessed by visual inspection of the MCMC trace plots for the model parameters and by monitoring the ratios of the Monte Carlo errors with respect to the corresponding standard deviation. The estimates of these ratios should be less than 0.05. In addition, MC error less than 0.05 also indicates that convergence may have been achieved [43]. The results of BLR showed training and validation accuracies of 70.30% and 58.30% respectively. In addition, the model had an average MC error of 0.047, Brooks-Gelman-Rubin (BGR) of 0.98, and model fitting performance (Deviance Information Criterion-DIC) of 322.

The comparative analysis showed that the proposed RNN model outperformed both MLP and BLR in terms of training and validation accuracies (Table 8). The BLR model performed better than MLP on the training dataset; however, its accuracy on the validation dataset was less than the MLP model. The MLP model uses only local contexts and therefore does not capture the spatial and temporal correlations in the dataset. The hidden units of the RNN model contain historical information from previous states, hence increasing the information about the data structure, which may be the main reason for its high validation accuracy over the other two methods. Building BLR models is more difficult than NN models, because they require expert domain knowledge and effective feature engineering processes. The NN model automatically captures the underlying structure of the dataset and extracts different levels of features abstractions. However, building deep learning models with various modules (e.g., fully connected networks, LSTM, convolutional layers) can be a challenging task. In addition, the BLR models are less prone to over-fitting of the training dataset than NN models, because they involve simpler relationships between the outcome and the explanatory variables. Complex networks with more hidden units and many modules often tend to over-fit more, because they detect almost any possible interaction so that the model becomes too specific to the training dataset. Therefore, optimization of network structures is very critical to avoid over-fitting and build practical prediction models. In computing the importance of an explanatory factor, the BLR models could easily calculate the factor importance and the confidence intervals of the predicted probabilities. In contrast, NN methods, which are not built primarily for statistical use, cannot easily calculate the factor importance or generate confidence intervals of the predicted probabilities unless extensive computations were done.

Table 8. Performance comparison of the proposed RNN model with Multilayer Perceptron (MLP) and Bayesian Logistic Regression (BLR) models.

Method	Training Accuracy (%)	Validation Accuracy (%)
MLP	68.90	65.48
BLR	70.30	58.30
RNN	71.79	71.77

In addition, based on the estimated weight of each accident related factor, the factors could be ranked by scaling the weights into the range of 1–100 and then giving them a rank from 1 to 9 as shown in Table 9. The three methods (MLP, BLR, and RNN) did not agree on the factors' ranking. For example, the RNN model ranked accident time as the most influential factor in injury severity, whereas accident time was ranked 2 and 7 by the BLR and MLP models, respectively. Both the RNN and BLR models agreed on several factors' ranking i.e., accident location (6), accident reported cause (5), and collision type (8). The correlation (R^2) between the RNN and BLR ranks was 0.72. In contrast, RNN and MLP did not agree on the ranking of the factors and their ranking correlation was the lowest (0.38).

Table 9. Calculated ranks of the accident related factors in the RNN, MLP, and BLR models.

Factor	MLP	BLR	RNN
Accident location	2	6	6
Road bound	8	4	2
Accident zone	5	3	4
Accident reported cause	3	5	5
Lighting condition	4	7	9
Surface condition	9	1	3
Collision type	6	8	8
Accident time	7	2	1
Vehicle type	1	9	7

$$\text{BLR:MLP: } R^2 = 0.51$$
$$\text{RNN:MLP: } R^2 = 0.38$$
$$\text{MLP:BLR: } R^2 = 0.51$$
$$\text{RNN:BLR: } R^2 = 0.72$$

4.8. Computational Complexity of the Model

The time complexity of the RNN model was measured in terms of training and testing time per iteration. Table 10 gives information about the average training time per iteration with batch size of 32 and the average testing time per prediction. It can be seen that the model on average spends around 150 milliseconds per iteration during training and only ~13 milliseconds per prediction for new unseen examples. Although this experiment shows the computational efficiency of the model, it is also worth to note that the training time can be increased by reducing the batch size or sequence length, and also by increasing the volume of the training data.

Table 10. Average training and testing time per iteration/ prediction of the proposed model.

Time (milliseconds per iteration)	RNN Model
Training Time	150.23
Testing Time	13.17

Additionally, the major computational problem of NNs is their scalability; sometimes they become unstable when applied to larger problems. However, recent advancements in hardware and computing performance such as Graphics Processing Units (GPUs), parallel computing, and cloud computing have decreased most of the limitations of NN models including RNN-based computations.

4.9. Applicability and Limitations of the Proposed Method

The proposed RNN model is a promising traffic accident forecasting tool that has several applications in practice. First, identifying the most risky road sections (i.e., site ranking) is a daily practice of transportation agencies in many cities around the world. Accident prediction models, such as the one we proposed in this research, are often an effective solution for identifying risky road sections and helping to conduct investigations into methods for improving the safety performance of the road systems. Second, the RNN model is able to explain the underlying relationships between several accidents' related factors, such as accident time and collision type, and the injury severity outcomes. Information about the effects of accident factors on the injury severity outcomes provides huge amount of information to the transportation agencies and stakeholders. Finally, estimating the expected number of traffic accidents in a road section can help road designers to optimize the geometric alignments of the roads based on the accident scenarios.

However, the proposed RNN model has some constraints and limitations. The major limitation of the model is that the input factors are prerequisite and if any of them is missing, the output probabilities cannot be accurately estimated. Another constraint of the model is the sequence length of the RNN model, which mainly depends on the number of accident records in the training dataset. To handle this limitation, the future works should develop RNN models that operate on input sequences of variable lengths via Tensorflow dynamic calculation.

5. Conclusions

In this paper, a Recurrent Neural Network (RNN) model was developed to predict the injury severity of traffic accidents on the NSE, Malaysia. An optimized network architecture was determined through a systematic grid search for the suboptimal network hyper-parameters. Several hyper-parameters of the RNN model were critical to achieve the highest validation accuracy. The best optimization algorithm was determined to be the SGD with learning rate, momentum, and weight decay of 0.01, 0.80, and 0.9, respectively. In addition, the dropout technique helped us to reduce the complexity of the model and also the chance of over-fitting the training dataset. The RNN model achieved the best validation accuracy of 71.77% when compared to the MLP and BLR models. This indicates that additional information about temporal and contextual correlations among accident records could help the RNN model to perform better than the other models. In the sensitivity analysis of RNN sequence length, the best accuracy was achieved with a sequence length of 2, whereas the accuracy was decreased when a sequence length of 10 or 5 was used. This means that the accident data has temporal and contextual structures that could not be used by the MLP and BLR models. The impact of each factor in the RNN model was calculated using the profile method. The results showed that the most influential factors in predicting injury severity of traffic accidents are accident time and road bound. In the study area, the model predicted that the southbound section is more dangerous than the northbound, with respct to injury severity. The model also predicted drivers had a greater risk of injury on a dry surface and with lighting conditions such as dark with and without street lights.

While the proposed RNN model outperformed the MLP and BLR models and could estimate the factors' impacts, further studies should focus on developing more flexible and optimized network structures to predict accident frequency and injury severity. Future theoretical studies are encouraged to focus on improving the ability of the RNN model to represent variables and data structures, and to store data over long timescales. In addition, more applied research is recommended to develop new techniques to make use of additional information in accident datasets, such as contextual structures, spatial and temporal interactions and underlying relationships between accident factors and injury severity outcomes.

Author Contributions: Maher Ibrahim Sameen and Biswajeet Pradhan conceived and designed the experiments; Maher Ibrahim Sameen performed the experiments and analyzed the data; Biswajeet Pradhan and Maher Ibrahim Sameen contributed reagents/materials/analysis tools; Maher Ibrahim Sameen and Biswajeet Pradhan wrote the paper.

References

1. Sameen, M.I.; Pradhan, B. Assessment of the effects of expressway geometric design features on the frequency of accident crash rates using high-resolution laser scanning data and gis. *Geomat. Nat. Hazards Risk* **2016**, 1–15. [CrossRef]
2. Pei, X.; Wong, S.; Sze, N.-N. A joint-probability approach to crash prediction models. *Accid. Anal. Prev.* **2011**, *43*, 1160–1166. [CrossRef] [PubMed]
3. Fogue, M.; Garrido, P.; Martinez, F.J.; Cano, J.-C.; Calafate, C.T.; Manzoni, P. A system for automatic notification and severity estimation of automotive accidents. *IEEE Trans. Mob. Comput.* **2014**, *13*, 948–963. [CrossRef]
4. Pawlus, W.; Reza, H.; Robbersmyr, K.G. Application of viscoelastic hybrid models to vehicle crash simulation. *Int. J. Crashworthiness* **2011**, *16*, 195–205. [CrossRef]
5. Pawlus, W.; Robbersmyr, K.G.; Karimi, H.R. Mathematical modeling and parameters estimation of a car crash using data-based regressive model approach. *Appl. Math. Model.* **2011**, *35*, 5091–5107. [CrossRef]
6. Pawlus, W.; Karimi, H.R.; Robbersmyr, K.G. Mathematical modeling of a vehicle crash test based on elasto-plastic unloading scenarios of spring-mass models. *Int. J. Adv. Manuf. Technol.* **2011**, *55*, 369–378. [CrossRef]
7. Karimi, H.R.; Pawlus, W.; Robbersmyr, K.G. Signal reconstruction, modeling and simulation of a vehicle full-scale crash test based on morlet wavelets. *Neurocomputing* **2012**, *93*, 88–99. [CrossRef]
8. Abdelwahab, H.; Abdel-Aty, M. Development of artificial neural network models to predict driver injury severity in traffic accidents at signalized intersections. *Transp. Res. Rec. J. Transp. Res. Board* **2001**, *1746*, 6–13. [CrossRef]
9. Lv, Y.; Duan, Y.; Kang, W.; Li, Z.; Wang, F.-Y. Traffic flow prediction with big data: A deep learning approach. *IEEE Trans. Intell. Transp. Syst.* **2015**, *16*, 865–873. [CrossRef]
10. Duan, Y.; Lv, Y.; Liu, Y.-L.; Wang, F.-Y. An efficient realization of deep learning for traffic data imputation. *Transp. Res. C* **2016**, *72*, 168–181. [CrossRef]
11. Alkheder, S.; Taamneh, M.; Taamneh, S. Severity prediction of traffic accident using an artificial neural network. *J. Forecast.* **2017**, *36*, 100–108. [CrossRef]
12. Yu, R.; Li, Y.; Shahabi, C.; Demiryurek, U.; Liu, Y. Deep learning: A generic approach for extreme condition traffic forecasting. In Proceedings of the 2017 SIAM International Conference on Data Mining (SDM), Houston, TE, USA, 27–29 April 2017.
13. Delen, D.; Sharda, R.; Bessonov, M. Identifying significant predictors of injury severity in traffic accidents using a series of artificial neural networks. *Accid. Anal. Prev.* **2006**, *38*, 434–444. [CrossRef] [PubMed]
14. Hashmienejad, S.H.-A.; Hasheminejad, S.M.H. Traffic accident severity prediction using a novel multi-objective genetic algorithm. *Int. J. Crashworthiness* **2017**, 1–16. [CrossRef]
15. Zeng, Q.; Huang, H. A stable and optimized neural network model for crash injury severity prediction. *Accid. Anal. Prev.* **2014**, *73*, 351–358. [CrossRef] [PubMed]
16. Dean, J.; Corrado, G.; Monga, R.; Chen, K.; Devin, M.; Mao, M.; Senior, A.; Tucker, P.; Yang, K.; Le, Q.V. Advances in neural information processing systems. In *Large Scale Distributed Deep Networks*; Association for Computing Machinery: New York, NY, USA, 2012; pp. 1223–1231.
17. Collobert, R.; Weston, J. A unified architecture for natural language processing: Deep neural networks with multitask learning. In Proceedings of the 25th International Conference on Machine learning, New York, NY, USA, 5–9 July 2008; pp. 160–167.
10. Huang, W.; Song, G.; Hong, H.; Xie, K. Deep architecture for traffic flow prediction: Deep belief networks with multitask learning. *IEEE Trans. Intell. Transp. Syst.* **2014**, *15*, 2191–2201. [CrossRef]

19. Xie, Z.; Yan, J. Detecting traffic accident clusters with network kernel density estimation and local spatial statistics: An integrated approach. *J. Transp. Geogr.* **2013**, *31*, 64–71. [CrossRef]

20. LeCun, Y.; Bengio, Y.; Hinton, G. Deep learning. *Nature* **2015**, *521*, 436–444. [CrossRef] [PubMed]

21. Sutskever, I.; Martens, J.; Hinton, G.E. Generating text with recurrent neural networks. In Proceedings of the 28th International Conference on Machine Learning (ICML-11), Bellevue, WA, USA, 28 June–2 July 2011; pp. 1017–1024.

22. Graves, A.; Mohamed, A.-r.; Hinton, G. Speech recognition with deep recurrent neural networks. In Proceedings of the 2013 IEEE International Conference on Acoustics, Speech and Signal Processing, Vancouver, BC, Canada, 26–31 May 2013; pp. 6645–6649.

23. Hochreiter, S.; Schmidhuber, J. Long short-term memory. *Neural Comput.* **1997**, *9*, 1735–1780. [CrossRef] [PubMed]

24. Donahue, J.; Anne Hendricks, L.; Guadarrama, S.; Rohrbach, M.; Venugopalan, S.; Saenko, K.; Darrell, T. Long-Term Recurrent Convolutional Networks for Visual Recognition and Description. In Proceedings of the IEEE Conference on Computer Vision and Pattern Recognition, Boston, MA, USA, 7–15 June 2015; pp. 2625–2634.

25. Wang, X.; Abdel-Aty, M. Temporal and spatial analyses of rear-end crashes at signalized intersections. *Accid. Anal. Prev.* **2006**, *38*, 1137–1150. [CrossRef] [PubMed]

26. Krizhevsky, A.; Sutskever, I.; Hinton, G.E. Imagenet classification with deep convolutional neural networks. In Proceedings of the Advances in Neural Information Processing Systems, Lake Tahoe, NV, USA, 3–6 December 2012; pp. 1097–1105.

27. Srivastava, N.; Hinton, G.E.; Krizhevsky, A.; Sutskever, I.; Salakhutdinov, R. Dropout: A simple way to prevent neural networks from overfitting. *J. Mach. Learn. Res.* **2014**, *15*, 1929–1958.

28. Hinton, G.; Rumelhart, D.; Williams, R. Learning internal representations by back-propagating errors. In *Parallel Distributed Processing: Explorations in the Microstructure of Cognition*; MIT Press: London, UK, 1985; Volume 1.

29. Hinton, G.E.; Srivastava, N.; Krizhevsky, A.; Sutskever, I.; Salakhutdinov, R.R. Improving neural networks by preventing co-adaptation of feature detectors. *arXiv* **2012**, arXiv:1207.0580.

30. Bousquet, O.; Bottou, L. The tradeoffs of large scale learning. In Proceedings of the Advances in Neural Information Processing Systems, Vancouver, BC, Canada; 2008; pp. 161–168.

31. Pascanu, R.; Mikolov, T.; Bengio, Y. On the difficulty of training recurrent neural networks. In Proceedings of the 30th International Conference on International Conference on Machine Learning, Atalnta, GA, USA, 16–21 June 2013; pp. 1310–1318.

32. Zur, R.M.; Jiang, Y.; Pesce, L.L.; Drukker, K. Noise injection for training artificial neural networks: A comparison with weight decay and early stopping. *Med. Phys.* **2009**, *36*, 4810–4818. [CrossRef] [PubMed]

33. Pedregosa, F.; Varoquaux, G.; Gramfort, A.; Michel, V.; Thirion, B.; Grisel, O.; Blondel, M.; Prettenhofer, P.; Weiss, R.; Dubourg, V. Scikit-learn: Machine learning in python. *J. Mach. Learn. Res.* **2011**, *12*, 2825–2830.

34. Abadi, M.; Agarwal, A.; Barham, P.; Brevdo, E.; Chen, Z.; Citro, C.; Corrado, G.S.; Davis, A.; Dean, J.; Devin, M. Tensorflow: Large-scale machine learning on heterogeneous distributed systems. *arXiv* **2016**, arXiv:1603.04467.

35. Duchi, J.; Hazan, E.; Singer, Y. Adaptive subgradient methods for online learning and stochastic optimization. *J. Mach. Learn. Res.* **2011**, *12*, 2121–2159.

36. Tieleman, T.; Hinton, G. Lecture 6.5-rmsprop: Divide the gradient by a running average of its recent magnitude. *COURSERA Neural Netw. Mach. Learn.* **2012**, *4*, 26–31.

37. Kingma, D.; Ba, J. Adam: A method for stochastic optimization. *arXiv* **2014**, arXiv:1412.6980.

38. Ruder, S. An overview of gradient descent optimization algorithms. *arXiv* **2016**, arXiv:1609.04747.

39. Lek, S.; Belaud, A.; Dimopoulos, I.; Lauga, J.; Moreau, J. Improved estimation, using neural networks, of the food consumption of fish populations. *Oceanogr. Lit. Rev.* **1996**, *9*, 929. [CrossRef]

40. Lek, S.; Belaud, A.; Dimopoulos, I.; Lauga, J.; Moreau, J. Improved estimation, using neural networks, of the food consumption of fish populations. *Mar. Freshw. Res.* **1995**, *46*, 1229–1236. [CrossRef]

41. Xu, C.; Wang, W.; Liu, P.; Guo, R.; Li, Z. Using the bayesian updating approach to improve the spatial and temporal transferability of real-time crash risk prediction models. *Transp. Res. C* **2014**, *38*, 167–176. [CrossRef]

42. El-Basyouny, K.; Sayed, T. Application of generalized link functions in developing accident prediction models. *Saf. Sci.* **2010**, *48*, 410–416. [CrossRef]

43. Wang, C.; Quddus, M.A.; Ison, S.G. Predicting accident frequency at their severity levels and its application in site ranking using a two-stage mixed multivariate model. *Accid. Anal. Prev.* **2011**, *43*, 1979–1990. [CrossRef] [PubMed]

12

Habitat Potential Mapping of Marten (*Martes flavigula*) and Leopard Cat (*Prionailurus bengalensis*) in South Korea using Artificial Neural Network Machine Learning

Saro Lee [1,2], Sunmin Lee [3,4], Wonkyong Song [5] and Moung-Jin Lee [4,*]

[1] Geological Research Division, Korea Institute of Geoscience and Mineral Resources (KIGAM), 124, Gwahak-ro Yuseong-gu, Daejeon 34132, Korea; leesaro@kigam.re.kr
[2] Korea University of Science and Technology, 217 Gajeong-ro Yuseong-gu, Daejeon 34113, Korea
[3] Department of Geoinformatics, University of Seoul, 163 Seoulsiripdaero, Dongdaemun-gu, Seoul 02504, Korea; smlee@kei.re.kr
[4] Center for Environmental Assessment Monitoring, Environmental Assessment Group, Korea Environment Institute (KEI), 370 Sicheong-daero, Sejong 30147, Korea
[5] Department of Landscape Architecture, Dankook University, 119 Dandae-ro, Dongnam-gu, Cheonan-si, Chungnam 31116, Korea; wksong@dankook.ac.kr
* Correspondence: leemj@kei.re.kr

Abstract: This study developed habitat potential maps for the marten (*Martes flavigula*) and leopard cat (*Prionailurus bengalensis*) in South Korea. Both species are registered on the Red List of the International Union for Conservation of Nature, which means that they need to be managed properly. Various factors influencing the habitat distributions of the marten and leopard were identified to create habitat potential maps, including elevation, slope, timber type and age, land cover, and distances from a forest stand, road, or drainage. A spatial database for each species was constructed by preprocessing Geographic Information System (GIS) data, and the spatial relationship between the distribution of leopard cats and environmental factors was analyzed using an artificial neural network (ANN) model. This process used half of the existing habitat location data for the marten and leopard cat for training. Habitat potential maps were then created considering the relationships. Using the remaining half of the habitat location data for each species, the model was validated. The results of the model were relatively successful, predicting approximately 85% for the marten and approximately 87% for the leopard cat. Therefore, the habitat potential maps can be used for monitoring the habitats of both species and managing these habitats effectively.

Keywords: habitat mapping; marten; leopard cat; ANN; South Korea

1. Introduction

Medium-sized predators control not only herbivorous animals but also intermediate predators, and have a combined effect on the overall food chain. Apex predators generally have low population densities and wide ranges of activities, and are vulnerable to local extinction due to habitat damage and disconnection [1]. In South Korea, members of the order Carnivora, such as tigers, leopards, and wolves, have become extinct or lost their ecological functions in natural ecosystems over the past century. As a result, the importance of the remaining carnivorous animals, such as martens, leopard cats, and otters, is increasing.

Martens (*Martes* sp.) are mammals with wide distribution worldwide, from subtropical to sub-Antarctic habitats. They are generally characterized by a larger sphere of activity than other mammals [2], and have roles as forest ecosystem indicator species sensitive to habitat disturbances and habitat fragmentation [3]. Therefore, the marten in Korea, the yellow-throated marten (*Martes flavigula*), is considered a second-grade endangered species of wild fauna and flora by the Ministry of Environment, South Korea, and belongs to Annex II of the Convention on International Trade in Endangered Species of Wild Fauna and Flora (CITES). Studies on marten habitats have been conducted mainly in tropical and subtropical areas such as Thailand [4], and there is insufficient information on marten habitats on the Korean Peninsula.

The leopard cat (*Prionailurus bengalensis*) is the only wild feline carnivore in South Korea, which also belongs to the second-grade endangered species of wild fauna and flora of the Ministry of Environment and the CITES Annex II list. Leopard cats are distributed within a wide range, from deep forests to rural and coastal areas in Korea. Moreover, they maintain relatively stable populations, due to their adaptability to inhabiting various habitats. It has been reported that leopard cats are distributed widely in forests and rural areas [5], but there have been few quantitative studies on the environmental spatial characteristics of their habitats. In particular, there is a lack of research on their distribution throughout South Korea and analysis of their habitat characteristics.

Machine learning is designed to process new data and predict results by learning patterns through training based on consistent patterns among variables in a dataset [6,7]. Recently, the amount of plant- and animal-related data has rapidly increased, and many studies using machine learning models have been conducted [8,9]. Depending on the approach applied, machine learning can use various algorithms, such as decision tree, neural network, and support vector machine, and it can be divided into supervised learning and unsupervised learning depending on the presence or absence of training data. Thus, many studies have been conducted to predict the distribution of plants and animals using machine learning techniques [10,11]. In addition, Geographic Information System (GIS) platforms have been used as a useful tool to model the spatial relationships between specific events and related factors [12–14].

Thus, recent studies have used GIS to indicate the distribution of habitats of various species. Studies on mapping and quantifying mammalian habitats have been conducted through GIS-based models [15,16]. Among mammals, studies on martens and leopard cats have been conducted mainly in Asia. Studies of the genetic structure and mitochondrial genome of the marten and the leopard cat have also been performed [17,18]. In particular, the complete mitochondrial genome has been analyzed for martens in Korea [19]. Meanwhile, studies on the movements and activity patterns of leopard cats have been conducted in China [20,21]. One study predicted the distribution of leopard cats in Borneo [22], and another similar study predicting the habitat potential of leopard cats was conducted in South Korea using GIS [23]. Habitat use and activity patterns of martens were also analyzed in central and northern Thailand [4]. Typically, the habitat distributions of badgers [24,25], leopard cats [26], and bears [27] have been mapped using logistic regression models. In recent years, research has been conducted on the creation of habitat distribution maps of leopard cats in South Korea through probabilistic and statistical models [23]. Supporting this, other probabilistic models have been used to map European bison habitats [28], and statistical models have been applied to the habitat distribution of bats [29,30]. Moreover, habitat mapping of water birds for conservation plans was conducted in the Hamun wetland [31]. In Canada, a high-resolution habitat map of Atlantic wolffish was also created [32]. However, artificial neural network (ANN) modeling has not been applied to analyzing the habitat of martens or leopard cats in South Korea. Therefore, the purposes of this study were to map the distribution of martens and leopard cats in South Korea using ANN modeling and to clarify the relationship between the habitat distributions and various environmental factors.

South Korea is located in East Asia between latitudes 33° and 39° N and longitudes 124° and 130° E, including all of its islands, and occupies the southern part of the Korean Peninsula (Figure 1); the total area of the country is approximately 100,032 km^2 [33]. The boundary of the study area in this study is marked with a red line in the Figure 1. South Korea can be divided into four regions: a western region of broad coastal plains, an eastern region of high mountain ranges, a southwestern mountainous region, and a southeastern region of a broad basin. Mountains cover about 70% of South Korea, and the country is surrounded by the Yellow Sea to the west, the East Sea to the east, and the Korea Strait and the East China Sea to the south. The north of the country is bordered by the Democratic People's Republic of Korea, via the demilitarized zone; therefore, it is difficult to support the return of extirpated terrestrial animals in South Korea, since their path is blocked in all directions. To support conservation efforts, habitat potential maps should be created to manage habitats before extirpation, including the selection of development areas. Therefore, in this study, ANN modeling was applied to generate habitat potential maps for martens and leopard cats. The weights of factors related to the habitat potentials were calculated with the model, and the results were validated to ensure the reliability of the maps.

Figure 1. Study area: (**a**) South Korea in East Asia; and (**b**) South Korea.

2. Data

2.1. Habitat Survey

This study used data from the Second National Survey on Natural Environment in South Korea, which included the identification of habitats of endangered species of wild animals and plants. From 1997 to 2003, the National Institute for Environmental Studies conducted a survey of species appearances and spatial distributions of wild animals by experts from various research institutes. Surveys were performed from February to October every year for mammals to consider changes in seasonal patterns. The survey methods were based on field observations, including direct observations, community surveys, tracking, feces, and footprints, to investigate the species occurrence and spatial distribution of wild animals. The locations observed or detected by the traces were geocoded via Global Positioning System and were composed of GIS data.

The marten has the second largest sphere of action among mammals in the Korean Peninsula after the Asiatic black bear. Martens prefer broadleaf forests and mainly inhabit mature forests, such as fourth-grade forests that have $\geq$50% occupancy rates of 31–40-year-old trees with diameters of $\geq$30 cm. Martens inhabiting the Jirisan area occur at a density of 1–1.6 per 10 km^2, and are diurnal animals with a wide range of behaviors, moving an average of 11.2 $\pm$ 5.4 km per day. Martens mainly hunt Eurasian red squirrels, rats, and other rodents as food, but they also hunt hares, young roe deer, and wild boars. They also consume tree fruit, such as those from the lotus persimmon tree. In winter, they often hunt for food in shrubs on the edge of forests and frequently cross two-lane roads in their radius of action [34].

Leopard cats have activity areas of 3.69 $\pm$ 1.34 km^2, and the core space of the species is estimated to be 0.64 $\pm$ 0.47 km^2 [5]. Leopard cats prefer forests, as well as adjacent grassland and agricultural land, as habitats. They have a high preference for inland wetlands, such as wild grassland along riverbanks. Owing to their behavioral radius and nocturnality of movement in a variety of regions, they suffer a high frequency of road kills on roads adjacent to forests and rivers. Rodents, birds, and small mammals are their main sources of food, and forest ridges and valleys are used as main transport routes by leopard cats.

In the Second National Survey on Natural Environment, martens were found at 156 points, and leopard cats were found at 630 points in the study area (Figure 2). In this study, the observed data were divided randomly in half for model training and validation. This confirmed that the distribution of martens is mainly limited to large forest areas, including the Mt. Baekdu range. Meanwhile, leopard cats inhabit a wide range of emergence sites, from major forests to rivers and agricultural areas.

Figure 2. Data obtained from field observations of (a) Martens; and (b) Leopard cats.

2.2. Habitat-Related Factors

Habitat distributions of wild animals are influenced by the combined impacts of various factors. Environmental factors such as topographic characteristics, forest properties [22], and distance from essential factors to support life, such as water, influence the habitat distributions of mammals, including martens and leopard cats [22]. In this study, ground elevation, slope gradient, slope aspect, timber type, timber age, land cover, and distance from road, water, and forest were selected as factors influencing the habitat distributions of marten and leopard cat (Table 1 and Figure 3). The attribute information of the timber type of forest map is shown separately in Table 2, due to the limitation of the figure size. The determination and collection of factors were a fundamental component of mapping the potential habitats of these species. The factors were collected from nationally generated thematic maps and field investigations, such as those described in Section 2.1.

Table 1. Data layers related to marten and leopard cat habitat.

Original Data	Factors	Data Type	Scale
Habitat	Marten Leopard cat	Point	-
Topographical map [a]	Ground elevation (m) Slope gradient (°) Slope aspect	GRID	1:5000
Forest map [b]	Timber type Timber age	Polygon	1:25,000
Land cover map [c]	Land cover Distance from road (m) Distance from water (m) Distance from forest (m)	Polygon	1:25,000

[a] The digital topographic map by National Geographic Information Institute (NGII); [b] The forest map published by Korea Forest Service (KFS); [c] The land use map offered by the Korea Ministry of Environment.

A digital elevation model (DEM) was produced by generating a triangulated irregular network and digitizing the contours in 100-m intervals from topographical maps published by the National Geographic Information Institute. The slope gradient and slope aspect calculations were performed in ArcGIS 10.3. Thematic maps of forest and land cover were prepared at a 1:25,000 scale in a polygon format. The forest map was provided by the Korea Forest Service, and the timber type and age were prepared from this map. The land cover map was generated and published by the Ministry of Environment, South Korea. Land cover, distance from road, water, and forest were also calculated using ArcGIS.

All factors were converted into a 100 m × 100 m grid format to apply the ANN model. The row and column sizes of the study area were 6056 and 3533 cells, and the study area consisted of a total of 9,952,165 cells, except in the no data region, based on the DEM. The marten and leopard cat habitat data were converted from point formats into grid formats. Half of both the marten and the leopard cat habitat data were selected via random sampling as input data. The other half of the data were used for validation. Ultimately, all factors were collated into one input dataset with randomly extracted habitat sample data for both marten and leopard cat.

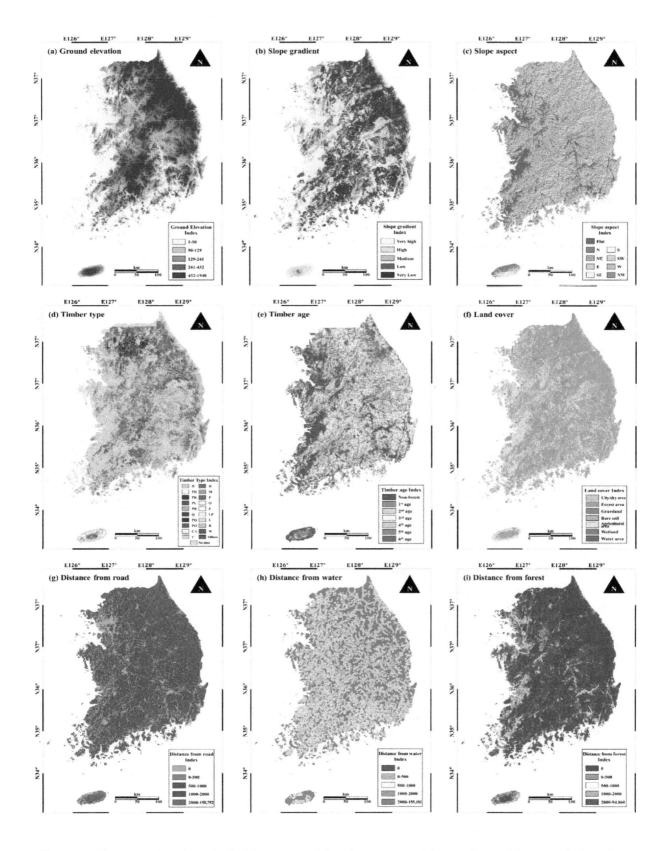

Figure 3. Factors related to the habitat potentials of marten and leopard cat: (**a**) ground elevation, (**b**) slope gradient, (**c**) slope aspect, (**d**) timber type, (**e**) timber age, (**f**) land cover, (**g**) distance from road, (**h**) distance from water, and (**i**) distance from forest.

Table 2. Attribute information of timber type from forest map.

	Code	Forest Type		Code	Forest Type
Forest species	D PD PK PL PR Q PQ PO CA	Pinus densiflora Forests Pinus densiflora artificial forest Pinus koraiensis forest Larch Pinus rigida forest Oak forest Oak artificial forest Poplar forest Chestnut artificial forest	Forest physiognomy	C H M	Conifer mixed forest Broadleaved forest Mixed forest of soft and hardwood
			Dentuded area	F O E LP L	Cut-over area Non-stocked forest land Dentuded land Grassland Farmland
			Left-over area	R W	Left-over area Water

3. Methods

The machine learning technique determines the algorithm patterns using the sampled training data and derives the results through learned patterns. In other words, it is possible to learn from results that are already known according to information from the samples. Habitats are influenced by various environmental factors based on the nature of animals; therefore, even if there are no animals at the present time, the possibility of migration into an area with a similar environment exists at any time. Therefore, the habitat pattern can be determined by training based on habitat data using the machine learning method, and the habitat potential of an entire study area can be created. Among various machine learning techniques, habitat potential mapping of martens and leopard cats was conducted using an ANN in this study.

Interest in ANNs has increased recently. ANN is a highly sophisticated modeling method that can model complex functions such as environmental problems or social issues with large numbers of variables. Mimicking the human biological neural system, an ANN adjusts the weight from the basic unit of perceptron between the input and output data, minimizing the error of the result. In addition to these strengths, ANNs have been used for prediction or classification in a remarkable range of fields, such as engineering, physics, geology, and environmental science. In particular, neural networks can be applied to nonlinear problems of dimensionality and can be used to identify spatial relationships between observation location data and influencing factors.

Multi-layer perceptron is one of the most popular ANN models, and was created by [35]. This network, which allows for the regression of nonlinear data, was used in this study with a hidden layer between the input layers of habitat-related factors and the output layer. In this study, a back-propagation algorithm was used for neural network training. In back propagation, the algorithm calculates the gradient vector of the error surface and distance from the current point. The error can be decreased when the point moves a short distance from the output layer to other layers. The process iterates through a number of epochs, submitting the training data and calculating the error by comparing the target and output. The weights can be corrected through the error from the iteration and surface gradient. The initial network is randomly set, and the stopping point is determined by a set number of epochs or a user-selected stopping point. The following equation indicates the weight-updating architecture of the back-propagation algorithm.

$$\omega_k^{up} = \omega_k - \eta * \frac{\partial E_{total}}{\omega_k} \tag{1}$$

where ω_{k_0} is the initial value of the weight ω_k, which is given randomly, and ω_k is the updated weight for connection k. η is the learning rate that determines the step size and is typically chosen experimentally. E_{total} is the error of the output.

$$E_{total} = \frac{1}{2}\sum_{j}^{n}\left(T_j - out_{o_j}\right)^2 \tag{2}$$

T is the expected output value according to the input data [36] and j is the number assigned to the node in each layer. n is the number of output layers. E_{total} is minimized through the updated weight from Equation (16). To detect the minimum point, differentiation of E_{total} is necessary. Thus, an activation function is needed for conversion from the result of real number x to a range of values between 0 and 1 when the value is passed through the next layer or when the final result is expected. The unipolar sigmoid function was used for the activation function in this study. The equation of the unipolar sigmoid function is as follows.

$$f(x) = \frac{1}{1 + e^{-x}} \tag{3}$$

In this study, the ANN was supported by MATLAB software. Using the training data of randomly extracted habitat sample data, the slope data of the flat area were used as non-habitat area for training, and the input data of the study area were rearranged between 0.1 and 0.9. The weights of the input layers were calculated using the back-propagation algorithm of the ANN with a $9 \times 18 \times 1$ structure of the networks. The initial weight was set randomly as described previously, and the learning rate and number of epochs were set to 0.01 and 1000, respectively. The root-mean-square error, the criterion of the stopping point of decreasing error, was set to 0.01.

The process of potential mapping for martens and leopard cats can be explained in three steps, as shown in Figure 4. First, a spatial database was constructed with nine potential habitat-influencing factors, including the observation points of martens and leopard cats. Half of the marten and leopard distribution data selected by random sampling were used as training sets. The locations of observations of martens and leopard cats were confirmed in field surveys for the application and validation of habitat potential models described in Section 2.2 (Figure 2). The observation location data of martens and leopard cats were set as dependent variables. Second, the ANN model was applied to a spatial database to map the habitat potentials. Nine factors that were considered to affect the marten and leopard cat habitats were set as independent variables. Elevation, slope, and aspect derived from topographic maps, timber type and timber age from forest maps, and land cover, distance from road, distance from water, and distance from forest from land cover map were included in the spatial database with half of the randomly sampled marten and leopard cat habitat data. The ANN model was applied after constructing the spatial database. To apply the model, MATLAB programming was used with the GIS data. The resulting maps of the habitat potentials were validated using a receiver operating characteristic (ROC) curve. Finally, validation of the marten and leopard cat habitat potential maps was performed using the remaining marten and leopard cat distribution data that were not included in the spatial database.

Figure 4. Flow chart of the steps used in this study

4. Results

4.1. Weight of Related Factors and Habitat Potential Mapping

The weights of the related factors between the layers acquired during the ANN model training process are shown in Table 3. The weight values show the contribution of each weight to the ANN model. Since the weights were initialized during the first training, the results could differ. This study attempted to obtain similar results by performing the calculation several times. The weight calculation was repeated for 10 epochs, with 100 cycles for each epoch, to identify the influence of the randomly extracted sample for a city/dry area (public area), non-habitat area, since most wild animals do not live in populated urban areas. No leopard cats were observed in public areas in the results of the frequency ratio analysis in a previous study [23]. In the iterations of each epoch, the weights were updated via back propagation. The standard deviation was 0.001–0.018 for martens and 0.002–0.021 for leopard cats. The ranges of standard deviations indicated that there were minimal effects of random sampling of non-habitat area on the results.

Table 3. Neural network weight between martens and leopard cats, and habitat-related factors.

	Marten			Leopard Cat		
	Average	Standard Deviation	Normalized Weight with Respect to Land Cover	Average	Standard Deviation	Normalized Weight with Respect to Land Cover
DEM	0.1657	0.0175	2.3877	0.1728	0.0208	2.3883
Slope gradient	0.1335	0.0102	1.9242	0.1939	0.0103	2.6790
Slope aspect	0.0901	0.0019	1.2986	0.0768	0.0019	1.0619
Timber type	0.1453	0.0107	2.0945	0.1159	0.0185	1.6014
Timber age	0.0704	0.0104	1.0146	0.1053	0.0059	1.4545
Land cover	0.0694	0.0015	1.0000	0.0724	0.0030	1.0000
Distance from road	0.1092	0.0023	1.5741	0.0817	0.0054	1.1285
Distance from water	0.1210	0.0032	1.7439	0.0854	0.0037	1.1803
Distance from forest	0.0953	0.0008	1.3735	0.0959	0.0044	1.3259

Regarding the average weights of the related factors for the habitat potential mapping of martens, DEM had the highest value (0.166), followed by timber type (0.145), while land cover showed the lowest value (0.069). The factor weights for leopard cats were similar. Slope gradient had the highest value (0.194), followed by DEM (0.173), and land cover had the lowest value (0.072).

However, for a comparison of the same standards between the two analyses, all average weight values were normalized by dividing by the smallest average weight value, land cover, for each factor for martens and leopard cats. The lowest values of 0.069 and 0.072 were normalized to 1.000. DEM showed the highest normalized weight value of 2.388 for martens and timber type was the second highest normalized factor at 2.095. Likewise, the weight analysis showed that the slope gradient and DEM presented the highest and second highest normalized weight values of 2.679 and 2.388 for leopard cats, respectively. The least important factors, land cover for both martens and leopard cats, were used as references for the normalized weights. Timber age (1.015) and slope aspect (1.299) for martens and slope aspect (1.062) and distance from road (1.129) for the leopard cats followed.

The weights calculated via iteration, shown above, were applied throughout the study area, and the final weights determined for martens and leopard cats were applied to the corresponding factor in each dataset. Figure 5 shows the habitat potential maps for martens and leopard cats. To simplify the interpretation of the habitat potential maps, the results were classified into five classes of potentials: very high (10%), high (10%), medium (20%), low (20%), and very low (40%). Based on the habitat potential maps of both martens and leopard cats, the eastern region of South Korea appeared to have high potential.

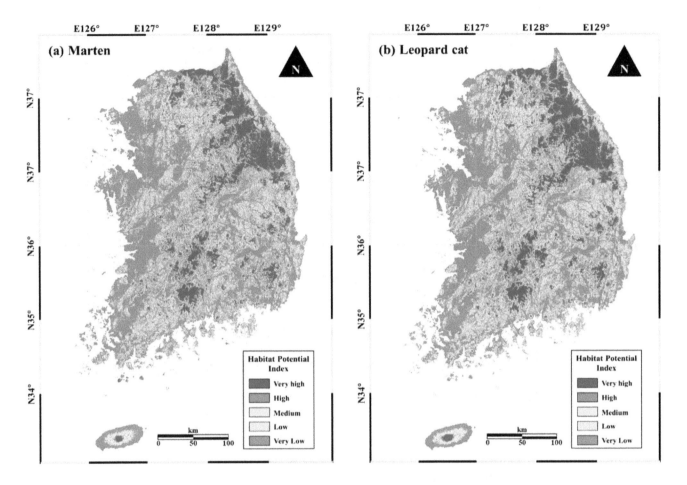

Figure 5. Habitat potential maps generated using ANN model of (**a**) martens; and (**b**) leopard cats.

4.2. Validation

Habitat data can be used to validate whether the habitat potential maps were effectively predicted. Therefore, the halves of marten and leopard cat habitat data not used for training were used as the validation data. ROC curves were used to validate the habitat potential maps of martens and leopard cats generated using the ANN model. The ROC curves were generated for the marten and leopard cat habitat potential maps to compare the habitat locations of martens and leopard cats, respectively.

The ROC curve is a graphical plot showing the diagnostic capabilities of classification models with various thresholds. It is a widely used method for validation [37–40]. ROC curves can show sensitivity and specificity on the x-axis and y-axis, respectively. In this study, specificity represented the percentage of area that martens and leopard cats could inhabit, while sensitivity represented the predicted potential marten and leopard cat habitat locations. For the ROC graph, the predicted habitat potential values after applying the ANN model were sorted in descending order. The ranks of the habitat potential values were identified and the values were equally classified into 100 classes of study area.

Figure 6 shows the ROC curves of the marten and leopard cat habitat potential maps; martens showed 85.01% (0.8501) accuracy, versus 87.03% (0.8703) accuracy for leopard cats. These results are depicted as a graph of the validation rate (Figure 6), where 65% of the study area included 100% of all habitat locations for both species.

Figure 6. Validation results of the habitat potential maps for martens and leopard cats using the ANN model.

5. Discussion

Machine learning focuses on processing new data and predicting outcomes using patterns learned through training [12,13]. In other words, the most noticeable advantage of ANN modeling is that the algorithm itself can learn to be more accurate, although it requires sufficient data to support this. In South Korea, a government-affiliated organization constructed detailed country-scale thematic maps. These previously constructed data were sufficient for generalizing and correctly processing new incoming untrained data. Therefore, a spatial database was constructed based on nationally constructed data, including topographical, forest, and land use maps. The spatial database built in this process could be used in conjunction with other methods. MATLAB was used to process large amounts of data quickly, and the spatial distribution of martens and leopard cats could be confirmed through a GIS analysis. The results for leopard cats using the ANN showed approximately 5% higher accuracy than those obtained using the frequency ratio (82.15%) and logistic model (81.48%) in a previous study [23], even though the input data differed. This confirmed that the ANN model had been applied appropriately to habitat potential mapping.

The results of machine learning models, such as the ANN model applied in this study, are strongly influenced by the input data. Therefore, in this study, an input spatial database was constructed with marten and leopard cat habitat point data based on the Second National Survey on Natural Environment, and data extracted and calculated from national theme maps, which reflected regional characteristics. The impacts of factors besides the input data, such as climate or sudden natural disasters, on martens and leopard cats were assumed to be negligible. In the case of the habitat data, the generation of errors may depend on the survey method. However, the behavior radii of martens and leopard cats were within 10 km, and the size of the entire study area was considerably larger by comparison; therefore, all data were considered to be reliable.

In both species, DEM, slope gradient and timber type were the three most important factors related to habitat. The relative weight of slope gradient was higher for leopard cats than for martens, indicating a greater influence of slope gradient in the result map for leopard cats (Figure 5). The smaller radius of action of leopard cats compared to martens may have influenced this result. Martens mainly inhabit forests; therefore, the habitat potential of martens was high in relatively dense forests. In contrast, the habitat potential of leopard cats tended to be evenly distributed in relatively low areas compared to martens, which was particularly evident on Jeju Island.

Considering the results of the habitat potential map of this study, elevation, slope gradient and timber type information of mountain areas should be considered as indices for prioritizing the protection and management of marten and leopard cat habitats. Moreover, the index should be managed with weighted indices for each factor. In the future, national natural environment survey data that are constructed and accumulated in a timely manner will enable the continuous construction and management of habitat maps for not only marten and leopard cat, but also other endangered species.

6. Conclusions

In this study, ANN modeling was used to predict the habitat potentials for martens and leopard cats. The factors affecting the habitat distribution of each species were selected and a spatial database was constructed with the collected data. Using the collected data and half of the field-surveyed habitat data, the habitat potentials of martens and leopard cats were predicted using the ANN model. The predicted maps were validated with the other half of the habitat data not used for training.

The first- to third-highest normalized weights with respect to the lowest average weight for both martens and leopard cats (land cover) were DEM, slope gradient and timber type. This indicates that factors related to DEM are positively correlated with the habitat location of martens and leopard cats. In contrast, land cover showed the lowest normalized weights for both martens and leopard cats. Subsequently, martens had lower weights for timber age and slope aspect, whereas leopard cats had lower weights for slope aspect and distance from road. Since both species are mammals with large activity radii, and steep slopes are considered an important factor, the weight value for distance from mountains was relatively low.

In South Korea, martens and leopard cats exhibited similar habitat patterns, with high habitat potentials in eastern mountainous areas, except along beach lines and southern areas. Likewise, coastal areas beyond the eastern range and western plains of the study area showed very low habitat potentials for both species. Nearly all areas with low habitat potentials were lowlands, coastal areas, and non-forest areas. These trends are likely similar even though the habitat characteristics of martens and leopard cats differ because of the large size of the study area.

The results of the marten and leopard cat habitat potential maps were validated with the half of the marten and leopard cat habitat data not used for modeling. The results of the validation showed accuracy of 85.01% for martens and 87.03% for leopard cats, both of which were satisfactory (>85%). These results could lead to more informed decisions for wildlife management planning and land development planning in areas with high habitat potentials.

Owing to the characteristics of the machine learning model, input data accuracy is very important. However, it is difficult to determine the exact habitat location of mammals with large activity radii, such as martens and leopard cats. Since inaccurate location data could result in difficulties in performing spatial analyses, reliable habitat data should be used when research is conducted for smaller administrative units. Quantitative evaluation of the ecological consequences of a wide range of spatial data could be performed using GIS. Integrating several characteristics representing the habitat potential of mammals is an important aspect of ecological management research.

This study identified habitat-related factors for martens and leopard cats. The methodology applied in this study could also be used to generate a time-series of habitat maps, and marten and leopard cat habitat data could be used, since the habitat-related data will be additionally and continuously constructed during the next national natural environment survey. In addition to martens and leopard cats, this method can also be applied to habitat mapping of other mammalian species examined in national natural environment surveys. Moreover, the habitat potential mapping results can be used as a basis for determining the locations of monitoring sites or creating protection plans for mammalian species. However, more studies are essential to generalize the habitat-related factors that affect the habitat characteristics for individual mammalian species.

Acknowledgments: This research was conducted by Korea Environment Institute (KEI) with support of a grant (16CTAP-C114629-01) from Technology Advancement Research Program (TARP) funded by Ministry of Land, Infrastructure and Transport of Korean government. This research was supported by the Basic Research Project of the Korea Institute of Geoscience and Mineral Resources (KIGAM) funded by the Ministry of Science and ICT. This research (NRF-2016K1A3A1A09915721) was supported by Science and Technology Internationalization Project through National Research Foundation of Korea (NRF) grant funded by the Ministry of Education, Science and Technology (MEST).

Author Contributions: Saro Lee organized the paperwork and constructed the input database. Sunmin Lee designed and performed the experiments. Wonkyong Song interpreted the input data and this resulted in the habitat aspect. Moung-Jin Lee suggested the idea and collected the data. All authors contributed to the writing of each part.

References

1. Ripple, W.J.; Estes, J.A.; Beschta, R.L.; Wilmers, C.C.; Ritchie, E.G.; Hebblewhite, M.; Berger, J.; Elmhagen, B.; Letnic, M.; Nelson, M.P. Status and ecological effects of the world's largest carnivores. *Science* **2014**, *343*, 1241484. [CrossRef] [PubMed]
2. Taylor, S.L.; Buskirk, S.W. Forest microenvironments and resting energetics of the American marten *Martes americana. Ecography* **1994**, *17*, 249–256. [CrossRef]
3. Harrison, D.J.; Fuller, A.K.; Proulx, G. *Martens and Fishers (Martes) in Human-Altered Environments: An International Perspective*; Springer Science & Business Media: Berlin, Germany, 2004.
4. Grassman, L.I., Jr.; Tewes, M.E.; Silvy, N.J. Ranging, habitat use and activity patterns of binturong arctictis binturong and yellow-throated marten *Martes flavigula* in north-central Thailand. *Wildl. Biol.* **2005**, *11*, 49–57. [CrossRef]
5. Park, C.-H.; Choi, T.-Y.; Kwon, H.-S.; Woo, D.-G. Habitat selection and management of the leopard cat (*Prionailurus bengalensis*) in a rural area of Korea. *Korean J. Environ. Ecol.* **2012**, *26*, 322–332.
6. Ramcharan, A.; Hengl, T.; Nauman, T.; Brungard, C.; Waltman, S.; Wills, S.; Thompson, J. Soil property and class maps of the conterminous us at 100 meter spatial resolution based on a compilation of national soil point observations and machine learning. *arXiv*, **2017**, arXiv:1705.08323.
7. Veronesi, F.; Korfiati, A.; Buffat, R.; Raubal, M. Assessing accuracy and geographical transferability of machine learning algorithms for wind speed modelling. In Proceedings of the International Conference on Geographic Information Science, Wageningen, The Netherlands, 9–12 May 2017; Springer: Berlin, Germany, 2017; pp. 297–310.
8. Folmer, E.O.; van Beusekom, J.E.; Dolch, T.; Gräwe, U.; Katwijk, M.M.; Kolbe, K.; Philippart, C.J. Consensus forecasting of intertidal seagrass habitat in the wadden sea. *J. Appl. Ecol.* **2016**, *53*, 1800–1813. [CrossRef]
9. Sahragard, H.P.; Chahouki, M.A.Z. Comparison of logistic regression and machine learning techniques in prediction of habitat distribution of plant species. *Range Manag. Agrofor.* **2016**, *37*, 21–26.
10. Fox, C.; Huettmann, F.; Harvey, G.; Morgan, K.; Robinson, J.; Williams, R.; Paquet, P. Predictions from machine learning ensembles: Marine bird distribution and density on Canada's pacific coast. *Mar. Ecol. Prog. Ser.* **2017**, *566*, 199–216. [CrossRef]
11. Wieland, R.; Kerkow, A.; Früh, L.; Kampen, H.; Walther, D. Automated feature selection for a machine learning approach toward modeling a mosquito distribution. *Ecol. Model.* **2017**, *352*, 108–112. [CrossRef]
12. Anderson, D.J.; Rojas, L.F.; Watson, S.; Knelson, L.P.; Pruitt, S.; Lewis, S.S.; Moehring, R.W.; Bennett, E.E.S.; Weber, D.J.; Chen, L.F. Identification of novel risk factors for community-acquired clostridium difficile infection using spatial statistics and geographic information system analyses. *PLoS ONE* **2017**, *12*, e0176285. [CrossRef] [PubMed]
13. Bui, D.T.; Bui, Q.-T.; Nguyen, Q.-P.; Pradhan, B.; Nampak, H.; Trinh, P.T. A hybrid artificial intelligence approach using gis-based neural-fuzzy inference system and particle swarm optimization for forest fire susceptibility modeling at a tropical area. *Agric. For. Meteorol.* **2017**, *233*, 32–44.
14. Rahimi, S.; Martin, M.J.; Obeysekere, E.; Hellmann, D.; Liu, X.; Andris, C. A geographic information system (gis)-based analysis of social capital data: Landscape factors that correlate with trust. *Sustainability* **2017**, *9*, 365. [CrossRef]

15. Ottaviani, D.; Panzacchi, M.; Lasinio, G.J.; Genovesi, P.; Boitani, L. Modelling semi-aquatic vertebrates' distribution at the drainage basin scale: The case of the otter lutra lutra in Italy. *Ecol. Model.* **2009**, *220*, 111–121. [CrossRef]

16. Poplar-Jeffers, I.O.; Petty, J.T.; Anderson, J.T.; Kite, S.J.; Strager, M.P.; Fortney, R.H. Culvert replacement and stream habitat restoration: Implications from brook trout management in an appalachian watershed, USA. *Restor. Ecol.* **2009**, *17*, 404–413. [CrossRef]

17. Patel, R.P.; Wutke, S.; Lenz, D.; Mukherjee, S.; Ramakrishnan, U.; Veron, G.; Fickel, J.; Wilting, A.; Förster, D.W. Genetic structure and phylogeography of the leopard cat (*Prionailurus bengalensis*) inferred from mitochondrial genomes. *J. Hered.* **2017**, *108*, 349–360. [CrossRef] [PubMed]

18. Tan, S.; Xu, J.-T.; Zou, F.-D.; Peng, Q.-K.; Peng, R. The complete mitochondrial genome of leopard cat, *Prionailurus bengalensis* chinensis (carnivora: Felidae). *Mitochondrial DNA A* **2016**, *27*, 3073–3074.

19. Jang, K.H.; Hwang, U.W. Complete mitochondrial genome of korean yellow-throated marten, *Martes flavigula* (carnivora, mustelidae). *Mitochondrial DNA A* **2016**, *27*, 1785–1786. [CrossRef] [PubMed]

20. Chen, M.-T.; Liang, Y.-J.; Kuo, C.-C.; Pei, K.J.-C. Home ranges, movements and activity patterns of leopard cats (*Prionailurus bengalensis*) and threats to them in Taiwan. *Mamm. Study* **2016**, *41*, 77–86. [CrossRef]

21. Vigne, J.-D.; Evin, A.; Cucchi, T.; Dai, L.; Yu, C.; Hu, S.; Soulages, N.; Wang, W.; Sun, Z.; Gao, J. Earliest "domestic" cats in China identified as leopard cat (*Prionailurus bengalensis*). *PLoS ONE* **2016**, *11*, e0147295. [CrossRef] [PubMed]

22. Mohamed, A.; Ross, J.; Hearn, A.J.; Cheyne, S.M.; Alfred, R.; Bernard, H.; Boonratana, R.; Samejima, H.; Heydon, M.; Augeri, D.M.; et al. Predicted distribution of the leopard cat *Prionailurus bengalensis* (mammalia: Carnivora: Felidae) on borneo. *Raffles Bull. Zool.* **2016**, *33*, 180–185.

23. Lee, M.-J.; Song, W.; Lee, S. Habitat mapping of the leopard cat (*Prionailurus bengalensis*) in South Korea using gis. *Sustainability* **2015**, *7*, 4668–4688. [CrossRef]

24. Bui, D.T.; Lofman, O.; Revhaug, I.; Dick, O. Landslide susceptibility analysis in the hoa binh province of vietnam using statistical index and logistic regression. *Nat. Hazards* **2011**, *59*, 1413. [CrossRef]

25. Huck, M.; Davison, J.; Roper, T.J. Predicting European badger meles meles sett distribution in urban environments. *Wildl. Biol.* **2008**, *14*, 188–198. [CrossRef]

26. Gavashelishvili, A.; Lukarevskiy, V. Modelling the habitat requirements of leopard panthera pardus in west and Central Asia. *J. Appl. Ecol.* **2008**, *45*, 579–588. [CrossRef]

27. Northrup, J.; Stenhouse, G.; Boyce, M. Agricultural lands as ecological traps for grizzly bears. *Anim. Conserv.* **2012**, *15*, 369–377. [CrossRef]

28. Kuemmerle, T.; Perzanowski, K.; Chaskovskyy, O.; Ostapowicz, K.; Halada, L.; Bashta, A.-T.; Kruhlov, I.; Hostert, P.; Waller, D.M.; Radeloff, V.C. European bison habitat in the Carpathian mountains. *Biol. Conserv.* **2010**, *143*, 908–916. [CrossRef]

29. Clement, M.J.; Castleberry, S.B. Estimating density of a forest-dwelling bat: A predictive model for rafinesque's big-eared bat. *Popul. Ecol.* **2013**, *55*, 205–215. [CrossRef]

30. Greaves, G.J.; Mathieu, R.; Seddon, P.J. Predictive modelling and ground validation of the spatial distribution of the New Zealand long-tailed bat (*Chalinolobus tuberculatus*). *Biol. Conserv.* **2006**, *132*, 211–221. [CrossRef]

31. Maleki, S.; Soffianian, A.R.; Koupaei, S.S.; Saatchi, S.; Pourmanafi, S.; Sheikholeslam, F. Habitat mapping as a tool for water birds conservation planning in an arid zone wetland: The case study hamun wetland. *Ecol. Eng.* **2016**, *95*, 594–603. [CrossRef]

32. Novaczek, E.; Devillers, R.; Edinger, E.; Mello, L. High resolution habitat mapping to describe coastal denning habitat of a Canadian species at risk, atlantic wolffish (*Anarhichas lupus*). *Can. J. Fish. Aquat. Sci.* **2017**. [CrossRef]

33. Park, J.-K.; Das, A.; Park, J.-H. A new approach to estimate the spatial distribution of solar radiation using topographic factor and sunshine duration in South Korea. *Energy Convers. Manag.* **2015**, *101*, 30–39. [CrossRef]

34. Woo, D.G. A Study on Ecological Characteristics and Conservation of Yellow-Throated Marten (*Martes flavigula*) in Temperate Forests of Korea. Ph.D. Thesis, Seoul National University, Seoul, Korea, 2014.

35. Rumelhart, D.E.; McClelland, J.L.; Group, P.R. *Parallel Distributed Processing*; MIT Press: Cambridge, MA, USA, 1987; Volume 1.

36. Hines, J.; Tsoukalas, L.H.; Uhrig, R.E. *Matlab Supplement to Fuzzy and Neural Approaches in Engineering*; John Wiley & Sons, Inc.: Hoboken, NJ, USA, 1997.

37. Althuwaynee, O.F.; Pradhan, B.; Lee, S. A novel integrated model for assessing landslide susceptibility mapping using chaid and ahp pair-wise comparison. *Int. J. Remote Sens.* **2016**, *37*, 1190–1209. [CrossRef]

38. Kim, J.-C.; Lee, S.; Jung, H.-S.; Lee, S. Landslide susceptibility mapping using random forest and boosted tree models in Pyeong-chang, Korea. *Geocarto Int.* **2017**, 1–16. [CrossRef]

39. Lee, S.; Kim, J.-C.; Jung, H.-S.; Lee, M.J.; Lee, S. Spatial prediction of flood susceptibility using random-forest and boosted-tree models in Seoul metropolitan city, Korea. *Geomat. Nat. Hazards Risk* **2017**, 1–19. [CrossRef]

40. Tahmassebipoor, N.; Rahmati, O.; Noormohamadi, F.; Lee, S. Spatial analysis of groundwater potential using weights-of-evidence and evidential belief function models and remote sensing. *Arab. J. Geosci.* **2016**, *9*, 79. [CrossRef]

13

Application of Deep Networks to Oil Spill Detection using Polarimetric Synthetic Aperture Radar Images

Guandong Chen [1], Yu Li [1,*], Guangmin Sun [1] and Yuanzhi Zhang [2,3,*]

[1]   Faculty of Information Technology, Beijing University of Technology, Beijing 100124, China;
     cgd@emails.bjut.edu.cn (G.C.); gmsun@bjut.edu.cn (G.S.)
[2]   National Astronomical Observatories, Chinese Academy of Sciences, Beijing 100012, China
[3]   Key Laboratory of Lunar Science and Deep-space Exploration, Chinese Academy of Sciences,
     Beijing 100012, China
*   Correspondence: yuli@bjut.edu.cn (Y.L.); yuanzhizhang@hotmail.com (Y.Z.)

Featured Application: Using polarimetric synthetic aperture radar (SAR) remote sensing to detect and classify sea surface oil spills, for the early warning and monitoring of marine oil spill pollution.

Abstract: Polarimetric synthetic aperture radar (SAR) remote sensing provides an outstanding tool in oil spill detection and classification, for its advantages in distinguishing mineral oil and biogenic lookalikes. Various features can be extracted from polarimetric SAR data. The large number and correlated nature of polarimetric SAR features make the selection and optimization of these features impact on the performance of oil spill classification algorithms. In this paper, deep learning algorithms such as the stacked autoencoder (SAE) and deep belief network (DBN) are applied to optimize the polarimetric feature sets and reduce the feature dimension through layer-wise unsupervised pre-training. An experiment was conducted on RADARSAT-2 quad-polarimetric SAR image acquired during the Norwegian oil-on-water exercise of 2011, in which verified mineral, emulsions, and biogenic slicks were analyzed. The results show that oil spill classification achieved by deep networks outperformed both support vector machine (SVM) and traditional artificial neural networks (ANN) with similar parameter settings, especially when the number of training data samples is limited.

Keywords: oil spill; polarimetric synthetic aperture radar (SAR); deep belief network; autoencoder; remote sensing

1. Introduction

As one of the most significant sources of marine pollution, oil spills have caused serious environmental and economic impacts to the ocean and coastal zone [1]. Oil spills near the coast can be caused by ship accidents, explosion of oil rig platforms, broken pipelines, and deliberate discharge of tank-cleaning wastewater from ships. The NEREIDs program, sponsored by the European Commission, was the first robust attempt to use shipping, geological and metocean data to characterize oil spills in one of the major oil exploration areas of the world, prior to any major oil spill accident. Based on this data, oil spill models were established to simulate the development and trajectories of oil spills and investigate the susceptibility of coastal zone and find suitable measures to alleviate its impacts to the environment [2–5].

Early warning and near-real-time monitoring of oil slicks plays a very important role in cleaning up operation of oil spill to alleviate its impact to coastal environment [2,3]. Synthetic aperture radar (SAR) is one of most promising remote sensing systems for oil spill monitoring, for it can provide valuable information about the position and size of the oil spill [1]. Moreover, the wide coverage and all-day, all-weather capabilities make SAR very suitable for large scale oil spill monitoring and early warning [6–8].

In their early stages, studies of oil spill detection are mainly based on single polarimetric SAR images [9–12]. The theoretical rationale of SAR oil spill detection is that the presence of oil slicks on the sea surface dampens short-gravity and capillary waves, so the Bragg scattering from the sea surface is largely weakened. The ideal sea surface wind speed for oil spills detection is 3–14 m/s [13]. As a result, oil spills can be detected as "dark" areas in SAR images. However, some other manmade or natural phenomena can result in very similar low scattering areas on the sea surface, e.g., biogenic slicks, waves, currents and low-wind areas, etc. Conventional oil spill detection procedures use intensity, morphological texture, and auxiliary information to distinguish mineral oil and its lookalikes, with its processing chain divided into three main steps [13]: (1) dark spot detection; (2) features extraction; and (3) classification between mineral and its lookalikes.

Single polarimetric SAR-based oil spill detection algorithms need auxiliary information and large number of data samples to classify mineral oil and its lookalikes. Sometimes the shape and texture of oil slicks may vary, affecting the robustness of intensity-based oil spill classification algorithms. Polarimetric observation capabilities provided by advanced SAR sensors have much stronger capabilities for oil spills detection [14]. For instance, biogenic slicks and mineral oil are difficult to distinguish by single polarimetric SAR images. Yet, their polarimetric scattering mechanisms are largely different: for oil-covered areas, Bragg scattering is largely suppressed, and high polarimetric entropy can be documented. In the case of a biogenic slick, Bragg scattering is still dominant, but with a low intensity. Thus, similar polarimetric behaviors as those of oil-free areas should be expected in the presence of biogenic films. Hence, polarimetric features can largely help the image classification between mineral and biogenic lookalikes [14].

Various polarimetric features have been proposed to classify oil spills. The standard deviation of copolarized phase difference (phase difference between Vertical transmit and Vertical receive-VV and Horizonal transmit and Horizonal receive-HH channel) has shown a strong oil classification capability on C-, X-, and L-band data [15]. Nunziata et al. (2011) proposed pedestal height to describe the different polarization signature between mineral oil and biogenic lookalikes [16]. Minchew et al. (2012) took the advantage of copolarization ratio to study the mixing status of crude oil and sea water [17]. Zhang et al. (2011) used the conformity coefficient as a binary classifier [18]. Other polarimetric features such as degree of polarization, entropy, alpha angle, and Bragg likelihood angle were also used to classify oil spills [19–21].

Some previous studies conducted automatic oil-spill classification algorithms. Marghany (2001) developed models to discriminate textures between oil and water by using co-occurrence textures [22]. Gambardella et al. (2008) proposed one-class classification with an optimized feature selection algorithm and obtained a promising oil spill classification [23]. Frate et al. (2000) proposed a semiautomatic detection of oil spills by neural network [24]. Garcia-Pineda et al. (2008) developed the Textural Classifier Neural Network Algorithm (TCNNA) to map an oil spill in the Gulf of Mexico Deepwater Horizon accident [11]. Marghany (2013) used a genetic algorithm (GA) for automatic detection of an oil spill from ENVISAT ASAR (Advanced Synthetic Aperture Radar) data [25]. Li et al. (2013) used a Support Vector Machine (SVM) to detect oil spills based on morphological features on very limited data samples [26].

Polarimetric SAR features contain massive complementary and redundancy information. The extraction and optimization of them are closely related to the performance of oil spill classification [27]. Deep learning algorithms have very strong capabilities of exploring complex correlation between features and achieve very promising fitting result on complicated problems. It has been a very popular technique for image processing, computer vision, and natural language processing. According to the authors, deep learning has not been used in features optimization for oil spills detection based on polarimetric SAR data, and it should be a very promising research topic.

Deep neural network with multilayer neuron has powerful capabilities in describing complex functions compared with shallow networks [28]. However, the traditional gradient descent technique works poorly on a deep neural network when the weights are initialized randomly. The reason is that when the derivative is calculated using the back propagation method, the magnitude of the gradient (from the output layer to the initial layer of the network) decreases dramatically as the network depth increases. As the result, the gradient of the overall loss function, with respect to the weights of the first few layers, is very small. Thus, when the gradient descent method is used, the weights of the first layers change very slowly, so that they cannot learn effectively from the samples. This problem is often referred to as "gradient dispersion". In 2006, Hinton et al. proposed the deep belief network (DBN), which is a belief network composed of Restricted Boltzmann Machine (RBM) one layer at a time, to take the advantage of complementary priors of the data. Inspired by DBN, Beigio et al. (2006) used a stacked autoencoder, which is a deep multilayer neural network that initialized its weights by a greedy layer-wise unsupervised training strategy [29].

Moreover, feature dimension reduction can be seen as an early fusion step. Fusion at different stages of classification procedures is a booming research field that has shown capabilities for improvement of classification results. For instance, Vergara et al. fused the output of nonindependent detectors to derive the optimum classification result [30]. Late fusion of scores of several classifiers could be adapted to the proposed problem as a future research work.

The aims of this paper are exploring the capabilities of deep learning algorithms on polarimetric SAR-based marine oil spill detection. In Section 2, research methods including the representation of polarimetric SAR data, feature extraction methods and deep learning algorithms including DBN and SAE will be introduced. In Section 3, experiments were conducted on RADARSAT-2 data containing verified oil spills and biogenic lookalikes. The performance of different algorithms on various sample sizes for oil spill classification will be compared. Finally, conclusions are drawn in Section 4, and the significance and future work of the study will be briefly presented.

2. Methods

2.1. Foudamentals of Polarimetric SAR

The scattering characteristics of the observed target can be described by matrix, S; which links the scattered and incident electromagnet field, in the backscattered coordinate system:

$$\mathbf{E}^S = \frac{e^{-jkr}}{r} \mathbf{S} \mathbf{E}^i \qquad (1)$$

where k is the wavenumber of the EM wave, r is the distance.

Fully polarimetric SAR observations can be achieved by quad-polarimetric mode, in which both horizontal and vertical polarized signals are transmitted alternatively and received coherently. The 2×2 scattering matrix is used to represent the single look complex quad-pol SAR data:

$$\mathbf{S} = \begin{pmatrix} S_{hh} & S_{hv} \\ S_{vh} & S_{vv} \end{pmatrix} \qquad (2)$$

where S_{ij} describes the transmitted and received polarization, respectively, with h denoting the horizontal direction and v denoting the vertical direction.

To take advantage of statistical properties and reduce the effect of speckle noise of SAR data, covariance matrix is often derived from the scattering matrix by multilook its second order products:

$$C = \begin{pmatrix} \langle S_{hh}^2 \rangle & \langle \sqrt{2} S_{hh} S_{hv}^* \rangle & \langle S_{hh} S_{vv}^* \rangle \\ \langle \sqrt{2} S_{hv} S_{hh}^* \rangle & \langle 2 S_{hv}^2 \rangle & \langle \sqrt{2} S_{hv} S_{vv}^* \rangle \\ \langle S_{vv} S_{hh}^* \rangle & \langle \sqrt{2} S_{vv} S_{hv}^* \rangle & \langle S_{vv}^2 \rangle \end{pmatrix} \qquad (3)$$

where "*" is the symbol of conjugate, and "< >" stands for multilook by using an averaging window. Multilook is applied as a standard procedure to obtain the second order statistics (covariance matrix, coherence matrix) of the SAR data, an average window of 5×5 is normally used for balancing the multilook result and maintaining the spatial resolution.

2.2. Features Extraction for Oil Spills Detection

Previous studies proved experimentally that various SAR features could assist oil spill detection and classifications [31]. In this study, ten features including single VV channel intensity, entropy, alpha angle, degree of polarization, ellipticity, pedestal height, copolarized phase difference (CPD), conformity coefficient, correlation coefficient and coherence coefficient are extracted from the covariance matrix (or coherence matrix and Stokes vector deriving from the covariance matrix) [32] of polarimetric SAR data. The ten features investigated in this study, and their behavior on clean sea surface and sea surface covered by different materials, are given in Table 1. Detailed definitions and their behavior on different targets are provided explicitly in [27].

Table 1. Features investigated in this study.

Feature	Definition	For Mineral Oil	For Biogenic Slicks	For Clean Sea Surface
VV intensity	S_{VV}^2	Lower [1]	low	High
Entropy (H)	$P_i = \dfrac{\lambda_i}{\sum\limits_{j=1}^{3} \lambda_j}$	High	Low	Lower
Alpha (α)	$\alpha = P_1\alpha_1 + P_2\alpha_2 + P_3\alpha_3$	High	Low	Lower
Degree of Polarization (DoP)	$P = \dfrac{\sqrt{g_{i1}^2+g_{i2}^2+g_{i3}^2}}{g_{i0}^2}$	Low	High	High
Ellipticity (χ)	$\sin(2\chi) = -\dfrac{s_3}{ms_0}$	Positive	Negative	Negative
Pedestal Height (PH)	$NPH = \dfrac{\min(\lambda_1,\lambda_2,\lambda_3)}{\max(\lambda_1,\lambda_2,\lambda_3)}$	High	Low	Lower
Standard Deviation of CPD	CPD: $\varphi_c = \arg(\langle S_{HH}S_{VV}^* \rangle)$	High	Low	Lower
Conformity Coefficient (Conf. Co.)	$\mu \cong \dfrac{2(\mathrm{Re}(S_{HH}S_{VV}^*) - \lvert S_{HV}\rvert^2)}{\lvert S_{HH}\rvert^2 + 2\lvert S_{HV}\rvert^2 + \lvert S_{VV}\rvert^2}$	Negative	Positive	Positive
Correlation Coefficient (Corr. Co.)	$\rho_{HH/VV} = \left\lvert \dfrac{\langle S_{HH}S_{VV}^* \rangle}{\langle S_{HH}^2 \rangle \langle S_{VV}^2 \rangle} \right\rvert$	Low	High	Higher
Coherence Coefficient (Conf. Co.)	$Coh = \dfrac{\lvert \langle T_{12} \rangle \rvert}{\sqrt{\langle T_{11} \rangle \langle T_{22} \rangle}}$	Low	High	Higher

[1] Note: "lower" and "higher" mean that the property of the feature on a certain type of surface is close to the other surface that has the property of "low" or "high", but slightly lower or higher. "Std. copolarized phase difference (CPD)" stands for the standard deviation of CPD.

2.3. Deep Belief Network (DBN)

2.3.1. Restricted Boltzmann Machine

RBM is a neural perceptron consisting of visible and hidden layers, and the neurons between the visible layer (v_i, $i = 1, \ldots, N_v$) and the hidden layer (h_j, $j = 1, \ldots, N_h$) are bidirectional and fully connected. The basic structure of RBM is shown in Figure 1:

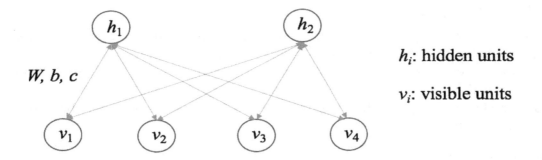

Figure 1. The illustration of a Restricted Boltzmann Machine (RBM) with two hidden units and four visible units.

In RBM, W represents the weight between any two connected neurons, in which each neuron has a bias coefficient b (of neurons) and c (of hidden neurons).

The energy of the RBM can be represented by:

$$E(v,h) = -\sum_{i=1}^{Nv} b_i v_i - \sum_{j=1}^{Nh} c_i j_i - \sum_{i,j=1}^{Nv,Nh} W_{i,j} v_i h_j \tag{4}$$

And the probability of the activation of the hidden layer neuron h_j is:

$$P(h_j|v) = \sigma(b_j + \sum_i W_{i,j} x_i) \tag{5}$$

Similarly, the neurons in the visible layer can also be activated by the bidirectional connected hidden neurons:

$$P(v_i|h) = \sigma(c_i + \sum_j W_{i,j} h_j) \tag{6}$$

where σ is the activation function, e.g., sigmoid function:

$$\sigma(x) = \frac{1}{1 + e^{-x}} \tag{7}$$

Since for RBM the neurons of the same layer is not connected, they are independent:

$$P(h|v) = \prod_{j=1}^{Nh} P(h_j|v) \tag{8}$$

$$P(v|h) = \prod_{i=1}^{Nv} P(v_i|h) \tag{9}$$

Based on the input data vector x, the possibility of the activation of each hidden layer neuron can be calculated. Similarly, based on the activation state of hidden layer neurons, the activation state of visible layers can be calculated. Through a contrastive divergence algorithm [28], the parameters of the RBM: (b, c, W) can be set based on the input data vector x iteratively by a Gibbs sampling technique. An RBM can be seen as a feature detector, which is often used for dimensional reduction of the data. The training process of RBM is to find a probability distribution that can best produce training samples.

2.3.2. The Structure of DBN

DBN is a generative model which establishes a joint distribution between a label and the data sample. It not only considers P (label/observation), but also P (observation/label). In a DBN, several RBMs are connected. The hidden layer of the previous RBM is the next RBM's visible layer, and the output of the previous RBM is the input of the next RBM. During the pre-training process, the upper layer of RBM is trained before the training of the current layer. Usually when the top RBM is trained, the label information is also considered as the visible units.

2.3.3. The Fine-Tuning of DBN

Contrastive Wake-Sleep algorithms are usually used to fine-tune the pre-trained DBN. In the wake stage, the status of nodes of each layer is generated by external features and cognitive weights (upward), and the generated weights (downward) are modified using gradient descent algorithm. In the sleep stage, the state of the bottom neurons is generated through the top-level representation (the states learned by waking) and the weights generated in previous stage, then the cognitive weights of each layer are modified.

2.4. Stacked Autoencoder

2.4.1. Autoencoder

As shown in Figure 2, to build an autoencoder, three layers, namely, an input layer, a hidden layer and an output layer have to be established. The explanations of symbols used in Figure 2 are listed below:

n: the size of the input and output layer.
m: the size of the hidden layer.
$x \in \mathbb{R}^n$, $h \in \mathbb{R}^m$, $y \in \mathbb{R}^n$ stand for the data vector of the input, hidden and output layers, respectively.
$b \in \mathbb{R}^m$, $c \in \mathbb{R}^n$ stand for the bias vector of the hidden and output layers, respectively.
$W \in \mathbb{R}^{m \times n}$ stands for weights matrix between the input and hidden layer.
$\widetilde{W} \in \mathbb{R}^{n \times m}$ stands for weights matrix between the hidden and input layer.

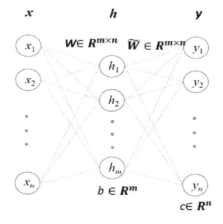

Figure 2. The structure of an autoencoder.

From the input layer to the output of hidden layer, the input signal is encoded. And from hidden layer to the output, the output of hidden layer is decoded by:

$$h = f(x) = s_f(Wx + b) \tag{10}$$

$$y = g(x) = s_g(\widetilde{W}x + c) \tag{11}$$

In Equations (10) and (11), $f()$ and $g()$ stand for the encoding and decoding functions, respectively. S_f and S_g are the corresponding activation functions of the encoder and decoder. sigmoid function can be chosen as the activation function and W^T can be taken as the weights $\widetilde{W}$ of the decoder.

Given input vectors, the autoencoder aims to minimize the difference between an input x and the output y. The reconstruction error can be described by the cross-entropy function:

$$L(x, y) = -\sum_{i=1}^{n} [x_i \log(y_i) + (1 - x_i) \log(1 - y_i)] \tag{12}$$

For the training set, S; the average reconstruction error can hence be established as:

$$\mathcal{L}(\theta) = \sum_{x \in S} L(x, g(f(x))) \tag{13}$$

By minimizing $\mathcal{L}(\theta)$, the parameter $\theta = \{W, b, c\}$ of the autoencoder can be fitted. The learning of an autoencoder does not need the label information, so it is an unsupervised procedure. The output of the hidden layer h can be seen as a representation of input x.

2.4.2. The Stacking of Autoencoders

In a SAE, autoencoders are stacked so that they take the output $h(k)$ of one hidden layer of the former autoencoder as the input for its successive autoencoder. Each layer is trained by a greedy unsupervised layer-wise training strategy, and the upper layers are the representations of relevant high-level abstractions (Figure 3). Stacked autoencoders can establish the deep neural network more efficiently by initializing its weights in a region near its local minimum.

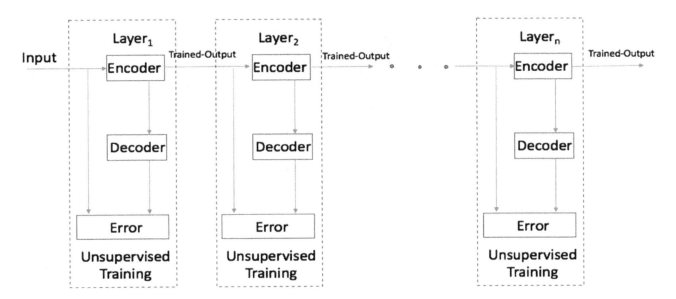

Figure 3. The demonstration of stacked autoencoder.

2.4.3. Fine-Tuning of the SAE

Normally the last layer of the SAE is connected to a classifier, it can be a neural network, Softmax classifier, SVM, etc. Finally, a fine-tuning process is also taken on either the whole network or only the classifier by taking the advantage of the label information through a supervised classification, using the back-propagation algorithm.

3. Experimental Results

3.1. The Experiment Data

In this study, RADARSAT-2 quad-pol SAR data acquired during the 2011 Norwegian oil-on-water experiment (59°59' N, 2°27' E) were used for analysis. The data was received at 17:27 of 8 June 2011 UTC in fine-quad polarimetric mode, with the spatial resolution of 4.7 × 4.8 m in range and azimuth directions. The incident angle of the image is 34.5–36.1° and the local wind speed is 1.6–3.3 m/s. For the convenience of processing and display, a data sample with 2000 × 2000 pixels was picked from the single look complex (SLC) data. The pseudo RGB image of the RADARSAT-2 data on the Pauli basis are provided in Figure 4. In the scene, three verified slicks were present; from left to right, they were: biogenic film, emulsions and mineral oil [33]. The biogenic film was simulated by Radiagreen plant oil. Emulsions were made of Oseberg blend crude oil mixed with 5% IFO380 (Intermediate Fuel Oil), released 5 h before the radar acquisition. Additionally, the Balder crude oil was released 9 h before the radar acquisition [34].

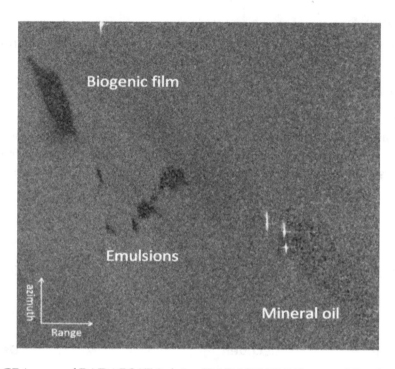

Figure 4. Pauli RGB image of RADARSAT-2 data. (RADARSAT-2 Data and Products © Macdonald, Dettwiler and Associates Ltd., Vancouver, BC, Canada, 2011—All Rights Reserved. RADARSAT is an official mark of the Canadian Space Agency).

3.2. The Experiment Procedure

The SLC quad-polarimetric SAR data was firstly multi-looked, and then the covariance matrix and coherency matrix of the data samples were generated. As mentioned before, 10 features are extracted and saved as a 10-dimension vector for each pixel.

As shown in Figure 5, the 24,000 data samples were picked up from the image, including 12,000 verified positive (mineral oil) and 12,000 negative (clean sea surface and biogenic slick) samples. The data samples were picked by squared boxes with the size of 20×20 for convenience and keeping the purity of the sample, and then their order was shuffled.

In order to test the performance of different algorithms and avoid over-fitting, a six-fold cross-validation was applied. We first divided the training set into six subsets equally. Then the five-sixths of the data samples were used as training set and the rest were taken as testing set. Sequentially, we repeat the classification and another one-sixth data sample were used as testing set. The experiment was conducted six times until each instance of the whole training set is predicted once. Finally, the cross-validation accuracy is the overall percentage of data which are correctly classified.

In order to test the performance of these algorithms on smaller sample sizes, the whole dataset was divided into smaller groups. All the 24,000 data samples were divided into 5 and 25 groups randomly. Then classifications were conducted on these groups, namely 4000 training, 800 testing and 800 training, and 160 testing samples respectively. In the experiment, the classification accuracy of smaller sample size was obtained by averaging the classification result on each group respectively.

In the experiment, two previously introduced deep learning algorithms (i.e., DBN and SAE) were tested on their performance of oil spill detection and classification. In addition, two traditional supervised classifiers including neural network (NN) and SVM were compared.

Figure 5. Demonstration of a selected area for analysis (taking VV^2 image as background); 24,000 pixels are picked as data samples.

The key parameters of these applied classifiers are shown in Tables 2–5:

Table 2. Parameter settings of the neural network.

Parameter	Value
Sizes of layers	[10, 8, 6, 2]
Activation function	Sigmoid
Learning rate	1
Number of epochs	100
Batch size	100

Table 3. Parameter settings of the support vector machine (SVM).

Parameter	Value
Type of SVM	C-SVC (n kind classification)
Type of kernel	Radial Basis Function (RBF): $\exp\left(-\gamma \times |u-v|^2\right)$
γ of the RBF	$1/k$ (k: number of features)
Cost	1
ε (termination criterion)	0.001
Weight w_i	1 (set the parameter C of class i to $w_i \times C$)
Shrinking h	1 (use the shrinking heuristics)

Table 4. Parameter settings of the stacked autoencoder (SAE).

Parameter	Value
Sizes of SAE layers	[8, 6]
Activation function SAE	Sigmoid
Learning rate of SAE	1
Input Zero Masked Fraction of SAE	0.5
Number of epochs when training SAE	10
Batch size when training SAE	100
Size of the whole network	[10, 8, 6, 2]
Activation function of the neural network	Sigmoid
Learning rate when making fine-tuning	3
Number of epochs when making fine-tuning	100
Batch size when making fine-tuning	100

Table 5. Parameter settings of the deep belief network (DBN).

Parameter	Value
Sizes of RBM layers	[8, 6]
Number of epochs when training RBM	10
Batch size when training RBM	100
Momentum for RBM	0
Learning rate alpha of RBM	1
Size of the whole network	[10, 8, 6, 2]
Activation function of the neural network	Sigmoid
Number of epochs when making fine-tuning	100
Batch size when making fine-tuning	100

LIBSVM-a library for Support Vector Machines [33] was used to implement the SVM algorithm. The parameters C and γ were derived by shrinking heuristics search technique. The neural network has ten input neurons, two hidden layers and two output neurons. The initialization of the former layers of deep learning algorithms SAE and DBN were carried out by unsupervised pretraining, and then the outputs were connected to a neural network with two output neurons.

3.3. Results and Discusion

To examine the feature dimension reduction capability of the deep neural networks, scatter plots of the main original features and the features derived by principal component analysis (PCA), DBN and SAE are shown in Figure 6. Two of the most discriminative features, conformity coefficient and degree of polarization (DoP) of HH and VV transition/receiving combinations, as two of the most effective feature in oil spill classification [31], are plotted in Figure 6a. Scatter plots of the first two components derived by PCA are shown in Figure 6b. In this paper, taking the advantage of DBN and SAE, the dimension of polarimetric features are reduced to six, then they are put into fully connected neural network. To show these features in a scatter plot, PCA was implemented on the six features, and then the first two components are shown in Figure 6c,d. It can be observed that deep neural network algorithms effectively extracted the information from high dimensional features and improved their separability to distinguish mineral oil and none mineral samples.

The classification results are shown in Table 6 statistically, some key findings and discussions are listed as follow:

- SAE achieved the highest classification accuracy (lowest testing error) among all the algorithms on different sample sizes. DBN achieved a close performance to SAE. SAE and DBN applied in the experiment had similar structures and both of them took the advantage of greedy unsupervised layer-wise pretraining, so very similar performances were achieved. The unsupervised pretraining worked as a feature optimizer, which can reveal the latent relationship and reduction of noise in features. It helps to improve the performance of the followed supervised classification procedure.
- On the small training data set, deep learning algorithms have much higher performance than neural networks. When the number of data set is reduced, the parameters of traditional NN cannot be sufficiently tuned. Based on unsupervised pretraining, deep learning algorithms such as SAE and DBN have much stronger capability to achieve the optimized solution of the learning problem.
- When the number of data sample size is reduced, the classification error will increase (i.e., the accuracy is reduced). When the number of data sets reduced, the characteristics of the studied object cannot be sufficiently expressed by the limited number of data samples, so the classification performance is reduced.
- On the large training data set, NN have a close performance to deep learning algorithms. With large number of training data, the parameters of NN can be sufficiently adjusted. In this experiment, the NN have a few hidden layers, the gradient of objective function could pass to the

layers in the front effectively. As the result, comparable classification performance was achieved by NN on large data set.

- SVM has better performance on small sample sizes than NN. SVM is based on structural risk minimization, which has superior performance on relative small data sets. It maximizes the classification margin, which is decided by a few support vectors and could successfully avoid the risk of the "curse of dimensionality". However, although the SVM has several advantages, it is equivalent to a NN with one hidden layer, so on learning complicated relationships its performance is no better than the other three more complex classifiers applied in the experiment.

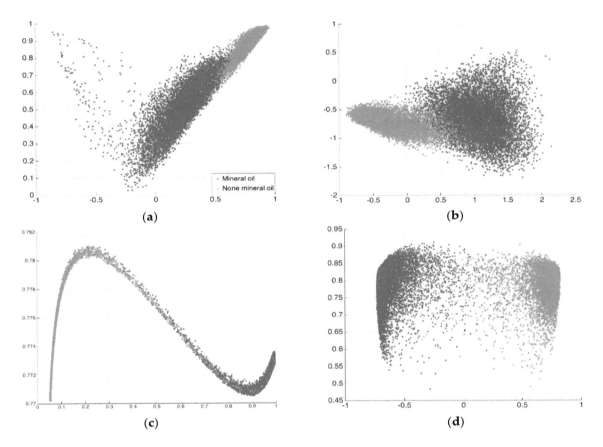

Figure 6. Scatter plots of the main original features (**a**) and the features derived by principal component analysis (PCA), deep believe network (DBN) and stacked autoencoder (SAE) (**b–d**).

Table 6. Testing error (1-accuracy) of classification achieved by different classifiers and on different sizes of data samples.

Classifier	Number of Data Samples (Training-Testing) Accuracy (Execution Time/Seconds)		
	a. 20,000–4000	**b. 4000–800**	**c. 800–160**
SVM (support vector Machine)	1.26% (1.16)	1.38% (0.07)	1.64% (0.01)
NN (nerual network)	1.16% (6.52)	1.68% (1.33)	2.13% (0.34)
PCA (principal component analysis)-SVM	1.20% (0.9)	1.33% (0.06)	1.58% (0.01)
PCA-NN	1.17% (6.2)	1.48% (1.2)	1.90% (0.35)
SAE (stacked autoencoder)	1.05% (6.33)	1.33% (1.24)	1.39% (0.32)
DBN (deep believe network)	1.16% (5.63)	1.42% (1.18)	1.53% (0.28)

The confusion matrix of the cross-validation testing result is shown in Tables 7–10. The best classification results were achieved by SAE on the largest data set: 20,000 training and 4000 testing samples. On the 24,000 testing set, 251 pixels were wrongly classified. 101 pixels of these 251 pixels were false positive (commission errors) and 150 pixels were false negative (omission errors). Similar false positive was achieved by DBN, with slightly higher false negative rate. In the confusion matrix achieved by NN and SAE, it can be discovered that compared with deep learning algorithms, they achieved lower false negative and higher false positive rates.

Table 7. Confusion Matrix (20,000 training, 4000 testing, six-fold) derived by SAE.

Confusion Matrix of SAE	Mineral Oil	Non Mineral Oil	Total
Mineral oil (truth)	11,844	150	12,000
Nonmineral oil (truth)	101	11,896	12,000
Total	11,948	12,052	24,000

Table 8. DBN Confusion Matrix (20,000 training, 4000 testing, six-fold) derived by DBN.

Confusion Matrix of DBN	Mineral Oil	Non Mineral Oil	Total
Mineral oil (truth)	11,819	181	12,000
Non mineral oil (truth)	101	11,899	12,000
Total	11,948	12,052	24,000

Table 9. SAE Confusion Matrix (20,000 training, 4000 testing, six-fold) derived by NN.

Confusion Matrix of NN	Mineral Oil	Non Mineral Oil	Total
Mineral oil (truth)	11,881	119	12,000
Non mineral oil (truth)	160	11,840	12,000
Total	11,948	12,052	24,000

Table 10. SAE Confusion Matrix (20,000 training, 4000 testing, six-fold) derived by SVM.

Confusion Matrix of SVM	Mineral Oil	Non Mineral Oil	Total
Mineral oil (truth)	11,893	107	12,000
Non mineral oil (truth)	201	11,799	12,000
Total	11,948	12,052	24,000

From the binary output that achieved by SAE (Figure 7), it can be observed that a few pixels in the area covered by the biogenic slick are classified as mineral oil. The possible reason of these "misclassifications" is the affection of signal noise on space-borne SAR data or the uniform distribution of the mineral oil and biogenic slicks. This misinterpretation can be further eliminated by a simple postprocessing step. Corrosion and swelling algorithms can be applied on the binary classification result to fix the small holes (missing alarm) in large oil-covered areas and isolated positive targets (false alarm) in the sea surface area.

The receiver operating characteristics (ROC) of these classifiers in oil spill detection are approached based on 20,000 training, 4000 testing (Figure 8a,b) and 4000 training, 800 testing samples (Figure 8c,d) respectively. All the ROC curves are very close to the upper-left corner of the ROC map (Figure 8a,c). In the zoomed-in map (Figure 8b,d), some minor differences can be observed. Compared with other classifiers, SVM achieved a lower true positive rate under low false positive rate requirements, and higher true positive rate under high false positive rate requirements. And for NN the situation is just opposite. Deep neural networks, SAE and DBN achieved a modest true positive rate in the whole false positive rate range.

Figure 7. Classification result achieved by SAE, 0 stands for nonoil and 1 stands for mineral oil.

Figure 8. Receiver operating characteristics (ROC) curves of the classifiers. (**a**) ROC curve achieved based on 20,000 training, 4000 testing samples; (**b**) Zoomed in map of (**a**); (**c**) ROC curve achieved based on 4,000 training, 800 testing samples; (**d**) Zoomed in map of (**c**).

4. Conclusions

In this paper, the capability of polarimetric SAR to detect and classify marine oil spills was investigated. Potential features were extracted from a covariance matrix, a coherence matrix and a Stokes vector of the original SLC quad-pol SAR data. Deep learning algorithms together with classic classifiers were compared and analyzed. A key discovery of this paper is that given insufficient number of data samples, deep learning algorithms such as SAE and DBN can achieve better performance than traditional algorithms by initializing their parameters from a position closer to the optimum solution. Polarimetric SAR data confirmed its strong capacity in distinguishing mineral oil and its biogenic lookalikes. This can be achieved by a one-step operation, with no need to firstly segment and then classify data samples based on auxiliary information. The advantages demonstrated by polarimetric SAR can greatly boost the efficiency and accuracy of marine oil spill detection. Further studies will be conducted on features extracted from compact polarimetric SAR modes, with wider swath width, to achieve larger monitoring areas and shorter revisit times: two of the prime requirements for marine surveillance through large areas.

Acknowledgments: The RADARSAT-2 data provided by CSA and MDA is highly appreciated. This research is jointly supported by the National Key Research and Development Program of China (2016YFB0501501) and the Natural Scientific Foundation of China (41471353 and 41706201). The authors would like to thank the anonymous reviewers for their valuable comments and suggestions that helped improve the quality of this manuscript.

Author Contributions: Yu Li and Yuanzhi Zhang conceived and designed the experiments. Guandong Chen performed the experiments and analyzed the data. Yu Li, Guangmin Sun and Yuanzhi Zhang wrote the paper.

References

1. Fingas, M. *The Basics of Oil Spill Cleanup*; Lewis Publisher: Boca Raton, FL, USA, 2001.
2. Alves, T.M.; Kokinou, E.; Zodiatis, G.; Lardner, R.; Panagiotakis, C.; Radhakrishnan, H. Modelling of oil spills in confined maritime basins: The case for early response in the Eastern Mediterranean Sea. *Environ. Pollut.* **2015**, *206*, 390–399. [CrossRef] [PubMed]
3. Lardner, R.; Zodiatis, G. Modelling oil plumes from subsurface spills. *Mar. Pollut. Bull.* **2017**. [CrossRef] [PubMed]
4. Soomere, T.; Döös, K.; Lehmann, A.; Meier, H.M.; Murawski, J.; Myrberg, K.; Stanev, E. The potential of current- and wind-driven transport for environmental management of the baltic sea. *Ambio* **2014**, *43*, 94–104. [CrossRef] [PubMed]
5. Chrastansky, A.; Callies, U. Model-based long-term reconstruction of weather-driven variations in chronic oil pollution along the German North Sea coast. *Mar. Pollut. Bull.* **2009**, *58*, 967–975. [CrossRef] [PubMed]
6. Alpers, W.H.; Espedal, A. Oils and Surfactants. In *Synthetic Aperture Radar Marine User's Manual*; Christopher, R.J., Apel, J.R., Eds.; US Department Commerce: Washington, DC, USA, 2004.
7. Gade, M.; Alpers, W. Using ERS-2 SAR for routine observation of marine pollution in European coastal waters. *Sci. Total Environ.* **1999**, *237*, 441–448. [CrossRef]
8. Ferraro, G.; Meyer-Roux, S.; Muellenhoff, O.; Pavliha, M.; Svetak, J.; Tarchi, D.; Topouzelis, K. Long term monitoring of oil spills in European seas. *Int. J. Remote Sens.* **2009**, *30*, 627–645. [CrossRef]
9. Migliaccio, M.; Tranfaglia, M.; Ermakov, S.A. A physical approach for the observation of oil spills in SAR images. *IEEE J. Ocean. Eng.* **2005**, *30*, 496–507. [CrossRef]
10. Topouzelis, K.; Karathanassi, V.; Pavlakis, P.; Rokos, D. Detection and discrimination between oil spills and look-alike phenomena through neural networks. *ISPRS J. Photogramm. Remote Sens.* **2007**, *62*, 264–270. [CrossRef]
11. Garcia-Pineda, O.; MacDonald, I.R.; Li, X.; Jackson, C.R.; Pichel, W.G. Oil spill mapping and measurement in the gulf of Mexico with Textural Classifier Neural Network Algorithm (TCNNA). *IEEE J. Sel. Top. Appl. Earth Obs. Remote Sens.* **2013**, *6*, 2517–2525. [CrossRef]
12. Pan, G.; Tang, D.; Zhang, Y. Satellite monitoring of phytoplankton in the East Mediterranean Sea after the 2006 Lebanon oil spill. *Int. J. Remote Sens.* **2012**, *33*, 7482–7490. [CrossRef]
13. Solberg, A.H.S. Remote sensing of ocean oil-spill pollution. *Proc. IEEE* **2012**, *100*, 2931–2945. [CrossRef]

14. Migliaccio, M.; Gambardella, A.; Tranfaglia, M. SAR polarimetry to observe oil spills. *IEEE Trans. Geosci. Remote Sens.* **2007**, *45*, 506–511. [CrossRef]

15. Migliaccio, M.; Nunziata, F.; Gambardella, A. On the co-polarized phase difference for oil spill observation. *Int. J. Remote Sens.* **2009**, *30*, 1587–1602. [CrossRef]

16. Nunziata, F.; Migliaccio, M.; Gambardella, A. Pedestal height for sea oil slick observation. *IET Radar Sonar Navig.* **2011**, *5*, 103–110. [CrossRef]

17. Minchew, B.; Jones, C.E.; Holt, B. polarimetric analysis of backscatter from the deepwater horizon oil spill using L-band synthetic aperture radar. *IEEE Trans. Geosci. Remote Sens.* **2012**, *50*, 3812–3830. [CrossRef]

18. Zhang, B.; Perrie, W.; Li, X.; Pichel, W. Mapping sea surface oil slicks using RADARSAT-2 quad-polarization SAR image. *Geophys. Res. Lett.* **2011**, *38*. [CrossRef]

19. Skrunes, S.; Brekke, C.; Eltoft, T. An experimental study on oil spill characterization by multi-polarization SAR. In Proceedings of the 9th European Conference on Synthetic Aperture Radar (2012 EUSAR), Nuremberg, Germany, 23–26 April 2012; pp. 139–142.

20. Shirvany, R.; Chabert, M.; Tourneret, J.-Y. Ship and oil-spill detection using the degree of polarization in linear and hybrid/compact dual-pol SAR. *IEEE J. Sel. Top. Appl. Earth Obs. Remote Sens.* **2012**, *5*, 885–892. [CrossRef]

21. Salberg, A.-B.; Rudjord, O.; Solberg, A.H.S. Model based oil spill detection using polarimetric SAR. In Proceedings of the IEEE International Geoscience and Remote Sensing Symposium (IGARSS), Munich, Germany, 22–27 July 2012; pp. 5884–5887.

22. Marghany, M. RADARSAT automatic algorithms for detecting coastal oil spill pollution. *Int. J. Appl. Earth Obs. Geoinf.* **2001**, *3*, 191–196. [CrossRef]

23. Gambardella, A.; Giacinto, G.; Migliaccio, M. On the Mathematical Formulation of the SAR Oil-Spill Observation Problem. In Proceedings of the IEEE International Geoscience and Remote Sensing Symposium (IGARSS), Boston, MA, USA, 7–11 July 2008; pp. III-1382–III-1385.

24. Frate, F.; Petrocchi, A.; Lichtenegger, J.; Calabresi, G. Neural networks for oil spill detection using ERS-SAR data. *IEEE Trans. Geosci. Remote Sens.* **2000**, *38*, 2282–2287. [CrossRef]

25. Marghany, M. Genetic Algorithm for Oil Spill Automatic Detection from Envisat Satellite Data. In Proceedings of the Computational Science and Its Applications—ICCSA, Ho Chi Minh City, Vietnam, 24–27 June 2013; Murgante, B., Misra, S., Carlini, M., Carmelo, M.T., Nguyen, H., Taniar, D., Apduhan, B.O., Gervasi, O., Eds.; Springer: Berlin/Heidelberg, Germany, 2013; pp. 587–598.

26. Li, Y.; Zhang, Y. Synthetic aperture radar oil spill detection based on morphological characteristics. *Geospat. Inf. Sci.* **2014**, *17*, 8–16. [CrossRef]

27. Zhang, Y.; Li, Y.; Liang, X.S.; Tsou, J. Comparison of oil spill classifications using fully and compact polarimetric SAR images. *Appl. Sci.* **2017**, *16*, 193. [CrossRef]

28. Hilton, G.; Salakhutdinov, R.R. Reducing the dimensionality of data with neural networks. *Science* **2006**, *313*, 504–507.

29. Bengio, Y.; Lamblin, P.; Popovici, D.; Larochelle, H. Greedy Layer-Wise Training of Deep Networks. In *Advances in Neural Information Processing Systems 19 (NIPS'06)*; MIT Press: Cambridge, MA, USA, 2006.

30. Vergara, L.; Soriano, A.; Safont, G.; Salazar, A. On the fusion of non-independent detectors. *Digit. Signal Process.* **2016**, *50*, 24–33. [CrossRef]

31. Li, Y.; LIN, H.; Chen, J.; Zhang, Y. Comparisons of circular transmit and linear receive compact polarimetric SAR features for oil slicks discrimination. *J. Sens.* **2015**, *2015*, 631561. [CrossRef]

32. Li, Y.; Zhang, Y.; Chen, J.; Zhang, H. Improved compact polarimetric SAR quad-pol reconstruction algorithm for oil spill detection. *IEEE Geosci. Remote Sens. Lett.* **2014**, *11*, 1139–1142. [CrossRef]

33. Chang, C.-C.; Lin, C.-J. LIBSVM: A library for support vector machines. *ACM Trans. Intell. Syst. Technol.* **2011**, *2*, 27. [CrossRef]

34. Skrunes, S.; Brekke, C.; Eltoft, T. Characterization of marine surface slicks by Radarsat-2 multipolarization features. *IEEE Trans. Geosci. Remote Sens.* **2014**, *52*, 5302–5319. [CrossRef]

14

Classification of Forest Vertical Structure in South Korea from Aerial Orthophoto and Lidar Data using an Artificial Neural Network

Soo-Kyung Kwon, Hyung-Sup Jung * , Won-Kyung Baek and Daeseong Kim

Department of Geoinformatics, University of Seoul, Seoul 02504, Korea; a99891902@uos.ac.kr (S.-K.K.); bekwkz@uos.ac.kr (W.-K.B.); kds2991@uos.ac.kr (D.K.)
* Correspondence: hsjung@uos.ac.kr

Abstract: Every vegetation colony has its own vertical structure. Forest vertical structure is considered as an important indicator of a forest's diversity and vitality. The vertical structure of a forest has typically been investigated by field survey, which is the traditional method of forest inventory. However, this method is very time- and cost-consuming due to poor accessibility. Remote sensing data such as satellite imagery, aerial photography, and lidar data can be a viable alternative to the traditional field-based forestry survey. In this study, we classified forest vertical structures from red-green-blue (RGB) aerial orthophotos and lidar data using an artificial neural network (ANN), which is a powerful machine learning technique. The test site was Gongju province in South Korea, which contains single-, double-, and triple-layered forest structures. The performance of the proposed method was evaluated by comparing the results with field survey data. The overall accuracy achieved was about 70%. It means that the proposed approach can classify the forest vertical structures from the aerial orthophotos and lidar data.

Keywords: forestry vertical structure; stratification; forest inventory; aerial orthophoto; lidar (light detection and ranging); ANN (Artificial Neural Network); machine learning

---

1. Introduction

Since forests are important for human life, forest inventories have been investigated for various purposes for centuries. In Europe, the first inventories were carried out in the 14th and 15th century for the purpose of intensive mine development. Since the 1910s, national forest inventories have been carried out in Norway, Sweden, and Finland, with an emphasis on timber production [1]. However, the demands of society have changed rapidly over recent decades. In this context, the principles for the conservation and sustainable management of forests have been newly added by the United Nations General Assembly [2]. This was in response to an increasing interest in non-timber aspects of forest structure and the demand for assessing these aspects [3]. In the Republic of Korea, in the 1970s, when the forest inventory was first investigated, it was aimed at reforestation and forest statistics. The purpose of the forest inventory has changed as the value of forest resources and impacts on the environment have evolved. Currently, forest inventories are developed to provide information that is useful to achieve the following goals: to maintain a healthy forest ecosystem, to preserve and protect the global environment and to promote sustainable development [4–6].

The vertical structure of a forest, which generally has four layers, is one of several elements representing forest vitality. In temperate zones, the forests are divided into layers of canopy, understory, shrub, and grass. The ability of the lower vegetation layer to grow under the canopy layer is determined by the condition of the soil, the species of vegetation, and the quantity of sunlight received by the lower vegetation layer due to the opening and closing rate of the crown [7–9].

In the case of artificial forests, they are usually single- or double-layered forests, as they are planned and managed for the purposes of wood production and agriculture. Natural forests however, have formed from a variety of vegetation communities through natural succession over lengthy periods of time, and have multiple layers. From an ecological point of view, natural forests with multiple-layers are highly resistant to pests, diseases, and environmental stress, and have high quality ecosystem services, such as providing habitats for wildlife. This is not the case with single-layered forests [10–14]. Therefore, vertical structure is a useful measure to evaluate the environmental aspects of a forest.

Typically, forest inventories have been investigated through terrestrial surveys. The Korea Forest Service currently uses aerial orthophotos to survey forests, but it is very difficult to understand forest structure because orthophotos only image the forest canopy. Thus, a field survey is required to understand the stratification of a forest. In Korea, more than 70% of the country is made up of mountainous areas, and a nationwide field survey would be very time- and cost-consuming. Since remote sensing data is advantageous for extensive regional studies [15], we attempt to develop a method of effective forest inventory using remotely sensed data. Such a method could reduce the time and cost of a forest inventory.

Multi-layered mature forests normally have a rough texture in remotely-sensed images, while single-layered young forests have a smoother texture [16–18]. Therefore, the structure of a forest can be estimated through the reflectance differences among the communities. The arrangement of the crown layer (canopy layer) in single-layered artificial forests can be considered consistent, but in natural forests it is inconsistent [17,18]. Since tree height is closely related to forest vertical structure, lidar data used for tree height measurements can be used to classify a forest's vertical structure [7,19–24]. Lidar is an abbreviation for light detection and ranging, and is a device that measures the distance of a target using laser light. The characteristics of remotely-sensed images and lidar data could enable us to classify the vertical structure of a forest.

The objective of this study is to classify forest vertical structure from aerial orthophotos and lidar data using an artificial neural network (ANN) approach. A total of five input layers are generated including: a median-filtered index map, a non-local (NL) means filtered index map, and reflectance texture map generated from the aerial orthophotos, and height difference and height texture maps generated from the lidar data. Since it is difficult to determine the presence of a grass layer in aerial images, it is omitted from our study and the forest vertical structure is classified into three groups for the ANN approach. The groups include: (i) single-layered forest that has only the canopy layer; (ii) double-layered forest that possesses the canopy layer and an understory or shrub layer; and (iii) triple-layered forest that is composed of the canopy, understory, and shrub layers. The red-green-blue (RGB) aerial orthophoto is used to obtain optical image information. The digital surface model (DSM) and digital terrain model (DTM) are extracted from the lidar point cloud. The height information is extracted by subtracting the DTM height from the DSM height. The accuracy of the classification is validated using field survey measurements.

2. Study Area and Dataset

The study area is located in Gongju province, South Korea, as shown in Figure 1. Gongju is 864.29 km^2 in area, about 0.95% of the total area of South Korea (99,407.9 km^2). The area of cultivated land is 185.82 km^2, accounting for 19.76% of the total area. Forests constitute 70% of the area and there are many hilly mountainous areas of 200–300 m above sea level. The average temperatures are around 11.8 °C in spring, 24.7 °C in summer, 14.0 °C autumn, and −0.9 °C in winter. The average annual precipitation is 1466 mm. The soils are loamy soils and clay loams and at the bottom it has sand soils. The image of the study area covers a 3.25 km^2 area that is 2344.8 m in length and 1385.76 m in width.

Figure 1. Location of the study area in Gongju province, South Korea. The red-green-blue (RGB) aerial orthophoto shows the spatial distribution of forests in the study area.

The RGB aerial orthophoto shows the spatial distribution of forests in the study area. Most forests are present on steep-sloped mountains and the types of forest include coniferous/broad-leaved forests and artificial/natural forests. Figure 2 shows the three types of forest classification map that are based on vertical structure, dominant species' leaf type, and whether the forest is natural or artificial. These maps were classified based on field survey measurements collected as a part of the Korean 3rd natural environment conservation master plan. The measurements were collected for the whole country over a period of six years from 2006 to 2012. The field survey of the study area was performed in 2009. The fourth masterplan started in 2014 and is due to be completed in 2018. The database construction for the fourth masterplan has not been completed. Figure 2a shows that triple-layered forests were dominant in the study area, and hence the natural forests are shown to be dominant in Figure 2c because the artificial forests are generally single- or double-layered. The study area includes a variety of different forest types including: (i) single-, double-, and triple-layered forests (Figure 2a); (ii) broad-leaved, coniferous, and mixed forests (Figure 2b); and, (iii) natural and artificial forests (Figure 2c).

Figure 3 shows examples of single-, double-, and triple-layered forests. Figure 3a shows an example of a single-layered forest. The forest is a chestnut plantation that the image shows to have been uniformly planted. Thus, it can be understood that it is an artificial forest. In Figure 3b, the image was acquired from a double-layered forest. It is a coniferous forest and the nut pine is dominant. The red reflectance value of coniferous trees was lower than that of broad-leaved trees based on analysis of the RGB image. This confirmed that the brightness is dependent upon species. Considering that coniferous forests are mainly single- or double-layered, species identification could be an important part of the classification of vertical structure. Figure 3c shows an example of a triple-layered forest. The forest is broad-leaved and oak is dominant tree type. Natural forests are a mixture of various species, and may contain coniferous and broad-leaved trees. Therefore, it is expected that it will be more useful to analyze height differences between trees rather than the identification of tree species in the classification of triple-layered forests. As mentioned above, single- and double-layered forests are mostly artificial forests. Artificial forests in the study area include broad-leaved forests, such as the chestnut tree forest, as well as coniferous forests such as the nut pine forest.

The RGB orthophoto was obtained using a DMC II airborne digital camera at an altitude of 1300 m (above sea level) with a redundancy rate of 63% in October 2015. The low-resolution RGB and high-resolution panchromatic (PAN) images were merged to create a high-resolution RGB image. The high-resolution RGB image was then orthorectified and mosaicked. The final mosaicked RGB orthophoto was used for this study. The ground sample distance (GSD) of the orthophoto was 12 cm in both the line and pixel. The lidar point cloud was acquired with an ALTM Gemini167 at an altitude of 1300 m (above sea level) in October 2015 with 2.5 points per square meter in the lowlands and 3~7 points per square meter in the mountain area. The difference in the point density occurred because the overlap rate of the lidar sensors is different between mountainous and non-mountainous areas. Points from terrain was classified from the Lidar point cloud, the DTM with the GSD of 1 m was created from the terrain-derived points. On the other hand, points from trees were extracted, and then the DSM with the GSD of 1 m was produced from the tree-derived points. The terrain- and tree-derived points were extracted by using the commercial software Global Mapper. The height difference map was generated by subtracting DTM from DSM. The height difference map is related with several tree parameters, including the height and density of trees, etc. That is, the pixel values in the height difference map are not tree height, but are closely related with tree height.

Figure 2. Forest classification maps of the study area produced from the field survey. (a) Forest vertical structure; (b) dominant species' leaf type; (c) natural and artificial forests.

Figure 3. Example images from the study area showing the different forest vertical structures. (**A**) A single-layered forest that is a broad-leaved forest. The chestnut tree (Castanea crenata var. dulcis) is dominant; (**B**) a double-layered forest, which is a coniferous forest in which the nut pine (Pinus koraiensis) is dominant; and (**C**) a triple-layered forest that is a broad-leaved forest. The sawtooth oak (Quercus acutissima) is dominant.

3. Methodology

In this study, we produced five input maps that were resampled to 0.48 m GSD from high-resolution RGB imagery and lidar DSM and DTM data. The three maps from the RGB image include: reflectance values from each tree, reflectance values from each colony, and the variance of reflectance values (i.e., a texture map of reflectance). The two maps from lidar data reflect the tree height of each individual species, and the variance of the height values. We classify forest vertical structure by applying an Artificial Neural Network (ANN) to the five maps. An ANN is a large network of extremely simple computational units that are massively interconnected with nodes and processed in parallel. In this study, a multi-layer artificial neural network was used. It contains three layers: an input layer, a hidden layer, and an output layer, and each layer is composed of nodes. Each node has a weight, which is adjusted from randomly generated initial values through the iterative experiment to obtain the most reasonable output [25,26]. Figure 4 shows the detailed workflow used to classify the forest vertical structure from RGB orthophoto and lidar data using a machine learning approach. The image processing steps such as median, NL-means filtering, texture calculation was implemented by using C language, the MATLAB software was used for the ANN processing, and ER-mapper was used to display input and output maps.

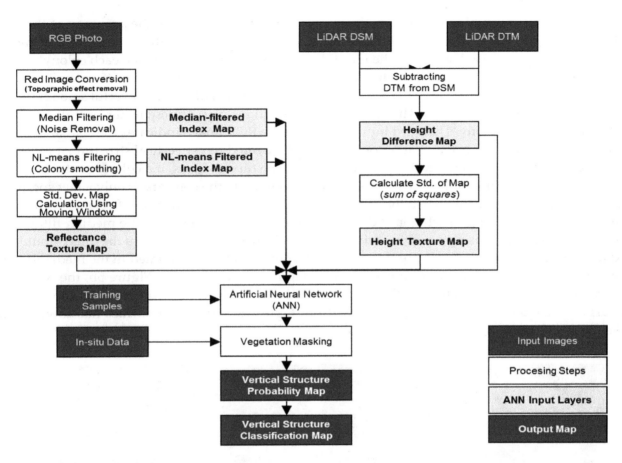

Figure 4. Detailed workflow used to classify the forest vertical structure from RGB orthophoto and lidar data using a machine learning approach.

3.1. Generation of Input Data from RGB Imagery

Vegetation has different reflectance values on remotely sensed imagery because different species have different leaf pigments. When these characteristics were determined from RGB imagery, the needle-leaf forest was brighter than the broad-leaf forest on the converted red image (Median-filtered Index Map). To analyze the optical imaging characteristics of each colony's vegetation, a normalized index map was produced. First, the initial RGB imagery (0.12 m GSD) has a deep shadow due to topographic effects that should be reduced. The shadow effect appears in both the red and green images. Thus, an index map was used instead of the red or green images. The index map is generated in two steps: (i) the red image is converted using the mean and standard deviation of the green image; and (ii) the ratio of the difference between the red and green images and the summation of the red and green images is calculated. This conversion reduces the topographic shadow artefact very effectively. The converted red image ($\overline{R}$) was calculated using:

$$\overline{R}(i,j) = \frac{\sigma_G}{\sigma_R} \times (R(i,j) - \mu_R) + \mu_G \tag{1}$$

where i and j are the line and pixel, respectively, R is the red image, σ_G and σ_R are the means of the green and red images, and μ_G are μ_R are the standard deviation of the green and red images. The normalized index map (NI) was defined using:

$$\mathrm{NI}(i,j) = (G(i,j) - \overline{R}(i,j)) / (G(i,j) + \overline{R}(i,j)) \tag{2}$$

where G is the green image.

The index map contains noise from the crown. Since it can degrade the accuracy of the ANN analysis, we applied a median filter with a kernel size of 3 × 3 pixels to reduce the noise. The median filtered index map was used as part of the input data for the ANN. To determine each colony's overall reflectance, we resampled the median-filtered index map to a spatial resolution of 48 cm and applied a NL-means filter with a kernel size of 21 × 21 pixels. The NL-means filter is a powerful smoothing filter that preserves the edges of objects. The NL-means filtered index map represents the overall reflectance of a forest vegetation colony. The third input map from the RGB image indicates the spatial texture of the reflectance values. If a colony has a double- or triple-layered forest vertical structure, it must have several species and the species' reflectance values must be mixed. Thus, if a colony contains various species, the spatial texture of the reflectance would be rough. On the contrary, if only a few species are present such as in an artificial forest, the spatial texture would be smooth. The reflectance texture map is generated by calculating the standard deviation of the difference between the median-filtered and NL-means filtered maps using a moving window of 5 × 5 pixels. Figure 5 shows the three input layers used for the ANN analysis. To ensure that the topographic effect was mitigated in the median-filtered index map from Figure 5a, it was compared with Figure 1. As shown in Figure 5b, the NL-means filtered index map was smoothed with well preserved object edges. The smooth or rough texture of the surface reflectance map can be recognized from Figure 5c. The water surface had lower values while the forest areas had higher texture values.

Figure 5. The five input layers used for the artificial neural network (ANN) analysis. (**a**) median-filtered index map; (**b**) non-local (NL)-means filtered index map; (**c**) reflectance texture map; (**d**) height difference map; and (**e**) height texture map.

3.2. Generation of Input Data from Lidar Data

Forest vertical structure is highly related to tree height. This is because the criteria for classification of canopy, understory, and shrub layers in a forest inventory are tree species and tree height. Two input layers were generated from the lidar data (initially 1 m GSD, but resampled to 0.48 m GSD): a height difference map and a height texture map. The height difference map was generated by subtracting the DTM from the DSM. The DSM is a kind of digital elevation model (DEM) that represents the Earth surface including features such as building, trees, and houses. The DTM is also a kind of DEM, but it only represents the Earth's terrain and excludes the natural and manmade features. Thus, the height difference map shows the height of the natural and manmade features, including the tree height Pixel values in the height difference map are not direct measurements of tree height, but a close approximation.

The height difference map was used for the ANN analysis. Since the triple- or double-layered forests have various tree species, the variance of the tree height in a colony would be uneven and the standard deviation of height difference in a colony can be large. The calculation of the height texture map from the lidar data is performed by two processing steps. In the first step, the height difference map is smoothed using a NL-means filter with the kernel size of 11×11 pixels. The second step involves calculating the standard deviation of the difference between the height difference map and the NL-means filtered height difference map using a moving window of 5×5 pixels. This map is used for the ANN classification processing. Figure 5d shows the height difference map. The relative height difference among trees can be recognized from Figure 5d. Figure 5e shows the height texture map that indicates the spatial variance in tree height.

Figure 6 shows the characteristics of the five input layers used for the ANN. The areas in the boxes labeled A, B, and C in Figure 6 are representative of single-, double-, and triple-layered forests, respectively. The dominant species in the area under box A is chestnut tree and the area is a broad-leaved and artificial forest. The area under box B includes nut pine trees as the dominant species and is a coniferous and artificial forest. The dominant species in the area under box C is oak tree and the area is a broad-leaved and natural forest. The nut pine (box B in Figure 6) was the brightest of the forests among the reflectance index maps. The reflectance texture maps in the single- and double-layered forests were smoother than the triple-layered forest. In box A of Figure 6d, there are some individual trees in the artificial forest that appear as bright patches. These trees show different reflectance values on the RGB and median-filtered index maps as compared with other trees in the same community and appear to be a different species. They show odd values on the maps from the lidar data as well. The brightest parts of the Height Texture Map (Figure 6e) are where the standard deviation was high.

Figure 6. *Cont.*

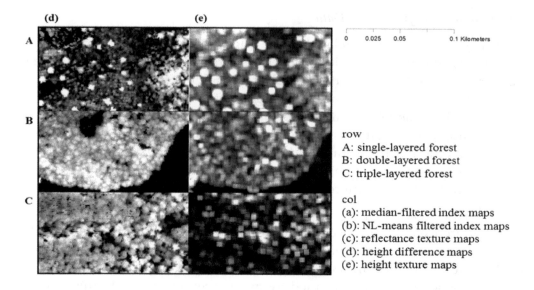

row
A: single-layered forest
B: double-layered forest
C: triple-layered forest

col
(a): median-filtered index maps
(b): NL-means filtered index maps
(c): reflectance texture maps
(d): height difference maps
(e): height texture maps

Figure 6. Characteristics of the five input layers used for the ANN approach in boxes A, B, and C of Figure 3: (**a**) median-filtered index maps; (**b**) NL-means filtered index maps; (**c**) reflectance texture maps; (**d**) height difference maps; and (**e**) height texture maps. The areas in boxes A, B, and C are the single-, double-, and triple-layered forests, respectively.

4. Results and Discussion

Figure 7 shows probability maps of the single-, double-, and triple-layered forests generated using the ANN and the five input layers. The ANN is a pixel-based analysis that calculates a probability of an event happening, and the brightness values represent probabilities. Training samples of ten thousand pixels were selected for the single-, double-, and triple-layered forests from field survey measurements. The non-vegetation areas and the road (red-dotted box in Figure 8) were masked out. The ANN determined the weighting factors from the five input layers. Table 1 summarizes the weighting factor for each input layer.

From the single-, double-, and triple-layered forests shown in Figure 7a–c, a final forest structure classification map was produced by selecting the highest value from the probability maps of the single-, double-, and triple-forests. Figure 8 shows the ANN-classified forest structure map and the field forest structure map for comparison. The red-dotted box in Figure 8b was classified as single-layered forest using the ANN approach, but the field measurements recorded it as triple-layered forest (e.g., compare Figure 8a,b). The field measurements have low spatial sampling because most forests in Korea occur in high mountainous regions. This discrepancy was caused by the spatial resolution differences. The forest vertical structure classes between the ANN approach and field measurement were compared and the results are summarized in Table 2. In Table 2, the number denotes the number of pixels. The classification accuracies of the single-, double-, and triple-layered forests were 82.84%, 56.56%, and 69.03%, respectively. The overall accuracy was 71.10%.

Table 1. Estimated weighting factor of the input layers.

Input Layers	Weighting Factor		
	Single-Layered	Double-Layered	Triple-Layered
Median-filtered Index Map	0.13	0.09	0.22
NL-means Filtered Index Map	0.17	0.47	0.23
Reflectance Texture Map	0.10	0.07	0.14
Height Difference Map	0.46	0.16	0.24
Height Texture Map	0.15	0.21	0.17

Summation of the weighting factors is 1.

Figure 7. Probability maps of the (**a**) single-; (**b**) double-; and (**c**) triple-layered forests.

Figure 8. (**a**) Field survey classification map of forest vertical structure (same as Figure 2a); (**b**) final classification map of forest vertical structure using the ANN with red box -road way-.

Table 2. Validation of forest vertical structure classified by using the ANN approach.

ANN \ Field	Single-Layered Forest	Double-Layered Forest	Triple-Layered Forest
Single-layered Forest	849,983 *	42,380	785,080
Double-layered Forest	50,920	105,795 *	611,421
Triple-layered Forest	125,118	38,862	3,112,900 *
total	1,026,021	187,037	4,509,401

* Correctly classified pixels.

For the sections with low classification accuracy, the reasons for the misclassification need to be investigated. However, the overall accuracy of the classification is relatively high, considering that it is difficult to classify the vertical structure of a mixed-species forest using remotely-sensed data [17]. Our results show that it is possible to classify the forest vertical structure using remotely-sensed data. Several discrepancies were identified and the causes determined. The triple-layered coniferous colonies on the ridge of the mountain were mostly classified as double-layered coniferous forest. This was due to the ANN classification being strongly influenced by the dominant species. The red-dotted box in Figure 8 is a roadway, which was classified as triple-layered forest in the field measurement data collected in 2009. However, a single-layered forest exists in the roadway when we checked it in the field in 2015 RGB imagery. The road was constructed before 2009. In this case, the ANN approach performed better than the field measurements due to a low spatial resolution of the field survey sampling. However, we should be aware that the time lag between airborne images and field survey data can result in differences such as boundary changes or forest succession. The discrepancy between the ANN and field measurement classifications was very high in the areas of double-layered forest (Table 2). This is because the double-layered forest is composed of a canopy layer and an understory or shrub layer, and it was very difficult to distinguish from the other vertical structures using the aerial orthophoto and lidar data. Even though using remotely-sensed data and an ANN has some limitations, the approach could be very useful to classify forest vertical structure, because it has a better spatial resolution and is more time- and cost-effective. One other issue was that the RGB image used for this study was created through image tiling by a vendor. The left side of the image (Figure 8b), is more comparable to the field survey data (Figure 8a) than the right side. The reason may be due to a slight difference in color when tiling several raw images to process one orthophoto. Therefore, in future studies, it may be better to use raw image data or high resolution satellite image, which have the same photographic conditions over a single image area.

There is much research on the internal structure of forests using dense lidar data [27], but this method is limited due to the high cost of investigating large areas. Since the forests of Korea comprise a variety of species, it has been difficult to apply the methods of previous studies that identified stratification in coniferous forests composed of similar species [28]. Therefore, the novelty of the approach used in this study is that it can estimate and classify the complex inner structure of a forest with simple remotely-sensed data.

5. Conclusions

Forest vertical structure one element that represents the vitality of a forest. In general, the forest inventory has been investigated through field surveys. Korea Forest Service currently uses the aerial orthophotos to survey the forests, but it is very difficult to understand the infrastructure of the forests using the remote sensing images because of the cost. Intensive lidar data and aerial orthophotos were used because the orthophotos can just image the forest canopy. Thus, the field survey is inevitable in order to understand the stratification of forest.

In this study, forest vertical structure was classified by using the ANN approach from aerial orthophotos, and lidar DTM and DSM. A total of five input layers were generated from these datasets: (i) a median-filtered index map; (ii) a NL-means filtered index map; (iii) a reflectance texture map;

(iv) a height difference map; and (v) a height texture map. Using these maps, a forest structure classification map was generated with the ANN. The classification accuracies of the single-, double-, and triple-layered forests were 82.84%, 56.56% and 69.03%, respectively. The overall accuracy was 71.10%. The accuracy seems good considering that it is not easy to detect the under layers of a mixed broad-leaf and conifer forest from the remotely sensed data. The ANN approach has a better spatial resolution, is a more time- and cost-effective procedure for mapping forest vertical structure. Future studies should consider the effect of time gaps between datasets, and include maps with more detailed information related plant growth such as soil, sunlight, and species characteristics.

Acknowledgments: This work was supported by "Public Technology Development Project based on Environmental Policy" (2016000210001) provided by Korea Environmental Industry and Technology Institute.

Author Contributions: H.-S.J. conceived and designed the experiments. S.-K.K. performed the experiments. H.-S.J. and S.-K.K. analyzed the data. W.-K.B. and D.K. helped with data processing for the imagery. S.-K.K. and H.-S.J. wrote the paper.

References

1.   Lund, H.G. *IUFRO Guidelines for Designing Multipurpose Resource Inventories: A Project of IUFRO Research Group 4.02.02*; International Union of Forest Research Organizations (IUFRO): Vienna, Austria, 1998.

2.   UN General Assembly. United Nations Sustainable Development. In Proceedings of the United Nations Conference on Environment and Development, Rio de Janeiro, Brazil, 3–14 June 1992; Available online: https://sustainabledevelopment.un.org/content/documents/Agenda21.pdf (accessed on 11 October 2017).

3.   Kenning, R.; Ducey, M.; Brissette, J.; Gove, J. Field efficiency and bias of snag inventory methods. *Can. J. For. Res.* **2005**, *35*, 2900–2910. [CrossRef]

4.   Kim, E.S.; Kim, C.M.; Kim, K.M.; Ryu, J.H.; Lim, J.S.; Kim, J.C. *The Change of Korean National Forest Inventory System (1971~2010)*; Korea Forest Institute: Seoul, Korea, 2015; ISBN 978-89-8176–158-5 93520.

5.   Kim, S.H.; Kim, J.C. *Guide Book for the Sixth Korean National Forest Inventory and Fieldwork for Forest Health and Vitality*; Korea Forest Institute: Seoul, Korea, 2011; ISBN 978-89-8176-805-8.

6.   Park, S.G.; Kang, H.M. Characteristics of Vegetation Structure in Chamaecyparis Obtusa Stands1. *Korean J. Environ. Ecol.* **2015**, *29*, 907–916. [CrossRef]

7.   Morsdorf, F.; Marell, A. Discrimination of vegetation strata in a multi-layered Mediterranean forest ecosystem using height and intensity information derived from airborne laser scanning. *Remote Sens. Environ.* **2010**, *114*, 1403–1415. [CrossRef]

8.   Lee, K.J.; Han, S.S. *Forest Ecology*, 2nd ed.; Hyangmunsa: Seoul, Korea, 1999; ISBN 8971871377.

9.   Kang, S.S. *Biology*, 3rd ed.; Academy Books: Seoul, Korea, 2000; ISBN 9788976161987.

10. Korea Forest Conservation Movement. Available online: http://www.kfca.re.kr/ (accessed on 16 August 2017).

11. Jeon, S.W.; Kim, J. A Study on the Forest Classification for Ecosystem Services Valuation. *Korean Environ. Res. Technol.* **2013**, *16*, 31–39. [CrossRef]

12. Isbell, F.; Calcagno, V. High plant diversity is needed to maintain ecosystem services. *Nature* **2011**, *477*, 199–202. [CrossRef] [PubMed]

13. Fraf, R.F.; Mathys, L. Habitat assessment for forest dwelling species using LiDAR remote sensing: Capercaillie in the Alps. *For. Ecol. Manag.* **2009**, *257*, 160–167. [CrossRef]

14. Onaindia, M.; Dominguez, I. Vegetation diversity and vertical structure as indicators of forest disturbance. *For. Ecol. Manag.* **2004**, *195*, 341–354. [CrossRef]

15. Seong, N.; Seo, M.; Lee, K.S.; Lee, C.; Kim, H.; Choi, S.; Han, K.S. A water stress evaluation over forest canopy using NDWI in Korean peninsula. *Korean J. Remote Sens.* **2015**, *31*, 77–83. [CrossRef]

16. Hay, G.J.; Niemann, K.O.; McLean, G.F. An object-specific image-texture analysis of H-resolution forest imagery. *Remote Sens. Environ.* **1996**, *55*, 108–122. [CrossRef]

17. National Forestry Cooperative Federation. Available online: http://iforest.nfcf.or.kr/ (accessed on 16 August 2017)

18. Korea Forest Service. Available online: http://www.forest.go.kr/ (accessed on 16 August 2017).

19. Joe, H.G.; Lee, K.S. Comparison between Hyperspectral and Multispectral Images for the Classification of Coniferous Species. *Korean J. Remote Sens.* **2014**, *30*, 25–36. [CrossRef]

20. Zimble, D.A.; Evans, D.L. Characterizing vertical forest structure using small-footprint airborne LiDAR. *Remote Sens. Environ.* **2003**, *87*, 171–182. [CrossRef]

21. Sun, G.; Ranson, K.J. Forest vertical structure form GLAS: An evaluation using LVIS and SRTM data. *Remote Sens. Environ.* **2008**, *112*, 107–117. [CrossRef]

22. Mund, J.P.; Wilke, R. Detecting multi-layered forest stands using high density airborne LiDAR data. *J. Geogr. Inf. Sci.* **2015**, *1*, 178–188. [CrossRef]

23. Fernadez-Ordonez, Y.; Soria-Ruiz, J.S.R. Forest Inventory using Optical and Radar Remote Sensing. In *Advances in Geoscience and Remote Sensing*; Jedlovec, G., Ed.; InTech: Rijeka, Yugoslavia, 2009; ISBN 978-953-307-005-6.

24. Yoon, J.S.; Lee, K.S.; Shin, J.I.; Woo, C.S. Characteristics of Airborne Lidar Data and Ground Points Separation in Forested Area. *Korean J. Remote Sens.* **2006**, *22*, 533–542.

25. LG CNS Blog 'Creative N Smart'-'what is the 'Artificial Neural Network?". Available online: http://www.blog.lgcns.com/1359/ (accessed on 20 September 2017).

26. Gopal, S.; Curtis, W. Remote Sensing of Forest Change Using Artificial Neural Networks. *IEEE Trans. Geosci. Remote Sens.* **1996**, *34*, 398–404. [CrossRef]

27. Jayathunga, S.; Owari, T.; Tsuyuki, S. Analysis of forest structural complexity using airborne LiDAR data and aerial photography in a mixed conifer–broadleaf forest in northern Japan. *J. For. Res.* **2017**, 1–15. [CrossRef]

28. Falkowski, M.J.; Evans, J.S.; Martinuzzi, S.; Gessler, P.E.; Hudak, A.T. Characterizing forest succession with lidar data: An evaluation for the Inland Northwest, USA. *Remote Sens. Environ.* **2009**, *113*, 946–956. [CrossRef]

PERMISSIONS

LIST OF CONTRIBUTORS

Ali Jozaghi and Babak Alizadeh
Department of Civil Engineering, University of Texas at Arlington, Arlington, TX 76019, USA

Mohsen Hatami and Ian Flood
M. E. Rinker, Sr. School of Construction Management, University of Florida, Gainesville, FL 32611, USA

Mohammad Khorrami
Department of Civil Engineering, Ferdowsi University of Mashhad, Mashhad 9177948974, Iran

Nastaran Khodaei
College of Architecture and Planning, University of Colorado Denver, Denver, CO 80217, USA

Erfan Ghasemi Tousi
Department of Civil and Architectural Engineering and Mechanics, University of Arizona, Tucson, AZ 85721, USA

Won-Kyung Baek and Hyung-Sup Jung
Department of Geoinformatics, University of Seoul, 163 Seoulsiripdaero, Dongdaemun-gu, Seoul 02504, Korea

Sunmin Lee
Department of Geoinformatics, University of Seoul, 163 Seoulsiripdaero, Dongdaemun-gu, Seoul 02504, Korea
Center for Environmental Assessment Monitoring, Environmental Assessment Group, Korea Environment Institute (KEI), 370 Sicheong-daero, Sejong 30147, Korea

Saro Lee
Geoscience Platform Division, Korea Institute of Geoscience and Mineral Resources (KIGAM), 124 Gwahak-ro, Yuseong-gu, Daejeon 34132, Korea
Department of Geophysical Exploration, Korea University of Science and Technology, 217 Gajeong-ro, Yuseong-gu, Daejeon 34113, Korea

Anna Kovacs-Györi
IDA Lab, University of Salzburg, 5020 Salzburg, Austria

Alina Ristea
Boston Area Research Initiative, School of Public Policy and Urban Affairs, Northeastern University, Boston, MA 02115, USA

Clemens Havas, Thomas Blaschke and Arthur Lehner
Department of Geoinformatics—Z_GIS, University of Salzburg, 5020 Salzburg, Austria

Michael Mehaffy
Centre for the Future of Places, KTH Royal Institute of Technology, 114 28 Stockholm, Sweden

Hartwig H. Hochmair
Fort Lauderdale Research and Education Center, University of Florida, Davie, FL 33314, USA

Bernd Resch
Department of Geoinformatics—Z_GIS, University of Salzburg, 5020 Salzburg, Austria
Center for Geographic Analysis, Harvard University, Cambridge, MA 02138, USA

Levente Juhasz
GIS Center, Florida International University, Miami, FL 33199, USA

Laxmi Ramasubramanian
Department of Urban & Regional Planning, San José State University, San Jose, CA 95192-0185, USA

Tom Brijs and Muhammad Aamir Basheer
Transportation Research Institute (IMOB), Hasselt University, Diepenbeek 3590, Belgium

Syyed Adnan Raheel Shah
Transportation Research Institute (IMOB), Hasselt University, Diepenbeek 3590, Belgium
Taxila Institute of Transportation Engineering, Department of Civil Engineering, University of Engineering & Technology, Taxila 47050, Pakistan

Naveed Ahmad
Taxila Institute of Transportation Engineering, Department of Civil Engineering, University of Engineering & Technology, Taxila 47050, Pakistan

Yongjun Shen
School of Transportation, Southeast University, Nanjing 210096, China

Ali Pirdavani
Faculty of Engineering Technology, Hasselt University, Diepenbeek 3590, Belgium

Alessandro Piscini, Vito Romaniello, Christian Bignami and Salvatore Stramondo
Istituto Nazionale di Geofisica e Vulcanologia, 00143 Roma, Italy

Giles M. Foody
School of Geography, University of Nottingham, University Park, Nottingham NG7 2RD, UK

Nádia F. Afonso and José C. M. Pires
Laboratório de Engenharia de Processos, Ambiente Biotecnologia e Energia (LEPABE), Departamento de Engenharia Química, Faculdade de Engenharia, Universidade do Porto, Rua Dr. Roberto Frias, 4200-465 Porto, Portugal

Prima Riza Kadavi and Chang-Wook Lee
Division of Science Education, Kangwon National University, 1 Kangwondaehak-gil, Chuncheon-si, Gangwon-do 24341, Korea

Won-Jin Lee
Earthquake Volcano Research Center, Korea Meteorological Administration, 61 16-Gil, Yeouidaebang-ro, Dongjak-Gu, Seoul 07062, Korea

Mustafa Ridha Mezaal, Helmi Zulhaidi Mohd Shafri and Zainuddin Md Yusoff
Department of Civil Engineering, Faculty of Engineering, Universiti Putra Malaysia, Serdang 43400, Malaysia

Biswajeet Pradhan
Department of Civil Engineering, Geospatial Information Science Research Center (GISRC), Faculty of Engineering, University Putra Malaysia, UPM, Serdang 43400, Malaysia

School of Systems, Management and Leadership, Faculty of Engineering and Information Technology, University of Technology Sydney, Building 11, Level 06, 81 Broadway, Ultimo NSW 2007, Australia

Maher Ibrahim Sameen
Department of Civil Engineering, Geospatial Information Science Research Center (GISRC), Faculty of Engineering, University Putra Malaysia, UPM, Serdang 43400, Malaysia

Moung-Jin Lee
Center for Environmental Assessment Monitoring, Environmental Assessment Group, Korea Environment Institute (KEI), 370 Sicheong-daero, Sejong 30147, Korea

Wonkyong Song
Department of Landscape Architecture, Dankook University, 119 Dandae-ro, Dongnam-gu, Cheonan-si, Chungnam 31116, Korea

Guandong Chen, Yu Li and Guangmin Sun
Faculty of Information Technology, Beijing University of Technology, Beijing 100124, China

Yuanzhi Zhang
National Astronomical Observatories, Chinese Academy of Sciences, Beijing 100012, China
Key Laboratory of Lunar Science and Deep-space Exploration, Chinese Academy of Sciences, Beijing 100012, China

Soo-Kyung Kwon and Daeseong Kim
Department of Geoinformatics, University of Seoul, Seoul 02504, Korea

Index

Printed in the USA
CPSIA information can be obtained
at www.ICGtesting.com
JSHW051403091023
49903JS00006B/253